Electric Power Systems

Electric Power Systems
Fifth Edition

B.M. Weedy, *University of Southampton, UK*
B.J. Cory, *Imperial College London, UK*
N. Jenkins, *Cardiff University, UK*
J.B. Ekanayake, *Cardiff University, UK*
G. Strbac, *Imperial College London, UK*

WILEY

A John Wiley & Sons, Ltd., Publication

This edition first published 2012
© 2012, John Wiley & Sons Ltd

Registered office

John Wiley & Sons Ltd, The Atrium, Southern Gate, Chichester, West Sussex, PO19 8SQ, United Kingdom

For details of our global editorial offices, for customer services and for information about how to apply for permission to reuse the copyright material in this book please see our website at www.wiley.com.

Library of Congress Cataloging-in-Publication Data

Electric power systems / Brian M. Weedy [...et al.]. – 5th ed.
    p. cm.
  Includes bibliographical references and index.
  ISBN 978-0-470-68268-5 (cloth)
  1. Electric power systems–Textbooks. 2. Electric power
transmission–Textbooks. I. Weedy, Brian M.
  TK1001.E4235 2012
  621.319′1–dc23

                      2012010322

A catalogue record for this book is available from the British Library.

Print ISBN: 9780470682685

Set in 10/12.5pt, Palatino-Roman by Thomson Digital, Noida, India

Contents

Preface to First Edition

In writing this book the author has been primarily concerned with the presentation of the basic essentials of power-system operation and analysis to students in the final year of first degree courses at universities and colleges of technology. The emphasis is on the consideration of the system as a whole rather than on the engineering details of its constituents, and the treatment presented is aimed at practical conditions and situations rather than theoretical nicety.

In recent years the contents of many undergraduate courses in electrical engineering have become more fundamental in nature with greater emphasis on electromagnetism, network analysis, and control theory. Students with this background will be familiar with much of the work on network theory and the inductance, capacitance, and resistance of lines and cables, which has in the past occupied large parts of textbooks on power supply. In this book these matters have been largely omitted, resulting in what is hoped is a concise account of the operation and analysis of electric power systems. It is the author's intention to present the power system as a system of interconnected elements which may be represented by models, either mathematically or by equivalent electrical circuits. The simplest models will be used consistently with acceptable accuracy and it is hoped that this will result in the wood being seen as well as the trees. In an introductory text such as this no apology is made for the absence of sophisticated models of plant (synchronous machines in particular) and involved mathematical treatments as these are well catered for in more advanced texts to which reference is made.

The book is divided into four main parts, as follows:

a. Introduction, including the establishment of equivalent circuits of the components of the system, the performance of which, when interconnected, forms the main theme.
b. Operation, the manner in which the system is operated and controlled to give secure and economic power supplies.
c. Analysis, the calculation of voltage, power, and reactive power in the system under normal and abnormal conditions. The use of computers is emphasised when dealing with large networks.

d. Limitations of transmittable power owing to the stability of the synchronous machine, voltage stability of loads, and the temperature rises of plant.

It is hoped that the final chapter will form a useful introduction to direct current transmission which promises to play a more and more important role in electricity supply.

The author would like to express his thanks to colleagues and friends for their helpful criticism and advice. To Mr J.P. Perkins for reading the complete draft, to Mr B.A. Carre on digital methods for load flow analysis, and to Mr A.M. Parker on direct current transmission. Finally, thanks are due to past students who for over several years have freely expressed their difficulties in this subject.

Birron M. Weedy
Southampton, 1967

Preface to Fourth Edition

As a university teacher for 40 years, I have always admired the way that Dr Birron Weedy's book has stood out from the numerous texts on the analysis and modelling of power systems, with its emphasis on practical systems rather than extensive theory or mathematics. Over the three previous editions and one revision, the text has been continually updated and honed to provide the essentials of electrical power systems sufficient not only for the final year of a first degree course, but also as a firm foundation for further study. As with all technology, progress produces new devices and understanding requiring revision and updating if a book is to be of continuing value to budding engineers. With power systems, there is another dimension in that changes in social climate and political thinking alter the way they are designed and operated, requiring consideration and understanding of new forms of infrastructure, pricing principles and service provision. Hence the need for an introduction to basic economics and market structures for electricity supply, which is given in a completely new Chapter 12.

In this edition, 10 years on from the last, a rewrite of Chapter 1 has brought in full consideration of CCGT plant, some new possibilities for energy storage, the latest thinking on electromagnetic fields and human health, and loss factor calculations. The major addition to system components and operation has been Flexible a.c. Transmission (FACT) devices using the latest semiconductor power switches and leading to better control of power and var flows. The use of optimisation techniques has been brought into Chapter 6 with powerflow calculations but the increasing availability and use of commercial packages has meant that detailed code writing is no longer quite so important. For stability (Chapter 8), it has been necessary to consider voltage collapse as a separate phenomenon requiring further research into modelling of loads at voltages below 95% or so of nominal. Increasingly, large systems require fast stability assessment through energy-like functions as explained in additions made to this chapter. Static-shunt variable compensators have been included in Chapter 9 with a revised look at h.v.d.c. transmission. Many d.c. schemes now exist around the world and are continually being added to so the description of an example scheme has been omitted. Chapter 11 now includes many new sections with updates on switchgear, and comprehensive introductions to digital (numerical) protection principles, monitoring and control with SCADA, state estimation, and the concept of Energy Management Systems (EMS) for system operation.

Readers who have been brought up on previous editions of this work will realise that detailed design of overhead and underground systems and components has been omitted from this edition. Fortunately, adequate textbooks on these topics are available, including an excellent book by Dr Weedy, and reference to these texts is recommended for detailed study if the principles given in Chapter 3 herein are insufficient. Many other texts (including some 'advanced' ones) are listed in a new organisation of the bibliography, together with a chapter-referencing key which I hope will enable the reader to quickly determine the appropriate texts to look up. In addition, mainly for historical purposes, a list of significant or 'milestone' papers and articles is provided for the interested student.

Finally, it has been an honour to be asked to update such a well-known book and I hope that it still retains much of the practical flavour pioneered by Dr Weedy. I am particularly indebted to my colleagues, Dr Donald Macdonald (for much help with a rewrite of the material about electrical generators) and Dr Alun Coonick for his prompting regarding the inclusion of new concepts. My thanks also go to the various reviewers of the previous editions for their helpful suggestions and comments which I have tried to include in this new edition. Any errors and omissions are entirely my responsibility and I look forward to receiving feedback from students and lecturers alike.

Brian J. Cory
Imperial College, London, 1998

Publisher's Note

Dr B. M. Weedy died in December 1997 during the production of this fourth edition.

Preface to Fifth Edition

We were delighted to be asked to revise this classic textbook. From the earlier editions we had gained much, both as undergraduate students and throughout our careers. Both Dr Weedy and Dr Cory can only be described as giants of power system education and the breadth of their vision and clarity of thought is evident throughout the text. Reading it carefully, for the purposes of revision, was a most rewarding experience and even after many years studying and teaching power systems we found new insights on almost every page.

We have attempted to stay true to the style and structure of the book while adding up-to-date material and including examples of computer based simulation. We were conscious that this book is intended to support a 3rd or 4th year undergraduate course and it is too easy when revising a book to continue to add material and so obscure rather than illuminate the fundamental principles. This we have attempted not to do. Chapter 1 has been brought up to date as many countries de-carbonise their power sector. Chapter 6 (load flow) has been substantially rewritten and voltage source converter HVDC added to Chapter 9. Chapter 10 has been revised to include modern switchgear and protection while recognising that the young engineer is likely to encounter much equipment that may be 30–40 years old. Chapter 12 has been comprehensively revised and now contains material suitable for teaching the fundamentals of the economics of operation and development of power systems. All chapters have been carefully revised and where we considered it would aid clarity the material rearranged. We have paid particular attention to the Examples and Problems and have created Solutions to the Problems that can be found on the Wiley website.

We are particularly indebted to Dave Thompson who created all the illustrations for this edition, Lewis Dale for his assistance with Chapter 12, and to IPSA Power for generously allowing us a license for their power system analysis software. Also we would like to thank: Chandima Ekanayake, Prabath Binduhewa, Predrag Djapic and Jelena Rebic for their assistance with the Solutions to the Problems. Bethany

Corcoran provided the data for Figure 1.1 while Alstom Grid, through Rose King, kindly made available information for some of the drawings of Chapter 11. Although, of course, responsibility for errors and omissions lies with us, we hope we have stayed true to the spirit of this important textbook.

For instructors and teachers, solutions to the problems set out in the book can be found on the companion website www.wiley.com/go/weedy_electric.

Nick Jenkins, Janaka Ekanayake, Goran Strbac
June 2012

Symbols

Throughout the text, symbols in bold type represent complex (phasor) quantities requiring complex arithmetic. Italic type is used for magnitude (scalar) quantities.

A,B,C,D Generalised circuit constants
a–b–c Phase rotation (alternatively R–Y–B)
a Operator $1 \angle 120°$
C Capacitance (farad)
D Diameter
E e.m.f. generated
F Cost function (units of money per hour)
f Frequency (Hz)
G Rating of machine
g Thermal resistivity (°C m/W)
H Inertia constant (seconds)
h Heat transfer coefficient (W/m² per °C)
I Current (A)
I* Conjugate of **I**
I_d In-phase current
I_q Quadrature current
j $1 \angle 90°$ operator
K Stiffness coefficient of a system (MW/Hz)
L Inductance (H)
ln Natural logarithm
M Angular momentum (J-s per rad or MJ-s per electrical degree)
N Rotational speed (rev/min, rev/s, rad/s)
P Propagation constant ($\alpha + j\beta$)
P Power (W)
$\dfrac{dP}{d\delta}$ Synchronising power coefficient
p.f. Power factor
p Iteration number
Q Reactive power (VAr)
q Loss dissipated as heat (W)
R Resistance (Ω); also thermal resistance (°C/W)
R–Y–B Phase rotation (British practice)
S Complex power $= P \pm jQ$
S Siemens
s Laplace operator
s Slip

SCR	Short-circuit ratio
T	Absolute temperature (K)
t	Time
t	Off-nominal transformer tap ratio
Δt	Interval of time
Ω^{-1}	Siemens
U	Velocity
$\mathbf{V}$	Voltage; ΔV scalar voltage difference
V	Voltage magnitude
W	Volumetric flow of coolant (m^3/s)
X'	Transient reactance of a synchronous machine
X''	Subtransient reactance of a synchronous machine
X_d	Direct axis synchronous reactance of a synchronous machine
X_q	Quadrature axis reactance of a synchronous machine
X_s	Synchronous reactance of a synchronous machine
$\mathbf{Y}$	Admittance (p.u. or Ω)
$\mathbf{Z}$	Impedance (p.u. or Ω)
$\mathbf{Z}_0$	Characteristic or surge impedance (Ω)
α	Delay angle in rectifiers and inverters–d.c. transmission
α	Attenuation constant of line
α	Reflection coefficient
β	Phase-shift constant of line
β	$(180 - \alpha)$ used in inverters
β	Refraction coefficient $(1 + \alpha)$
γ	Commutation angle used in converters
δ	Load angle of synchronous machine or transmission angle across a system (electrical degrees)
δ_0	Recovery angle of semiconductor valve
ε	Permittivity
η	Viscosity (g/(cm-s))
θ	Temperature rise (°C) above reference or ambient
λ	Lagrange multiplier
ρ	Electrical resistivity (Ω-m)
ρ	Density (kg/m^3)
τ	Time constant
ϕ	Angle between voltage and current phasors (power factor angle)
ω	Angular frequency (rad/s)

Subscripts 1, 2, and 0 refer to positive, negative, and zero symmetrical components, respectively.

1

Introduction

1.1 History

In 1882 Edison inaugurated the first central generating station in the USA. This fed a load of 400 lamps, each consuming 83 W. At about the same time the Holborn Viaduct Generating Station in London was the first in Britain to cater for consumers generally, as opposed to specialized loads. This scheme used a 60 kW generator driven by a horizontal steam engine; the voltage of generation was 100 V direct current.

The first major alternating current station in Great Britain was at Deptford, where power was generated by machines of 10 000 h.p. and transmitted at 10 kV to consumers in London. During this period the battle between the advocates of alternating current and direct current was at its most intense with a similar controversy raging in the USA and elsewhere. Owing mainly to the invention of the transformer the supporters of alternating current prevailed and a steady development of local electricity generating stations commenced with each large town or load centre operating its own station.

In 1926, in Britain, an Act of Parliament set up the Central Electricity Board with the object of interconnecting the best of the 500 generating stations then in operation with a high-voltage network known as the Grid. In 1948 the British supply industry was nationalized and two organizations were set up: (1) the Area Boards, which were mainly concerned with distribution and consumer service; and (2) the Generating Boards, which were responsible for generation and the operation of the high-voltage transmission network or grid.

All of this changed radically in 1990 when the British Electricity Supply Industry was privatized. Separate companies were formed to provide competition in the supply of electrical energy (sometimes known as electricity retail businesses) and in power generation. The transmission and distribution networks are natural monopolies, owned and operated by a Transmission System Operator and Distribution Network Operators. The Office of Gas and Electricity Markets (OFGEM) was

Electric Power Systems, Fifth Edition. B.M. Weedy, B.J. Cory, N. Jenkins, J.B. Ekanayake and G. Strbac.
© 2012 John Wiley & Sons, Ltd. Published 2012 by John Wiley & Sons, Ltd.

established as the Regulator to ensure the market in electricity generation and energy supply worked effectively and to fix the returns that the Transmission and Distribution Companies should earn on their monopoly businesses.

For the first 80 years of electricity supply, growth of the load was rapid at around 7% per year, implying a doubling of electricity use every 10 years and this type of increase continues today in rapidly.industrializing countries. However in the USA and in other industrialized countries there has been a tendency, since the oil shock of 1973, for the rate of increase to slow with economic growth no longer coupled closely to the use of energy. In the UK, growth in electricity consumption has been under 1% per year for a number of years.

A traditional objective of energy policy has been to provide secure, reliable and affordable supplies of electrical energy to customers. This is now supplemented by the requirement to limit greenhouse gas emissions, particularly of CO_2, and so mitigate climate change. Hence there is increasing emphasis on the generation of electricity from low-carbon sources that include renewable, nuclear and fossil fuel plants fitted with carbon capture and storage equipment. The obvious way to control the environmental impact of electricity generation is to reduce the electrical demand and increase the efficiency with which electrical energy is used. Therefore conservation of energy and demand reduction measures are important aspects of any contemporary energy policy.

1.2 Characteristics Influencing Generation and Transmission

There are three main characteristics of electricity supply that, however obvious, have a profound effect on the manner in which the system is engineered. They are as follows:

Electricity, unlike gas and water, cannot be stored and the system operator traditionally has had limited control over the load. The control engineers endeavour to keep the output from the generators equal to the connected load at the specified voltage and frequency; the difficulty of this task will be apparent from a study of the load curves in Figure 1.1. It will be seen that the load consists of a steady component known as the base load, plus peaks that depend on the time of day and days of the week as well as factors such as popular television programmes.

The electricity sector creates major environmental impacts that increasingly determine how plant is installed and operated. Coal burnt in steam plant produces sulphur dioxide that causes acid rain. Thus, in Europe, it is now mandatory to fit flue gas desulphurisation plant to coal fired generation. All fossil fuel (coal, oil and gas) produce CO_2 (see Table 1.1) which leads to climate change and so its use will be discouraged increasingly with preference given to generation by low-carbon energy sources.

The generating stations are often located away from the load resulting in transmission over considerable distances. Large hydro stations are usually remote from urban centres and it has often been cost-effective to burn coal close to where it is mined and transport the electricity rather than move the coal. In many countries, good sites for wind energy are remote from centres of population and,

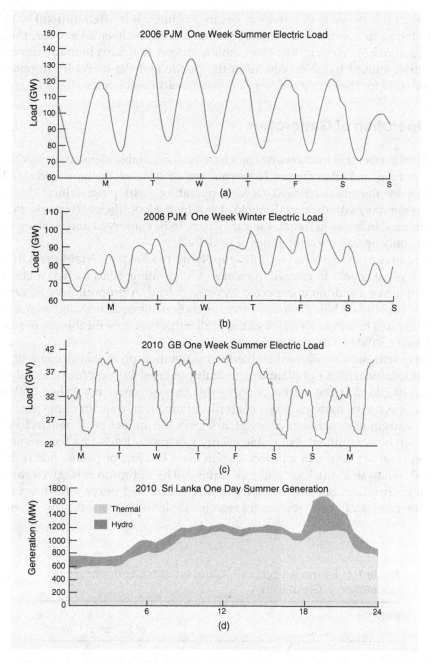

Figure 1.1 Load curves. (a) PJM (Pennsylvania, Jersey, Maryland) control area in the east of the USA over a summer week. The base load is 70 GW with a peak of 140 GW. This is a very large interconnected power system. (b) PJM control area over a winter week. Note the morning and evening peaks in the winter with the maximum demand in the summer. (c) Great Britain over a summer week. The base load is around 25 GW with a daily increase/decrease of 15 GW. GB is effectively an isolated power system. (d) Sri Lanka over 1 day. Note the base load thermal generation with hydro used to accommodate the rapid increase of 500 MW at dusk

although it is possible to transport gas in pipelines, it is often difficult to obtain permission to construct generating stations close to cities. Moreover, the construction of new electrical transmission is subject to delays in many developed countries caused by objections from the public and the difficulty in obtaining permission for the construction of new overhead line circuits.

1.3 Operation of Generators

The national electrical load consists of a base plus a variable element, depending on the time of day and other factors. In thermal power systems, the base load should be supplied by the most efficient (lowest operating cost) plant which then runs 24 hours per day, with the remaining load met by the less efficient (but lower capital cost) stations. In hydro systems water may have to be conserved and so some generators are only operated during times of peak load.

In addition to the generating units supplying the load, a certain proportion of available plant is held in reserve to meet sudden contingencies such as a generator unit tripping or a sudden unexpected increase in load. A proportion of this reserve must be capable of being brought into operation immediately and hence some machines must be run at, say, 75% of their full output to allow for this spare generating capacity, called spinning reserve.

Reserve margins are allowed in the total generation plant that is constructed to cope with unavailability of plant due to faults, outages for maintenance and errors in predicting load or the output of renewable energy generators. When traditional national electricity systems were centrally planned, it was common practice to allow a margin of generation of about 20% over the annual peak demand. A high proportion of intermittent renewable energy generation leads to a requirement for a higher reserve margin. In a power system there is a mix of plants, that is, hydro, coal, oil, renewable, nuclear, and gas turbine. The optimum mix gives the most economic operation, but this is highly dependent on fuel prices which can fluctuate with time and from region to region. Table 1.2 shows typical plant and

Table 1.1 Estimated carbon dioxide emissions from electricity generation in Great Britain

Fuel	Tonnes of CO_2/GWh of Electrical Output
Coal	915
Oil	633
Gas	405
Great Britain generation portfolio (including nuclear and renewables)	452

Data from the Digest of UK Energy Statistics, 2010, published by the Department of Energy and Climate Change.

Table 1.2 Example of costs of electricity generation

Generating Technology	Capital Cost of Plant £/MW	Cost of electricity £/MWh
Combined Cycle Gas Turbine	720	80
Coal	1800	105
Onshore wind	1520	94
Nuclear	2910	99

Data from UK Electricity Generating Costs Update, 2010, Mott MacDonald, reproduced with permission

generating costs for the UK. It is clear some technologies have a high capital cost (for example, nuclear and wind) but low fuel costs.

1.4 Energy Conversion

1.4.1 Energy Conversion Using Steam

The combustion of coal, gas or oil in boilers produces steam, at high temperatures and pressures, which is passed through steam turbines. Nuclear fission can also provide energy to produce steam for turbines. Axial-flow turbines are generally used with several cylinders, containing steam of reducing pressure, on the same shaft.

A steam power-station operates on the Rankine cycle, modified to include superheating, feed-water heating, and steam reheating. High efficiency is achieved by the use of steam at the maximum possible pressure and temperature. Also, for turbines to be constructed economically, the larger the size the less the capital cost per unit of power output. As a result, turbo-generator sets of 500 MW and more have been used. With steam turbines above 100 MW, the efficiency is increased by reheating the steam, using an external heater, after it has been partially expanded. The reheated steam is then returned to the turbine where it is expanded through the final stages of blading.

A schematic diagram of a coal fired station is shown in Figure 1.2. In Figure 1.3 the flow of energy in a modern steam station is shown.

In coal-fired stations, coal is conveyed to a mill and crushed into fine powder, that is pulverized. The pulverized fuel is blown into the boiler where it mixes with a supply of air for combustion. The exhaust steam from the low pressure (L.P.) turbine is cooled to form condensate by the passage through the condenser of large quantities of sea- or river-water. Cooling towers are used where the station is located inland or if there is concern over the environmental effects of raising the temperature of the sea- or river-water.

Despite continual advances in the design of boilers and in the development of improved materials, the nature of the steam cycle is such that vast quantities of heat are lost in the condensate cooling system and to the atmosphere. Advances in design and materials in the last few years have increased the thermal

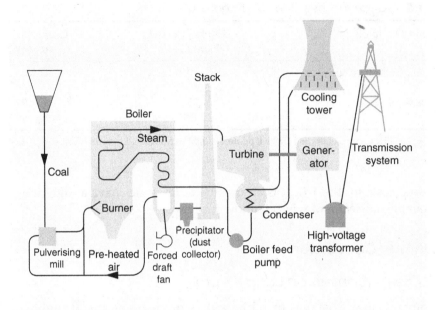

Figure 1.2 Schematic view of coal fired generating station

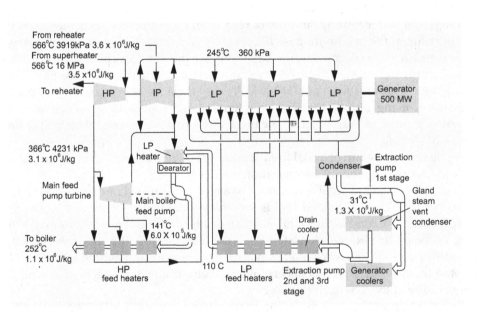

Figure 1.3 Energy flow diagram for a 500 MW turbine generator (Figure adapted from Electrical Review)

efficiencies of new coal stations to approaching 40%. If a use can be found for the remaining 60% of energy rejected as heat, fairly close to the power station, forming a Combined Heat and Power (or Co-generation) system then this is clearly desirable.

1.4.2 Energy Conversion Using Water

Perhaps the oldest form of energy conversion is by the use of water power. In a hydroelectric station the energy is obtained free of cost. This attractive feature has always been somewhat offset by the very high capital cost of construction, especially of the civil engineering works. Unfortunately, the geographical conditions necessary for hydro-generation are not commonly found, especially in Britain. In most developed countries, all the suitable hydroelectric sites are already fully utilized. There still exists great hydroelectric potential in many developing countries but large hydro schemes, particularly those with large reservoirs, have a significant impact on the environment and the local population.

The difference in height between the upper reservoir and the level of the turbines or outflow is known as the head. The water falling through this head gains energy which it then imparts to the turbine blades. Impulse turbines use a jet of water at atmospheric pressure while in reaction turbines the pressure drops across the runner imparts significant energy.

A schematic diagram of a hydro generation scheme is shown in Figure 1.4.

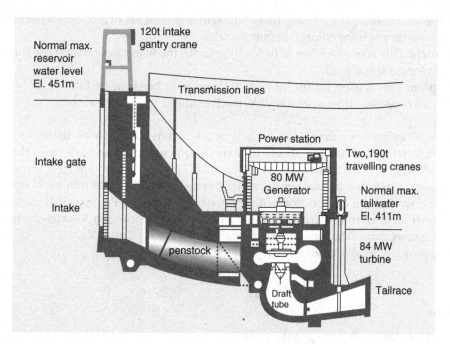

Figure 1.4 Schematic view of a hydro generator (Figure adapted from Engineering)

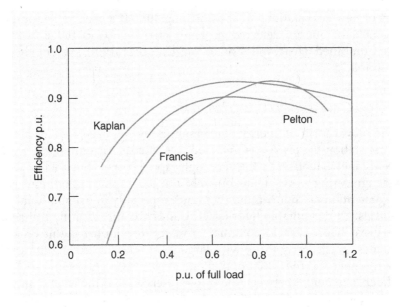

Figure 1.5 Typical efficiency curves of hydraulic turbines (1 per unit (p.u.). = 100%)

Particular types of turbine are associated with the various heights or heads of water level above the turbines. These are:

1. **Pelton:** This is used for heads of 150–1500 m and consists of a bucket wheel rotor with water jets from adjustable flow nozzles.
2. **Francis:** This is used for heads of 50–500 m with the water flow within the turbine following a spiral path.
3. **Kaplan:** This is used for run-of-river stations with heads of up to 60 m. This type has an axial-flow rotor with variable-pitch blades.

Typical efficiency curves for each type of turbine are shown in Figure 1.5. Hydroelectric plant has the ability to start up quickly and the advantage that no energy losses are incurred when at a standstill. It has great advantages, therefore, for power generation because of this ability to meet peak loads at minimum operating cost, working in conjunction with thermal stations – see Figure 1.1(d). By using remote control of the hydro sets, the time from the instruction to start up to the actual connection to the power network can be as short as 3 minutes.

The power available from a hydro scheme is given by

$$P = \rho g Q H \qquad [\text{W}]$$

where

Q = flow rate (m^3/s) through the turbine;
ρ = density of water (1000 kg/m^3);

g = acceleration due to gravity (9.81 m/s²);
H = head, that is height of upper water level above the lower (m).

Substituting,

$$P = 9.81QH \qquad [kW]$$

1.4.3 Gas Turbines

With the increasing availability of natural gas (methane) and its low emissions and competitive price, prime movers based on the gas turbine cycle are being used increasingly. This thermodynamic cycle involves burning the fuel in the compressed working fluid (air) and is used in aircraft with kerosene as the fuel and for electricity generation with natural gas (methane). Because of the high temperatures obtained, the efficiency of a gas turbine is comparable to that of a steam turbine, with the additional advantage that there is still sufficient heat in the gas-turbine exhaust to raise steam in a conventional boiler to drive a steam turbine coupled to another electricity generator. This is known as a combined-cycle gas-turbine (CCGT) plant, a schematic layout of which is shown in Figure 1.6. Combined efficiencies of new CCGT generators now approach 60%.

The advantages of CCGT plant are the high efficiency possible with large units and, for smaller units, the fast start up and shut down (2–3 min for the gas turbine, 20 min for the steam turbine), the flexibility possible for load following, the comparative speed of installation because of its modular nature and factory-supplied units,

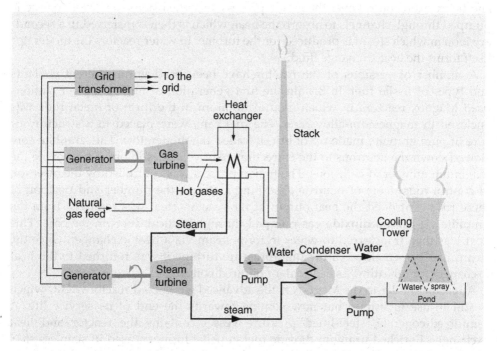

Figure 1.6 Schematic diagram of a combined-cycle gas-turbine power station

and its ability to run on light oil (from local storage tanks) if the gas supply is interrupted. Modern installations are fully automated and require only a few operators to maintain 24 hour running or to supply peak load, if needed.

1.4.4 Nuclear Power

Energy is obtained from the fission reaction which involves the splitting of the nuclei of uranium atoms. Compared with chemical reactions, very large amounts of energy are released per atomic event. Uranium metal extracted from the base ore consists mainly of two isotopes, ^{238}U (99.3% by weight) and ^{235}U (0.7%). Only ^{235}U is fissile, that is when struck by slow-moving neutrons its nucleus splits into two substantial fragments plus several neutrons and 3×10^{-11} J of kinetic energy. The fast moving fragments hit surrounding atoms producing heat before coming to rest. The neutrons travel further, hitting atoms and producing further fissions. Hence the number of neutrons increases, causing, under the correct conditions, a chain reaction. In conventional reactors the core or moderator slows down the moving neutrons to achieve more effective splitting of the nuclei.

Fuels used in reactors have some component of ^{235}U. Natural uranium is sometimes used although the energy density is considerably less than for enriched uranium. The basic reactor consists of the fuel in the form of rods or pellets situated in an environment (moderator) which will slow down the neutrons and fission products and in which the heat is evolved. The moderator can be light or heavy water or graphite. Also situated in the moderator are movable rods which absorb neutrons and hence exert control over the fission process. In some reactors the cooling fluid is pumped through channels to absorb the heat, which is then transferred to a secondary loop in which steam is produced for the turbine. In water reactors the moderator itself forms the heat-exchange fluid.

A number of versions of the reactor have been used with different coolants and types of fissile fuel. In Britain the first generation of nuclear power stations used Magnox reactors in which natural uranium in the form of metal rods was enclosed in magnesium-alloy cans. The fuel cans were placed in a structure or core of pure graphite made up of bricks (called the moderator). This graphite core slowed down the neutrons to the correct range of velocities in order to provide the maximum number of collisions. The fission process was controlled by the insertion of control rods made of neutron-absorbing material; the number and position of these rods controlled the heat output of the reactor. Heat was removed from the graphite via carbon dioxide gas pumped through vertical ducts in the core. This heat was then transferred to water to form steam via a heat exchanger. Once the steam had passed through the high-pressure turbine it was returned to the heat exchanger for reheating, as in a coal- or oil-fired boiler.

A reactor similar to the Magnox is the advanced gas-cooled reactor (AGR) which is still in use in Britain but now coming towards the end of its service life. A reinforced-concrete, steel-lined pressure vessel contains the reactor and heat exchanger. Enriched uranium dioxide fuel in pellet form, encased in stainless steel cans, is used; a number of cans are fitted into steel fitments within a graphite tube to

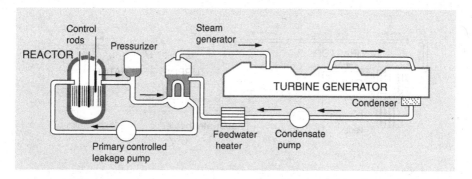

Figure 1.7 Schematic diagram of a pressurized-water reactor (PWR)

form a cylindrical fuel element which is placed in a vertical channel in the core. Depending on reactor station up to eight fuel elements are held in place one above the other by a tie bar. Carbon dioxide gas, at a higher pressure than in the Magnox type, removes the heat. The control rods are made of boron steel. Spent fuel elements when removed from the core are stored in a special chamber and lowered into a pond of water where they remain until the level of radioactivity has decreased sufficiently for them to be removed from the station and disassembled.

In the USA and many other countries pressurized-water and boiling-water reactors are used. In the pressurized-water type the water is pumped through the reactor and acts as a coolant and moderator, the water being heated to 315 °C at around 150 bar pressure. At this temperature and pressure the water leaves the reactor at below boiling point to a heat exchanger where a second hydraulic circuit feeds steam to the turbine. The fuel is in the form of pellets of uranium dioxide in bundles of zirconium alloy.

The boiling-water reactor was developed later than the pressurized-water type. Inside the reactor, heat is transferred to boiling water at a pressure of 75 bar (1100 p.s.i.). Schematic diagrams of these reactors are shown in Figures 1.7 and 1.8. The ratio of pressurized-water reactors to boiling-water reactors throughout the world is around 60/40%.

Both pressurized- and boiling-water reactors use light water.[1] The practical pressure limit for the pressurized-water reactor is about 160 bar (2300 p.s.i.), which limits its efficiency to about 30%. However, the design is relatively straightforward and experience has shown this type of reactor to be stable and dependable. In the boiling-water reactor the efficiency of heat removal is improved by use of the latent heat of evaporation. The steam produced flows directly to the turbine, causing possible problems of radioactivity in the turbine. The fuel for both light-water reactors is uranium enriched to 3–4% ^{235}U. Boiling-water reactors are probably the cheapest to construct; however, they have a more complicated fuel make up with different enrichment levels within each pin. The steam produced is saturated and requires wet-steam turbines. A further type of water reactor is the heavy-water

---

[1] Light water refers to conventional H_2O while heavy water describes deuterium oxide (D_2O).

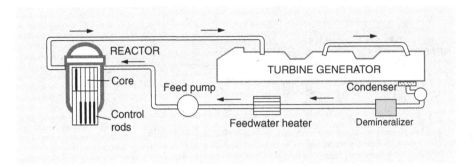

Figure 1.8 Schematic diagram of a boiling-water reactor (BWR)

CANDU type developed by Canada. Its operation and construction are similar to the light-water variety but this design uses naturally occurring, un-enriched or slightly enriched uranium.

Concerns over the availability of future supplies of uranium led to the construction of a number of prototype breeder reactors. In addition to heat, these reactors produce significant new fissile material. However, their cost, together with the technical and environmental challenges of breeder reactors, led to most of these programmes being abandoned and it is now generally considered that supplies of uranium are adequate for the foreseeable future.

Over the past years there has been considerable controversy regarding the safety of reactors and the management of nuclear waste. Experience is still relatively small and human error is always a possibility, such as happened at Three Mile Island in 1979 and Chernobyl in 1986 or a natural event such as the earthquake and tsunami in Fukishima in 2011. However, neglecting these incidents, the safety record of power reactors has been good and now a number of countries (including Britain) are starting to construct new nuclear generating stations using Light Water Reactors. The decommissioning of nuclear power stations and the long term disposal of spent fuel remains controversial.

1.5 Renewable Energy Sources

There is considerable international effort put into the development of renewable energy sources. Many of these energy sources come from the sun, for example wind, waves, tides and, of course, solar energy itself. The average peak solar energy received on the earth's surface is about $600 \, \text{W}/\text{m}^2$, but the actual value, of course, varies considerably with time of day and cloud conditions.

1.5.1 Solar Energy–Thermal Conversion

There is increasing interest in the use of solar energy for generating electricity through thermal energy conversion. In large-scale (central station) installations the sun's rays are concentrated by lenses or mirrors. Both require accurately curved surfaces and steering mechanisms to follow the motion of the sun. Concentrators may

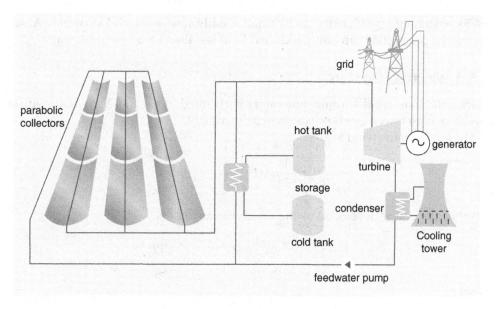

Figure 1.9 Solar thermal generator

be designed to follow the sun's seasonal movement, or additionally to track the sun throughout the day. The former is less expensive and concentration of the sun up to 30 times has been obtained. However, in the French solar furnace in the Pyrenees, two-axis mirrors were used and a concentration of 16 000 was achieved. The reflectors concentrated the rays on to a single receiver (boiler), hence raising steam.

An alternative to this scheme (with lower temperatures) is the use of many individual parabolic trough absorbers tracking the sun in one direction only (Figure 1.9), the thermal energy being transferred by a fluid to a central boiler. In the arid regions of the world where direct solar radiation is strong and hence solar thermal generation effective the limited supply of water for the steam cycle and for cooling can be an important consideration. In solar thermal schemes, heat energy storage can be used to mitigate the fluctuating nature of the sun's energy.

1.5.2 Solar Energy-Photovoltaic Conversion

Photovoltaic conversion occurs in a thin layer of suitable material, typically silicon, when hole-electron pairs are created by incident solar photons and the separation of these holes and electrons at a discontinuity in electrochemical potential creates a potential difference. Whereas theoretical efficiencies are about 25%, practical values are lower. Single-crystal silicon solar cells have been constructed with efficiencies of the complete module approaching 20%. The cost of fabricating and interconnecting cells is high. Polycrystalline silicon films having large-area grains with efficiencies of over 16% have been made. Although photovoltaic devices do not pollute they occupy large areas if MWs of output are required. It has been estimated that to produce 10^{12} kWh per year (about 65% of the 1970 US generation output) the necessary cells would occupy about 0.1% of the US land area (highways occupied 1.5% in

1975), assuming an efficiency of 10% and a daily insolation of $4\,\text{kWh/m}^2$. Automated cell production can now produce cells at less than US $3 per peak watt.

1.5.3 Wind Generators

Horizontal axis wind turbine generators each rated at up to $5\,\text{MW}$ mounted on 90–100 m high towers are now commercially available.

The power in the wind is given by

$$P_w = \frac{1}{2}\rho A U^3 \qquad [W]$$

While the power developed by the aerodynamic rotor is

$$P = C_p P_w = C_p \frac{1}{2}\rho A U^3 \qquad [W]$$

where

ρ = density of air ($1.25\,\text{kg/m}^3$);
U = wind velocity (m/s);
A = swept area of rotor (m^2).
C_p = power coefficient of the rotor

The operation of a wind turbine depends upon the wind speed and is shown in Figure 1.10.

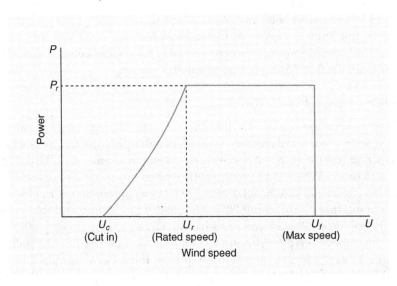

Figure 1.10 Wind turbine power curve

At low wind speeds, there is insufficient energy to operate the turbine and no power is produced. At the cut-in U_c speed, between 3 and 5 m/s, power starts to be generated until rated power P_r is produced at rated wind speed U_r. At higher wind speeds, the turbine is controlled, usually by altering the blade pitch angle, to give rated output up to a maximum wind speed U_f. After this the blades are 'furled' and the unit is shut down to avoid excessive wind loading. Typically, wind turbines with rotors of 80 m diameter, rotate at 15–20 rpm, and are geared up to a generator speed of around 1000 r.p.m. All modern large wind turbines operate at variable speed using power electronic converters to connect the generator to the 50/60 Hz electrical network. This is in order to reduce mechanical loads and to allow the aerodynamic rotor to run at its most effective speed.

Example 1.1

Calculate the number of wind generators required to produce the equivalent energy of a 600 MW CCGT operating at 80% load factor. Assume the average wind speed is 8 m/s, rotor diameter is 80 m, and conversion efficiency (coefficient of performance, C_p) is 0.45.

Calculation
Power in the wind:

$$P_{wind} = \frac{1}{2}\rho A U^3$$

$$= \frac{1}{2} \times 1.25 \times \pi \times 40^2 \times 8^3 = 1.6 \, \text{MW}$$

Power in generator:

$$P_{generator} = 1.6 \times 0.45 = 724 \, \text{kW}$$

Number of turbines required $= 600 \times 0.8/0.724 = 663$.
A regular spacing of turbines at 5 times rotor diameter (400 m), gives 6.25 turbines/km^2. Thus a total area of 106 km^2 is required.
From this calculation, it is apparent that wind generators spread over a wide area (e.g. 11 km $\times$ 10 km) would be required although the ground beneath them could be used for grazing. The saving in CO_2 emissions would be approximately:

Daily electrical energy generated $= 600 \times 0.8 \times 24 = 11.5$ GWh.
CO_2 emissions saved (see Table 1.1) $= 11.5 \times 405 = 4657$ tonnes/day

---

1.5.4 Biofuels

Biofuels are derived from vegetable matter produced by agriculture or forestry operations or from waste materials collected from industry, commerce and residential households. As an energy resource, biomass used as a source of heat by burning wood, dung, and so on, in developing countries is very important and contributes about 14% of the world's energy requirements. Biomass can be used to produce electricity in two ways:

1. by burning in a furnace to produce steam to drive turbines; or
2. by fermentation in landfill sites or in special anaerobic tanks, both of which produce a methane-rich gas which can fuel a spark ignition engine or gas turbine.

It can also be co-fired with coal in large steam power stations.

If crops are cultivated for combustion, either as a primary source of heat or as a by-product of some other operation; they can be considered as CO_2 neutral, in that their growing cycle absorbs as much CO_2 as is produced by their combustion. In industrialized countries, biomass has the potential to produce up to 5% of electricity requirements if all possible forms are exploited, including household and industrial waste, sewage sludge (for digestion), agricultural waste (chicken litter, straw, sugar cane, and so on). The use of good farmland to grow energy crops is controversial as it obviously reduces the area of land available to grow food.

1.5.5 Geothermal Energy

In most parts of the world the vast amount of heat in the earth's interior is too deep to be tapped. In some areas, however, hot springs or geysers and molten lava streams are close enough to the surface to be used. Thermal energy from hot springs has been used for many years for producing electricity, starting in 1904 in Italy. In the USA the major geothermal power plants are located in northern California on a natural steam field called the Geysers. Steam from a number of wells is passed through turbines. The present utilization is about 900 MW and the total estimated capacity is about 2000 MW. Because of the lower pressure and temperatures the efficiency is less than with fossil-fuelled plants, but the capital costs are less and, of course, the fuel is free. New Zealand and Iceland also exploit their geothermal energy resources.

1.5.6 Other Renewable Resources

1.5.6.1 Tides

An effective method of utilizing the tides is to allow the incoming tide to flow into a basin, thus operating a set of turbines, and then at low tide to release the stored water, again operating the turbines. If the tidal range from high to low water is h (m) and the area of water enclosed in the basin is A (m^2), then the energy in the full basin with the tide outside at its lowest level is:

$$E = \rho g A \int_0^h x\, dx$$

$$= \frac{1}{2} \rho g h^2 A \qquad [J]$$

Table 1.3 Sites that have been studied for tidal range generation

Site	Tidal Range(m)	Area (km²)	Generators (MW)
Passamaquoddy Bay (N. America)	5.5	262	1800
Minas-Cohequid (N. America)	10.7	777	19 900
San Jose (S. America)	5.9	750	5870
Severn (U.K.)	9.8	700	8000

The maximum total energy for both flows is therefore twice this value, and the maximum average power is $pgAh^2/T$, where T is the period of tidal cycle, normally 12 h 44 min. In practice not all this energy can be utilized. The number of sites with good potential for tidal range generation is small. Typical examples of those which have been studied are listed in Table 1.3 together with the size of generating plant considered.

A 200 MW installation using tidal flow has been constructed on the La Rance Estuary in northern France, where the tidal height range is 9.2 m (30 ft) and the tidal flow is estimated at 18 000 m³/s. Proposals for a 8000 MW tidal barrage in the Severn Estuary (UK) were first discussed in the nineteenth century and are still awaiting funding.

The utilization of the energy in tidal flows has long been the subject of attention and now a number of prototype devices are undergoing trials, Figure 1.11. The technical and economic difficulties are considerable and there are only a limited number of locations where such schemes are feasible.

1.5.6.2 Wave Power

The energy content of sea waves is very high. The Atlantic waves along the northwest coast of Britain have an average energy value of 80 kW/m of wave crest length. The energy is obviously very variable, ranging from greater than 1 MW/m for 1% of the year to near zero for a further 1%. Over several hundreds of kilometres a vast source of energy is available.

The sea motion can be converted into mechanical energy in several ways with a number of innovative solutions being trialled, Figure 1.12. An essential attribute of any wave power device is its survivability against the extreme loads encountered during storms.

1.6 Energy Storage

The tremendous difficulty in storing electricity in any large quantity has shaped the architecture of power systems as they stand today. Various options exist for the large-scale storage of energy to ease operation and affect overall economies. However, energy storage of any kind is expensive and incurs significant power losses. Care must be taken in its economic evaluation.

The options available are as follows: pumped storage, compressed air, heat, hydrogen gas, secondary batteries, flywheels and superconducting coils.

1.6.1 Pumped Storage

Very rapid changes in load may occur (for example 1300 MW/min at the end of some programmes on British TV) or the outage of lines or generators. An instantaneous loss of 1320 MW of generation (two 660 MW generating units) is considered when planning the operation of the Great Britain system. Hence a considerable amount of conventional steam plant must operate partially loaded to respond to these events. This is very expensive because there is a fixed heat loss for a steam turbogenerator regardless of output, and the efficiency of a thermal generating unit is reduced at part load. Therefore a significant amount of energy storage capable of instantaneous use would be an effective method of meeting such loadings, and by far the most important method to date is that of pumped storage.

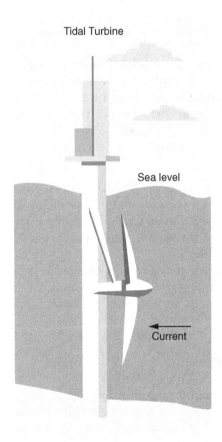

Figure 1.11 Tidal stream energy (Figure adapted from Marine Current Turbines)

A pumped storage scheme consists of an upper and a lower reservoir and turbine-generators which can be used as both turbines and pumps. The upper reservoir typically has sufficient storage for 4–6 hours of full-load generation.

The sequence of operation is as follows. During times of peak load on the power system the turbines are driven by water from the upper reservoir and the electrical machines generate in the normal manner. During the night, when only base load stations are in operation and electricity is being produced at its cheapest, the water in the lower reservoir is pumped back into the higher one ready for the next day's peak load. At these times of low network load, each generator changes to synchronous motor action and, being supplied from the general power network, drives its turbine which now acts as a pump.

Typical operating efficiencies attained are:

- Motor and generator 96%
- Pump and turbine 77%
- Pipeline and tunnel 97%
- Transmission 95%

giving an overall efficiency of 68%. A further advantage is that the synchronous machines can be easily used as synchronous compensators to control reactive power if required.

A large pumped hydro scheme in Britain uses six 330 MVA pump-turbine (Francis-type reversible) generator-motor units generating at 18 kV. The flow of water and hence power output is controlled by guide vanes associated with the turbine. The maximum pumping power is 1830 MW. The machines are 92.5% efficient as turbines and 91.7% efficient as pumps giving an exceptionally high round trip efficient of 85%. The operating speed of the 12-pole electrical machines is 500 r.p.m. Such a plant can be used to provide fine frequency control for the whole British system. The machines will be expected to start and stop about 40 times a day as well as

Figure 1.12 Wave power generation (Figure adapted from Pelamis)

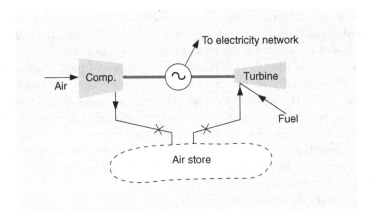

Figure 1.13 Storage using compressed air in conjunction with a gas turbine generator

provide frequency response in the event of a sudden load pick up or tripping of other generators.

1.6.2 Compressed-Air Storage

Air is pumped into large receptacles (e.g. underground caverns or old mines) at night and used to drive gas turbines for peak, day loads. The energy stored is equal to the product of the air pressure and volume. The compressed air allows fuel to be burnt in the gas turbines at twice the normal efficiency. The general scheme is illustrated in Figure 1.13. A German utility has installed a 290 MW scheme. In one discharge/charge cycle it generated 580 MWh of on-peak electricity and consumed 930 MWh of fuel plus 480 MWh of off-peak electricity. A similar plant has been installed in the USA. One disadvantage of these schemes is that much of the input energy to the compressed air manifests itself as heat and is wasted. Heat could be retained after compression, but there would be possible complications with the store walls rising to a temperature of 450 °C at 20 bar pressure. A solution would be to have a separate heat store that could comprise stacks of stones or pebbles which store heat cheaply and effectively. This would enable more air to be stored because it would now be cool. At 100 bar pressure, approximately 30 m^3 of air is stored per MWh output.

1.6.3 Secondary Batteries

Although demonstrated in a number of pilot projects (for example, a 3 MW battery storage plant was installed in Berlin for frequency control in emergencies and a 35 MW battery system is used to smooth the output of a wind farm in Japan) the large-scale use of battery storage remains expensive and the key area where the use of secondary batteries is likely to have impact is in electric vehicles. The popular lead-acid cell, although reasonable in price, has a low energy density (15 Wh/kg). Nickel-cadmium cells are better (40 Wh/kg) but more

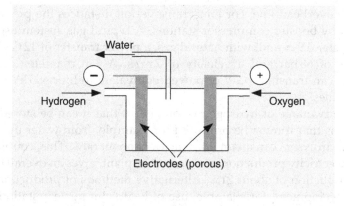

Figure 1.14 Hydrogen-oxygen fuel cell

expensive. Still under intensive development and demonstration is the sodium-sulphur battery (200 Wh/kg), which has a solid electrolyte and liquid electrodes and operates at a temperature of 300 °C. Modern electric vehicles use Lithium ion batteries (100–200 Wh/kg) but these remain expensive. Other combinations of materials are under active development in attempts to increase output and storage per unit weight and cost.

1.6.4 Fuel Cells

A fuel cell converts chemical energy to electrical energy by electrochemical reactions. Fuel is continuously supplied to one electrode and an oxidant (usually oxygen) to the other electrode. Figure 1.14 shows a simple hydrogen-oxygen fuel cell, in which hydrogen gas diffuses through a porous metal electrode (nickel). A catalyst in the electrode allows the absorption of H_2 on the electrode surface as hydrogen ions which react with the hydroxyl ions in the electrolyte to form water $(2H_2 + O_2 \rightarrow 2H_2O)$. A theoretical e.m.f. of 1.2 V at 25 °C is obtained. Other fuels for use with oxygen are carbon monoxide (1.33 V at 25 °C), methanol (1.21 V at 25 °C), and methane (1.05 V at 25 °C). In practical cells, conversion efficiencies of 80% have been attained. A major use of the fuel cell could be in conjunction with a future hydrogen energy system.

Intensive research and development is still proceeding on various types of fuel cell – the most successful to date for power generation being the phosphoric fuel cell. A demonstration unit used methane as the input fuel and operated at about 200–300 °C to produce 200 kW of electrical power plus 200 kW of heat energy, with overall efficiency of around 80%. Compared with other forms of energy conversion, fuel cells have the potential of being up to 20% more efficient. Much attention is now being given to the high-temperature molten carbonate cell which has a high efficiency.

1.6.5 Hydrogen Energy Systems

The transmission capacity of a pipe carrying natural gas (methane) is high compared with electrical links, the installed cost being about one tenth of an equivalent capacity H.V. overhead line. For long transmission distances the pressure drop is compensated by booster compressor stations. A typical gas system uses a pipe of internal diameter 0.9 m and, with natural gas, a power transfer of 12 GW is possible at a pressure of 68 bar and a velocity of 7 m/s. A 1 m diameter pipe carrying hydrogen gas can transmit 8 GW of power, equivalent to four 400 kV, three phase transmission lines.

The major advantage of hydrogen is, of course, that it can be stored; the major disadvantage is that it must be produced for example, from water by electrolysis. Very large electrolysers can attain efficiencies of about 60%. This, coupled with the efficiency of electricity production from a nuclear plant, gives an overall efficiency of hydrogen production of about 21%. Alternative methods of production are under laboratory development, for example, use of heat from nuclear stations to 'crack' water and so release hydrogen; however, temperatures of 3000 °C are required.

1.6.6 Superconducting Magnetic Energy Stores (SMES)

Continuing development of high-temperature superconductors, where the transition temperature can be around 60–80 K (K is degrees Kelvin where 0 K is absolute zero and 273 K is 0 °C) has led to the possibility of storing energy in the magnetic field produced by circulating a large current (over 100 kA) in an inductance. For a coil of inductance L in air, the stored energy is given by

$$E = \frac{1}{2}LI^2 \qquad [J]$$

A big advantage of the high-temperature superconductor is that cooling by liquid nitrogen can be used, which is far cheaper than using helium to reach temperatures closer to absolute zero. Initially, it is expected that commercial units will be used to provide an uninterruptible supply for sensitive loads to guard against voltage sags or to provide continuity whilst emergency generators are started. Another use in transmission networks would be to provide fast response for enhanced transient stability and improved power quality.

1.6.7 Flywheels

The most compact energy store known is that of utilizing high-speed flywheels. Such devices coupled to an electrical generator/motor have been employed in buses on an experimental basis and also in special industrial applications. For power systems, very large flywheels constructed of composite high-tensile resisting materials have been proposed, but their cost and maintenance problems have so far ruled them out of economic contention compared with alternative forms of energy supply.

1.6.8 Supercapacitors

The interface between an anode and cathode immersed in an electrolyte has a very high permittivity. This property can be exploited in a capacitor to produce a 25 V capsule with a capacitance of 0.1 F. Many units in series and parallel would have the capability of storing many MWh of energy, which can be quickly released for transient control purposes. To date, higher voltage forms of a commercially useful device have not got beyond their employment for pulse or actuator applications.

1.7 Environmental Aspects of Electrical Energy

Increasingly, environmental considerations influence the development of energy resources, especially those involving electricity production, transmission, and distribution. Conversion of one form of energy to another produces unwanted side effects and, often, pollutants which need to be controlled and disposed of. In addition, safety and health are subject to increasing legislation by national and international bodies, thereby requiring all engineers to be aware of the laws and regulations governing the practice of their profession.

It must be appreciated that the extraction of fossil fuels from the earth is not only a hazardous business but also, nowadays, one controlled through licensing by governments and state authorities. Hydro plants require careful study and investigation through modelling, widespread surveys, and environmental impact statements to gain acceptance. Large installations of all kinds require, often lengthy, planning enquiries which are both time consuming and expensive, thereby delaying the start up of energy extraction and production. As a consequence, methods of producing electrical energy which avoid or reduce the enquiry process are to be favoured over those needing considerable consultation before receiving the go ahead. This is likely to favour small-scale projects or the redevelopment of existing sites where industry or production facilities are already operating.

In recent years, considerable emphasis has been placed on 'sustainable development', by which is meant the use of technologies that do not harm the environment, particularly in the long term. It also implies that anything we do now to affect the environment should be recoverable by future generations. Irreversible damage, for example, damage to the ozone layer or increase in CO_2 in the atmosphere, should be avoided.

1.7.1 Global Emissions from Fossil Fuelled Power Stations

It is generally accepted that the burning of fossil fuels and the subsequent emission of greenhouse gases, particularly CO_2, is leading to climate change and potentially catastrophic increase in the earth's temperature. Hence concern over the emission of greenhouse gases is a key element of energy policy and, in Europe, is recognized through the EU Emissions Trading Scheme which requires major emitters of CO_2, such as power stations, to purchase permits to emit CO_2. This has the effect of making high carbon generation (particularly from coal) increasingly expensive.

1.7.2 Regional and Local Emissions from Fossil Fuelled Power Stations

Fossil fuelled power plants produce sulphur oxides, particulate matter, and nitrogen oxides. Of the former, sulphur dioxide accounts for about 95% and is a by-product of the combustion of coal or oil. The sulphur content of coal varies from 0.3 to 5%. Coal can only be used for generation in some US states if it is below a certain percentage sulphur.

In the eastern USA this has led to the widespread use of coal from western states because of its lower sulphur content or the use of gas as an alternative fuel. Sulphur dioxide forms sulphuric acid (H_2SO_4) in the air which causes damage to buildings and vegetation. Sulphate concentrations of 9–10 $\mu g/m^3$ of air aggravate asthma and lung and heart disease. This level has been frequently exceeded in the past, a notorious episode being the London fog of 1952 (caused by domestic coal burning). It should be noted that although sulphur does not accumulate in the air it does so in the soil.

Sulphur oxide emission can be controlled by:

- the use of fuel with less than, say, 1% sulphur;
- the use of chemical reactions to remove the sulphur, in the form of sulphuric acid, from the combustion products, for example limestone scrubbers, or fluidized bed combustion;
- removing the sulphur from the coal by gasification or flotation processes.

European legislation limits the amount of SO_2, NO_x, and particulate emission, as in the USA. This has led to the retrofitting of flue gas desulphurization (FGD) scrubbers to coal burning plants. Without such equipment coal fired power stations must be retired. Emissions of NO_x can be controlled by fitting advanced technology burners which can ensure a more complete combustion process, thereby reducing the oxides going up the stack (chimney).

Particulate matter, particles in the air, is injurious to the respiratory system, in sufficient concentration, and by weakening resistance to infection may well affect the whole body. Apart from settling on the ground or buildings to produce dirt, a further effect is the reduction of the solar radiation entering the polluted area. Reported densities (particulate mass in $1\,m^3$ of air) are 10 $\mu g/m^3$ in rural areas rising to 2000 $\mu g/m^3$ in polluted areas. The average value in US cities is about 100 $\mu g/m^3$.

About one-half of the oxides of nitrogen in the air in populated areas are due to power plants and originate in high-temperature combustion processes. At levels of 25–100 parts per million they can cause acute bronchitis and pneumonia. Increasingly, city pollutants are due to cars and lorries and not power plants.

A 1000 MW(e) coal plant burns approximately 9000t of coal per day. If this has a sulphur content of 3% the amount of SO_2 emitted per year is 2×10^5t. Such a plant produces the following pollutants per hour (in kg): CO_2 8.5×10^5, CO 0.12×10^5, sulphur oxides 0.15×10^5, nitrogen oxides 3.4×10^3, and ash.

Both SO_2 and NO_x are reduced considerably by the use of FGD, but at considerable cost and reduction in the efficiency of the generating unit caused by the power used by the scrubber. Gas-fired CCGT plants produce very little NO_x or SO_2 and their CO_2 output is about 55% of an equivalent size coal-fired generator.

The concentration of pollutants can be reduced by dispersal over a wider area by the use of high stacks. If, in the stack, a vertical wire is held at a high negative potential relative to the wall, the expelled electrons from the wire are captured by the gas molecules moving up the stack. Negative ions are formed which accelerate to the wall, collecting particles on the way. When a particle hits the wall the charge is neutralized and the particle drops down the stack and is collected. Precipitators have particle-removing (by weight) efficiencies of up to 99%, but this is misleading as performance is poor for small particles; of, say, less than 0.1 μm in diameter. The efficiency based on number of particles removed is therefore less. Disposal of the resulting fly-ash is expensive, but the ash can be used for industrial purposes, for example, building blocks. Unfortunately, the efficiency of precipitators is enhanced by reasonable sulphur content in the gases. For a given collecting area the efficiency decreases from 99% with 3% sulphur to 83% with 0.5% sulphur at 150 °C. This results in much larger and more expensive precipitator units with low-sulphur coal or the use of fabric filters in 'bag houses' situated before the flue gas enters the stack.

1.7.3 Thermal Pollution from Power Stations

Steam from the low-pressure turbine is liquefied in the condenser at the lowest possible temperatures to maximize the steam-cycle efficiency. Where copious supplies of water exist the condenser is cooled by 'once-through' circulation of sea- or river-water. Where water is more restricted in availability, for example, away from the coasts, the condensate is circulated in cooling towers in which it is sprayed in nozzles into a rising volume of air. Some of the water is evaporated, providing cooling. The latent heat of water is 2×10^6 J/kg compared with a sensible heat of 4200 J/kg per degree C. A disadvantage of such towers is the increase in humidity produced in the local atmosphere.

Dry cooling towers in which the water flows through enclosed channels (similar to a car radiator), past which air is blown, avoid local humidity problems, but at a much higher cost than 'wet towers'. Cooling towers emit evaporated water to the atmosphere in the order of 75 000 litres/min for a 1000 MW(e) plant.

A crucial aspect of once-through cooling in which the water flows directly into the sea or river is the increased temperature of the natural environment due to the large volume per minute (typically 360 m³/s for a coolant rise of 2.4 °C for a 2.4 GW nuclear station) of heated coolant. Because of their lower thermal efficiency, nuclear power stations require more cooling water than fossil-fuelled plants. Extreme care must be taken to safeguard marine life, although the higher temperatures can be used effectively for marine farming if conditions can be controlled.

1.7.4 Electromagnetic Radiation from Overhead Lines, Cables and Equipment

The biological effects of electromagnetic radiation have been a cause of considerable concern amongst the general public as to the possible hazards in the home and the

Table 1.4 Typical electric and magnetic field strengths directly under overhead lines[a]. Note that the magnetic field depends upon the current carried

Line Voltage (kV)	Electric Field Strength (V/m)	Magnetic Flux Density (μT)
400 and 275 tower lines	3000–5000	5–10
132 tower line	1000–2000	0.5–2
33 and 11 wood pole	200	0.2–0.5
UK recommendations on exposure limits (to the public)	9000	360
Earth's magnetic field	—	40–50

Data from http://www.emfs.info/. This site is maintained by the National Grid Company, reproduced with permission.

workplace. Proximity of dwellings to overhead lines and even buried cables has also led to concerns of possible cancer-inducing effects, with the consequence that research effort has been undertaken to allay such fears. In general it is considered that the power frequencies used (50 or 60 Hz) are not harmful. The electric field and magnetic field strengths below typical HV transmission lines are given in Table 1.4.

Considerable international research and cooperative investigation has now been proceeding for over 40 years into low-frequency electric and magnetic field exposure produced by household appliances, video display terminals, and local power lines. To date there is no firm evidence that they pose any demonstrable health hazards. Epidemiologic findings of an association between electric and magnetic fields and childhood leukaemia or other childhood or adult cancers are inconsistent and inconclusive; the same is likely to be true of birth defects or other reproductive problems.

1.7.5 Visual and Audible Noise Impacts

The presence of overhead lines constitutes an environmental problem (perhaps the most obvious one within a power system) on several counts.

1. Space is used which could be used for other purposes. The land allocated for the line is known as the right of way (or wayleave in Britain). The area used for this purpose is already very appreciable.
2. Lines are considered by many to mar the landscape. This is, of course, a subjective matter, but it cannot be denied that several tower lines converging on a substation or power plant, especially from different directions, is offensive to the eye.
3. Radio interference, audible noise, and safety considerations must also be considered.

Although most of the above objections could be overcome by the use of underground cables, these are not free of drawbacks. The limitation to cable transmitting current because of temperature-rise considerations coupled with high manufacture and installation costs results in the ratio of the cost of transmission underground to that for overhead transmission being between 10 and 20 at very high operating voltages. With novel cables, such as superconducting, still under development it is hoped to reduce this disadvantage. However, it may be expected that in future a larger proportion of circuits will be placed underground, especially in suburban areas. In large urban areas circuits are invariably underground, thereby posing increasing problems as load densities increase.

Reduction of radio interference can be achieved in the same way that electromagnetic effects are reduced, but audible noise is a function of the line design. Careful attention to the tightness of joints, the avoidance of sharp or rough edges, and the use of earth screen shielding can reduce audible noise to acceptable levels at a distance dependent upon voltage.

Safety clearances dependent upon International Standards are, of course, extremely important and must be maintained in adverse weather conditions. The most important of these is the increased sag due to ice and snow build up in winter and heavy loading of the circuits in summer.

1.8 Transmission and Distribution Systems

1.8.1 Representation

Modern electricity supply systems are invariably three-phase. The design of transmission and distribution networks is such that normal operation is reasonably close to balanced three-phase working, and often a study of the electrical conditions in one phase is sufficient to give a complete analysis. Equal loading on all three phases of a network is ensured by allotting, as far as possible, equal domestic loads to each phase of the low-voltage distribution feeders; industrial loads usually take three-phase supplies.

A very useful and simple way of graphically representing a network is the schematic or line diagram in which three-phase circuits are represented by single lines. Certain conventions for representing items of plant are used and these are shown in Figure 1.15.

A typical line or schematic diagram of a part of a power system is shown in Figure 1.16. In this, the generator is star connected, with the star point connected to earth through a resistance. The nature of the connection of the star point of rotating machines and transformers to earth is of vital importance when considering faults which produce electrical imbalance in the three phases. The generator feeds two three-phase circuits (overhead or underground). The line voltage is increased from that at the generator terminals by transformers connected as shown. At the end of the lines the voltage is reduced for the secondary distribution of power. Two lines are provided to improve the security of the supply that is, if one line develops a fault and has to be switched out the remaining one still delivers power to the receiving end. It is not necessary in straightforward current and voltage calculations to indicate the presence of switches on the diagrams, but in some cases, such as stability calculations, marking the location of switches, current transformers, and protection is very useful.

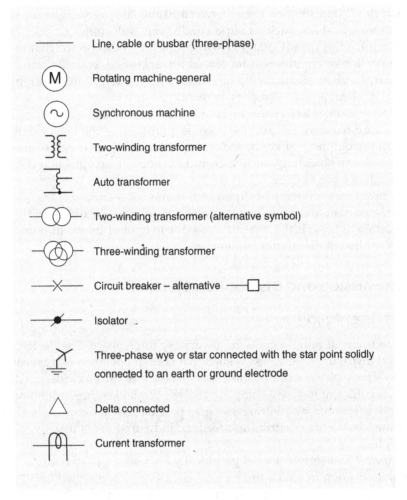

————	Line, cable or busbar (three-phase)
Ⓜ	Rotating machine-general
(∿)	Synchronous machine
ƷE	Two-winding transformer
	Auto transformer
—ⵛ—	Two-winding transformer (alternative symbol)
—ⵥ—	Three-winding transformer
—✕—	Circuit breaker – alternative —▢—
—•—	Isolator
	Three-phase wye or star connected with the star point solidly connected to an earth or ground electrode
△	Delta connected
—ⱱ—	Current transformer

Figure 1.15 Symbols for representing the components of a three-phase power system

A short list of terms used to describe power systems follows, with explanations.

System: This is used to describe the complete electrical network: generators, circuits, loads, and prime movers.

Load: This may be used in a number of ways: to indicate a device or collection of devices which consume electricity; to indicate the power required from a given supply circuit; to indicate the power or current being passed through a line or machine.

Busbar: This is an electrical connection of zero impedance (or node) joining several items, such as lines, generators or loads. Often this takes the form of actual busbars of copper or aluminium.

Earthing (Grounding): This is connection of a conductor or frame of a device to the main body of the earth. This must be done in such a manner that the resistance

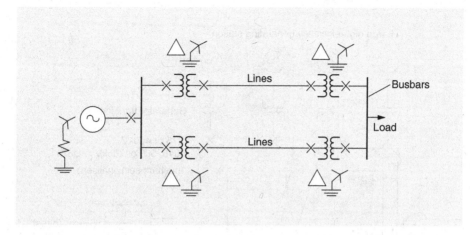

Figure 1.16 Line diagram of a simple system

between the item and the earth is below prescribed limits. This often entails the burying of the large assemblies of conducting rods in the earth and the use of connectors of a large cross sectional area. It is usual to earth the neutral point of 3-phase circuits at least once at each voltage level.

Fault: This is a malfunctioning of the network, usually due to the short-circuiting of conductors phase-phase or phase-earth.

Outage: Removal of a circuit either deliberately or inadvertently.

Security of Supply: Provision must be made to ensure continuity of supply to consumers, even with certain items of plant out of action. Usually, two circuits in parallel are used and a system is said to be secure when continuity is assured. This is obviously the item of first priority in design and operation.

1.8.2 Transmission

Transmission refers to the bulk transfer of power by high-voltage links between central generation and load centres. Distribution, on the other hand, describes the conveyance of this power to consumers by means of lower voltage networks.

Generators usually produce voltages in the range 11–25 kV, which is increased by transformers to the main transmission voltage. At substations the connections between the various components of the system, such as lines and transformers, are made and the switching of these components is carried out. Large amounts of power are transmitted from the generating stations to the load-centre substations at 400 kV and 275 kV in Britain, and at 765, 500 and 345 kV in the USA. The network formed by these very high-voltage lines is sometimes referred to as the Supergrid. Most of the large and efficient generating stations feed through transformers directly into this network. This grid, in turn, feeds a sub-transmission network operating at 132 kV in Britain and 115 kV in the USA. In Britain the lower voltage networks operate at 33, 11, or 6.6 kV and supply the final consumer feeders at 400 V three-phase, giving 230 V between phase and neutral. Other voltages exist

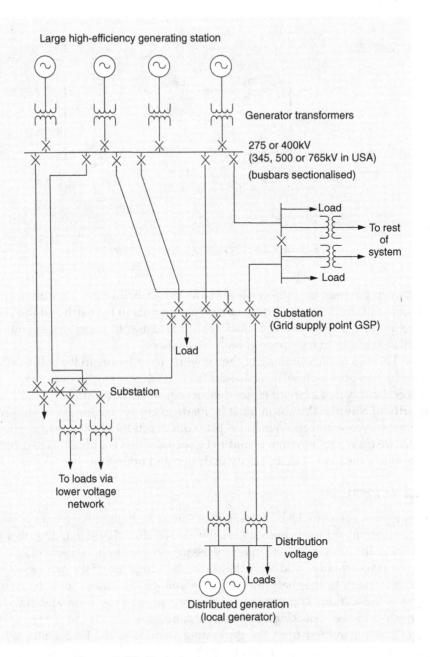

Large high-efficiency generating station

Generator transformers

275 or 400kV
(345, 500 or 765kV in USA)
(busbars sectionalised)

Load

To rest
of
system

Load

Substation
(Grid supply point GSP)

Load

Substation

To loads via
lower voltage
network

Distribution
voltage

Loads

Distributed generation
(local generator)

Figure 1.17 Part of a typical power system

in isolation in various places, for example the 66 and 22 kV London cable systems. A typical part of a supply network is shown schematically in Figure 1.17. The power system is thus made up of networks at various voltages. There exist, in effect, voltage tiers as represented in Figure 1.18.

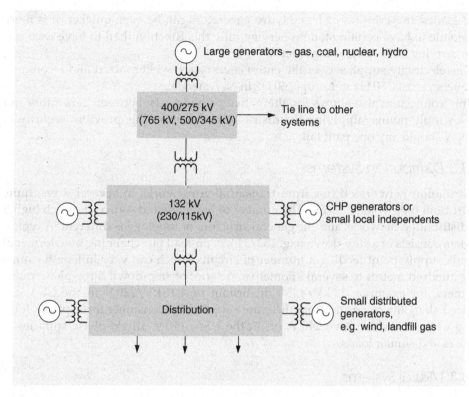

Figure 1.18 Schematic diagram of the constituent networks of a supply system. USA voltages in parentheses

Summarizing, transmission networks deliver to wholesale outlets at 132 kV and above; sub-transmission networks deliver to retail outlets at voltages from 115 or 132 kV, and distribution networks deliver to final step-down transformers at voltages below 132 kV, usually operated as radial systems.

1.8.2.1 Reasons for Interconnection

Many generating sets are large and power stations of more than 2000 MW are used to provide base load power. With CCGT units, their high efficiency and cheap long-term gas contracts mean that it is often more economic to use these efficient stations to full capacity 24 h a day and transmit energy considerable distances than to use less efficient more local stations. The main base load therefore is met by these high-efficiency stations which must be interconnected so that they feed into the general system and not into a particular load.

To meet sudden increases in load a certain amount of generating capacity, known as the spinning reserve, is required. This consists of part-loaded generators synchronized with the system and ready to supply power instantaneously. If the machines are stationary a reasonable time is required (especially for steam turbo-alternators) to run up to speed; this can approach 6 h, although small gas turbines can be started and loaded in 3 minutes or less. Hydro generators can be even quicker. It is more economic to have certain stations serving only this function than to have each station carrying its own spinning reserve.

The electricity supplies over the entire country are synchronized and a common frequency exists: 50 Hz in Europe, 60 Hz in N. America.

Interconnection also allows for alternative paths to exist between generators and bulk supply points supplying the distribution systems. This provides security of supply should any one path fail.

1.8.3 Distribution Systems

Distribution networks differ from transmission networks in several ways, quite apart from their voltage levels. The number of branches and sources is much higher in distribution networks and the general structure or topology is different. A typical system consists of a step-down (e.g. 132/11 kV) on-load tap-changing transformer at a bulk supply point feeding a number of circuits which can vary in length from a few hundred metres to several kilometres. A series of step-down three-phase transformers, for example, 11 kV/433 V in Britain or 4.16 kV/220 V in the USA, are spaced along the route and from these are supplied the consumer three-phase, four-wire networks which give 240 V, or, in the USA, 110 V, single-phase supplies to houses and similar loads.

1.8.3.1 Rural Systems

In rural systems, loads are relatively small and widely dispersed (5–50 kVA per consumer group is usual). In Great Britain a predominantly overhead line system at 11 kV, three-phase, with no neutral or single phase for spurs from the main system is used. Pole-mounted transformers (5–200 kVA) are installed, protected by fuses which require manual replacement after operation; hence rapid access is desirable by being situated as close to roads as possible. Essentially, a radial system is supplied from one step-down point; distances up to 10–15 miles (16–24 km) are feasible with total loads of 500 kVA or so (see Figure 1.19), although in sparsely populated areas, distances of 50 miles may be fed by 11 kV. Single-phase earth-return systems operating at 20 kV are used in some developing countries.

Over 80% of faults on overhead distribution systems are transitory due to flashover following some natural or man-made cause. This produces unnecessary fuse-blowing unless auto-reclosers are employed on the main supply; these have been used with great success in either single- or three-phase form. The principle, as shown in Figure 1.20, is to open on fault before the fuse has time to operate and to reclose after 1–2 s. If the fault still persists, a second attempt is made to clear, followed by another reclose. Should the fault still not be cleared, the recloser remains closed for a longer period to blow the appropriate protective fuse (for example, on a spur line or a transformer). If the fault is still not cleared, then the recloser opens and locks out to await manual isolation of the faulty section. This process requires the careful coordination of recloser operation and

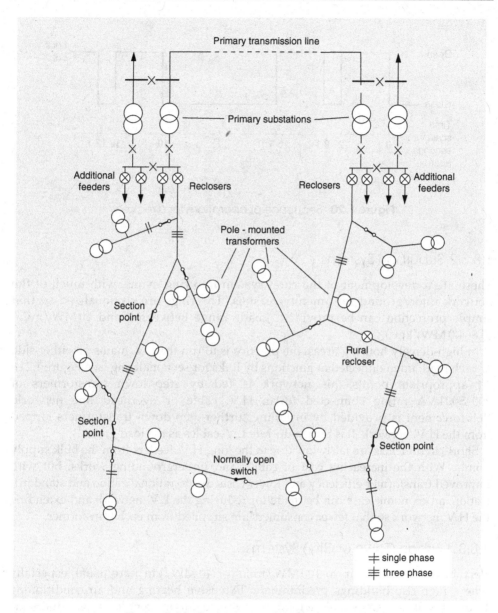

Figure 1.19 Typical rural distribution system

fuse-blowing characteristics where time grading is important. Section switches operated by radio or tele-command enable quick resupply routes to be established following a faulty section isolation.

Good earthing at transformer star points is required to prevent overvoltages at consumers' premises. Surge protection using diverters or arcing horns is essential in lightning-prone areas.

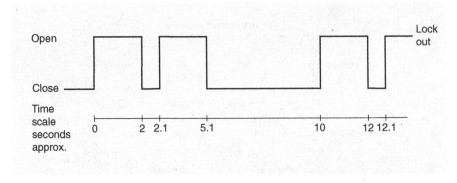

Figure 1.20 Sequence of operations for a recloser

1.8.3.2 Suburban Systems

These are a development of the rural system into ring mains, with much of the network underground for amenity reasons. The rings are sectionalized so that simple protection can be provided. Loads range between 2 and 10 MW/mile² (0.8–4.0 MW/km²).

In high-density housing areas, the practice is to run the L.V. mains on either side of each road, interconnected at junctions by links for sectionalizing (see Figure 1.21). At appropriate points this network is fed by step-down transformers of 200–500 kVA rating connected to an H.V. cable or overhead line network. Reinforcement is provided by installing further step-down transformers tapped from the H.V. network. It is rare to up-rate L.V. cables as the load grows.

Short-circuit levels are fairly low due to the long H.V. feeders from the bulk supply points. With the increasing cost of cables and undergrounding works, but with improved transformer efficiency and lower costs due to rationalization and standardization, an economic case can be made for reducing the L.V. network and extending the H.V. network so that fewer consumers are supplied from each transformer.

1.8.3.3 Urban (Town or City) Systems

Very heavy loadings (up to 100 MW/mile² or 40 MW/km²) are usual, especially where high-rise buildings predominate. Extensive heating and air-conditioning loads as well as many small motors predominate. Fluorescent lighting reduces the power factor and leads to some waveform distortion, but computer and TV loads and power electronic motor drives now cause considerable harmonics on all types of network.

Again, a basic L.V. grid, reinforced by extensions to the H.V. network, as required, produces minimum costs overall. The H.V. network is usually in the form of a ring main fed from two separate sections of a double busbar substation where 10–60 MVA transformers provide main supply from the transmission system (see Figure 1.22). The H.V. network is sectionalized to contain short-circuit levels and to ease protection grading. A high security of supply is possible by

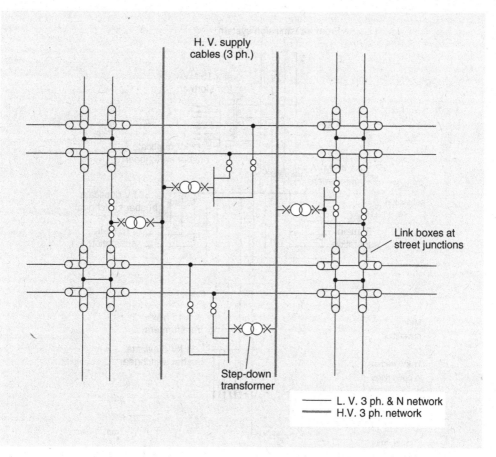

Figure 1.21 Principle of suburban distribution system

overlapping H.V. rings so that the same L.V. grid is fed from several transformers supplied over different routes. Failure of one portion of the H.V. system does not affect consumers, who are then supported by the L.V. network from adjacent H.V. supplies.

In the UK transformers of 500 or 1000 kVA rating are now standard, with one H.V. circuit breaker or high rupturing capacity (HRC) fuse-switch and two isolators either side to enable the associated H.V. cable to be isolated manually in the event of failure. The average H.V. feeder length is less than 1 mile and restoration of H.V. supplies is usually obtainable in under 1 h. Problems may arise due to back-feeding of faults on the H.V. system by the L.V. system, and in some instances reverse power relay protection is necessary.

In new urban developments it is essential to acquire space for transformer chambers and cable access before plans are finalized. High-rise buildings may require substations situated on convenient floors as well as in the basement.

Apart from the supply of new industrial and housing estates or the electrification of towns and villages, a large part of the work of a planning engineer

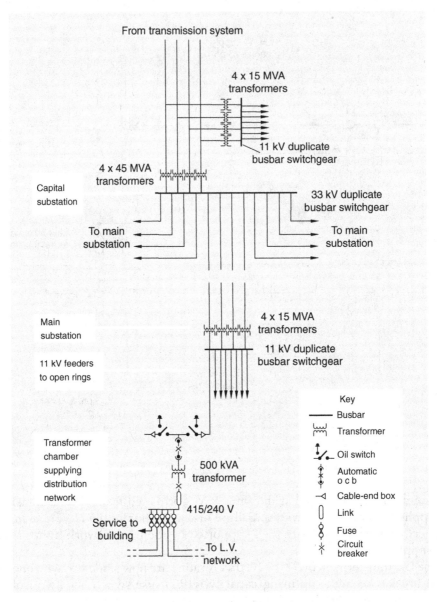

Figure 1.22 Typical arrangement of supply to an urban network–British practice

is involved with the up-rating of existing supplies. This requires good load forecasting over a period of 2–3 years to enable equipment to be ordered and access to sites to be established. One method of forecasting is to survey the demand on transformers by means of a maximum-demand indicator and to treat any transformer which has an average load factor of 70% or above as requiring up-rating over the next planning period. Another method is to analyze consumers' bills, sectionalized into areas and distributors and, by surveys and computer analysis, to relate energy consumed to maximum demand. This method can be particularly useful and quite economic where computerized billing is used. The development of the Smart Grid is leading to much greater monitoring being installed on the 400 V network and control on the 11 kV system.

In practice, good planning requires sufficient data on load demands, energy growth, equipment characteristics, and protection settings. All this information can be stored and updated from computer files at periodic intervals and provides the basis for the installation of adequate equipment to meet credible future demands without unnecessary load shedding or dangerous overloading.

1.8.4 Typical Power Systems

Throughout the world the general form of power systems follows the same pattern. Voltage levels vary from country to country, the differences originating mainly from geographical and historical reasons.

Several frequencies have existed, although now only two values – 50 and 60 Hz – remain: 60 Hz is used on the American continent, whilst most of the rest of the world uses 50 Hz, although Japan still has 50 Hz for the main island and 60 Hz for the northern islands. The value of frequency is a compromise between higher generator speeds (and hence higher output per unit of machine volume) and the disadvantage of high system reactance at higher frequencies. Historically, the lower limit was set by the need to avoid visual discomfort caused by flicker from incandescent electric lamps.

The distance that a.c. transmission lines can transfer power is limited by the maximum permissible peak voltage between conductor and ground. As voltages increase, more and more clearance must be allowed in air to prevent the possibility of flashover or danger to people or animals on the ground. Unfortunately, the critical flashover voltage increases non-linearly with clearance, such that, with long clearances, proportionally lower peak voltages can be used safely. Figure 1.23 illustrates this effect.

The surge impedance of a line, described for a lossless line where the resistance and conductance are zero, is given by $Z_0 = \sqrt{L/C}$, where L is the series inductance and C is the shunt capacitance per unit length of the line. When a line is terminated in a resistive load equal to its surge impedance then the reactive power $I^2 X_L$ drawn by the line is balanced by the reactive power generated by the line capacitance V^2/X_C. The surge impedance loading is V^2/Z_0, typical values are shown in Table 1.5.

The curve given in Figure 1.24 shows that, in practice, short lines are usually loaded above their surge impedance loading (SIL) but, to ensure stability is maintained, long lines are generally loaded below their SIL. A useful rule-of-thumb to ensure stability is to restrict the phase difference between the sending and receiving end voltages of an EHV transmission circuit to 30°. Lower voltage circuits will have much smaller phase angle differences.

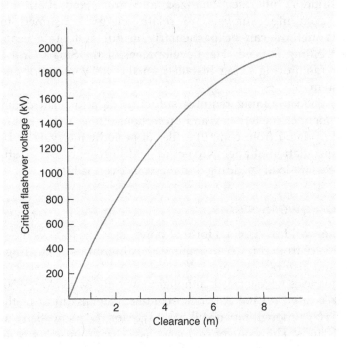

Figure 1.23 Critical flashover voltage for V-string insulators in a window tower (Figure adapted from Edison Electric Institute)

The highest a.c. voltage used for transmission is around 750 kV, although experimental lines above 1000 kV have been built.

The topology of the power system and the voltage magnitudes used are greatly influenced by geography. Very long lines are to be found in North and South American nations and Russia. This has resulted in higher transmission voltages, for example, 765 kV, with the possibility of voltages in the range 1000–1500 kV. In some South American countries, for example, Brazil, large hydroelectric resources have been developed, resulting in very long transmission links. In highly developed countries the available hydro resources have been utilized and a considerable proportion of new generation is from wind. In geographically smaller countries, as exist in Europe, the degree of interconnection is much higher, with shorter transmission distances, the upper voltage being about 420 kV.

Systems are universally a.c. with the use of high-voltage d.c. links for specialist purposes, for example, very long circuits, submarine cable connections, and back-to-back converters to connect different a.c. areas. The use of d.c. has been limited by the high cost of the conversion equipment. This requires overhead line lengths of a few hundred kilometres or cables of 30–100 km to enable the reduced circuit costs to offset the conversion costs.

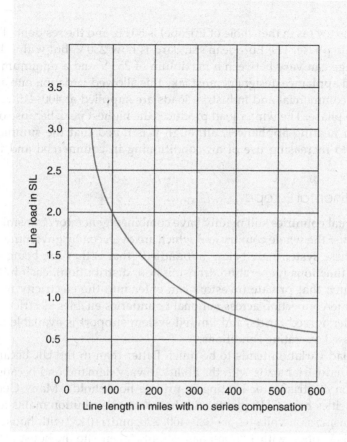

Figure 1.24 Practical transmission-line capability in terms of surge-impedance loading (SIL) (Figure adapted from Edison Electric Institute)

1.8.4.1 USA

The power system in the USA is based on a comparatively few investor owned generation/transmission utilities responsible for bulk transmission as well as operating and constructing new generation facilities when needed. Delivery is to smaller municipal or cooperative rural distribution companies or to the distribution arm of the transmission utility. The industry is heavily regulated by Federal and State Agencies and the investor profits are carefully controlled.

The loads differ seasonally from one part of the country to another and load diversity occurs because of the time zones. Generally, the summer load is the highest due to extensive use of air-conditioning (Figure 1.1(a) and (b)). Consumers are supplied at either 110 V, 60 Hz, single phase for lighting and small-consumption appliances but at 220 V for loads above about 3 kW (usually cookers, water-heaters and air-conditioners). This type of connection normally uses a 220 V centre-tapped transformer secondary to provide the 110 V supply.

1.8.4.2 UK

Here, the frequency (as in the whole of Europe) is 50 Hz and the residential supply is 230–240 V single phase. The European standard is now 230 V, but with ±10% tolerance this voltage can vary between a maximum of 253 V and a minimum of 207 V; equipment and appliance designers must take this allowed variation into account or beware! Most commercial and industrial loads are supplied at 400–415 V or higher voltages, three phase. The winter load produces the highest peak because of the preponderance of heating appliances, although it is noted that the summer load is growing due to increasing use of air-conditioning in commercial and industrial premises.

1.8.4.3 Continental Europe

Many continental countries still mainly have combined generator/transmission utilities, which cover the whole country and which are overseen by government control. Increasingly these systems are being 'unbundled', that is they are being separated into different functions (generation, transmission, distribution), each individually accountable such that private investors can enter into the electricity market (see Chapter 12). Interconnection across national boundaries enables electrical energy to be traded under agreed tariffs, and limited system support is available under disturbed or stressed operating conditions.

The daily load variation tends to be much flatter than in the UK because of the dominance of industrial loads with the ability to vary demand and because there is less reliance on electricity for heating in private households. Many German and Scandinavian cities have CHP plants with hot water distribution mains for heating purposes. Transmission voltages are 380–400, 220, and 110 kV with household supplies at 220–230 V, often with a three-phase supply taken into the house.

1.8.4.4 China and the Pacific Rim

The fastest-growing systems are generally found in the Far East, particularly China, Indonesia, the Philippines and Malaysia. Voltages up to 500/750 kV are used for transmission and 220–240 V are employed for households. Hydroelectric potential is still quite large (particularly in China) but with gas and oil still being discovered and exploited, CCGT plant developments are underway. Increasing interconnection, including by direct current, is being developed.

1.9 Utilization

1.9.1 Loads

The major consumption groups are industrial, residential (domestic) and commercial. Industrial consumption accounts for up to 40% of the total in many industrialized countries and a significant item is the induction motor. The percentage of electricity in the total industrial use of energy is expected to continue to increase due to greater mechanization and the growth of energy intensive industries, such as chemicals and aluminium. In the USA the following six industries account for over 70% of the industrial electricity consumption: metals (25%), chemicals (20%), paper and products (10%), foods (6%), petroleum products (5%) and transportation equipment (5%). Over the past 25 years the amount of electricity per unit of industrial output has increased annually by 1.5%, but this is dependent on the economic cycle. Increase in consumption of electricity in industrialized countries since about 1980 has been no more than 2% per year due largely to the contraction of energy-intensive industries (e.g. steel manufacturing) combined with efforts to load manage and to make better use of electricity. Residential loads are largely made up of refrigerators, fridge-freezers, freezers, cookers (including microwave ovens), space heating, water heating, lighting, and (increasingly in Europe) air-conditioning. Together these loads in the UK amount to around 40% of total load.

The commercial sector comprises offices, shops, schools, and so on. The consumption here is related to personal consumption for services, traditionally a relatively high-growth quantity. In this area, however, conservation of energy measures are particularly effective and so modify the growth rate.

Quantities used in measurement of loads are defined as follows:

Maximum Load: The average load over the half hour of maximum output.
Load Factor: The units of electricity exported by the generators in a given period divided by the product of the maximum load in this period and the length of the period in hours. The load factor should be high; if it is unity, all the plant is being used over all of the period. It varies with the type of load, being poor for lighting (about 12%) and high for industrial loads (e.g. 100% for pumping stations).
Diversity Factor: This is defined as the sum of individual maximum demands of the consumers, divided by the maximum load on the system. This factor measures the diversification of the load and is concerned with the installation of sufficient generating and transmission plant. If all the demands occurred simultaneously, that is, unity diversity factor, many more generators would have to be installed. Fortunately, the factor is much higher than unity, especially for domestic loads.

Table 1.5 Surge impedance loading and charging MVA for EHV overhead lines. The range of values at each line voltage is due to variations in line construction

Line Voltage (kV)	Surge Impedance Loading (MW)	Charging MVAr per 100 miles
230	132–138	27–28
345	320–390	65–81
500	830–910	170–190
700	2150	445
750	2165	450

Data from Edison Electric Institute.

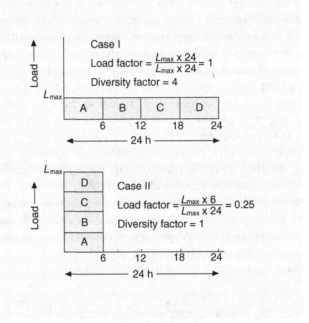

Figure 1.25 Two extremes of load factor and diversity factor in a system with four consumers

A high diversity factor could be obtained with four consumers by compelling them to take load as shown in Case I of Figure 1.25. Although compulsion obviously cannot be used, encouragement can be provided in the form of tariffs. An example is the two-part tariff in which the consumer has to pay an amount dependent on the maximum demand required, plus a charge for each unit of energy consumed. Sometimes the charge is based on kilovoltamperes instead of power to penalize loads of low power factor.

1.9.1.1 Load Management

Attempts to modify the shape of the load curve to produce economy of operation have already been mentioned. These have included tariffs, pumped storage, and the use of seasonal or daily diversity between interconnected systems. A more direct method is the control of the load either through tariff structure or direct electrical control of appliances, the latter, say, in the form of remote on/off control of electric water-heaters where inconvenience to the consumer is least. For many years this has been achieved with domestic time switches, but some schemes use switches radio-controlled from the utility to give greater flexibility. This permits load reductions almost instantaneously and defers hot-water and air-conditioning load until after system peaks.

1.9.1.2 Load Forecasting

It is evident that load forecasting is a crucial activity in electricity supply. Forecasts are based on the previous year's loading for the period in question, updated by factors such as general load increases, major new loads, and weather trends. Both power demand (kW) and energy (kWh) forecasts are used, the latter often being the more readily obtained. Demand values may be determined from energy forecasts. Energy trends tend to be less erratic than peak power demands and are considered better growth indicators; however, load factors are also erratic in nature.

As weather has a much greater influence on residential than on industrial demands it may be preferable to assemble the load forecast in constituent parts to obtain the total. In many cases the seasonal variations in peak demand are caused by weather-sensitive domestic appliances, for example, heaters and air-conditioning. A knowledge of the increasing use of such appliances is therefore essential. Several techniques are available for forecasting. These range from simple curve fitting and extrapolation to stochastic modelling. The many physical factors affecting loads, for example, weather, national economic heath, popular TV programmes, public holidays, and so on, make forecasting a complex process demanding experience and high analytical ability using probabilistic techniques.

Problems

1.1 In the U.S.A. in 1971 the total area of right of ways for H.V. overhead lines was 16 000 km². Assuming a growth rate for the supply of electricity of 7% per annum calculate what year the whole of the USA will be covered with transmission systems (assume area to approximate 4800×1600 km). Justify any assumptions made and discuss critically why the result is meaningless.

(Answer: 91.25 years)

1.2 The calorific value of natural gas at atmospheric pressure and temperature is 40 MJ/m³. Calculate the power transfer in a pipe of lm diameter with gas at 60 atm (gauge) flowing at 5 m/s. If hydrogen is transferred at the same velocity and pressure, calculate the power transfer. The calorific value of hydrogen is 13 MJ/m³ at atmospheric temperature and pressure.

(Answer: 9.4 GW, 3.1 GW)

1.3 a. An electric car has a steady output of 10 kW over its range of 100 km when running at a steady 40 km/h. The efficiency of the car (including batteries) is 65%. At the end of the car's range the batteries are recharged over a period of 10 h. Calculate the average charging power if the efficiency of the battery charger is 90%.

   b. The calorific value of gasoline (petrol) is roughly 16 500 kJ/gallon. By assuming an average filling rate at a pump of 10 gallon/minute, estimate the rate of energy transfer on filling a gasoline-driven car. What range and what cost/km would the same car as (a) above produce if driven by gasoline with a 7 gallon tank? (Assume internal combustion engine efficiency is 60% and gasoline costs £3 per gallon)

(Answer: (a) 4.3 kW; (b) 2.75 MW, 77 km, 27p/km)

1.4 The variation of load (P) with time (t) in a power supply system is given by the expression,

$$P(kW) = 4000 + 8t - 0.00091t^2$$

where t is in hours over a total period of one year.

This load is supplied by three 10 MW generators and it is advantageous to fully load a machine before connecting the others. Determine:

a. the load factor on the system as a whole;
b. the total magnitude of installed load if the diversity factor is equal to 3;
c. the minimum number of hours each machine is in operation;
d. the approximate peak magnitude of installed load capacity to be cut off to enable only two generators to be used.

(Answer: (a) 0.73 (b) 65 MW (c) 8760, 7209, 2637 h (d) 4.2 MW)

1.5 a. Explain why economic storage of electrical energy would be of great benefit to power systems.
b. List the technologies for the storage of electrical energy which are available now and discuss, briefly, their disadvantages.
c. Why is hydro power a very useful component in a power system?
d. Explain the action of pumped storage and describe its limitations.
e. A pumped storage unit has an efficiency of 78% when pumping and 82% when generating. If pumping can be scheduled using energy costing 2.0 p/kWh, plot the gross loss/profit in p/kWh when it generates into the system with a marginal cost between 2p and 6p/kWh.
f. Explain why out-of-merit generation is sometimes scheduled.

(Answer: (e) Overall efficiency 64%; Profit max. 2.87p/kWh; Loss max. 1.25 p/kWh)

(From Engineering Council Examination, 1996)

2

Basic Concepts[1]

2.1 Three-Phase Systems

The rotor flux of an alternating current generator induces sinusoidal e.m.f.s in the conductors forming the stator winding. In a single-phase machine these stator conductors occupy slots over most of the circumference of the stator core. The e.m.f.s that are induced in the conductors are not in phase and the net winding voltage is less than the arithmetic sum of the individual conductor voltages. If this winding is replaced by three separate identical windings, as shown in Figure 2.1(a), each occupying one-third of the available slots, then the effective contribution of all the conductors is greatly increased, yielding a greatly enhanced power capability for a given machine size. Additional reasons why three phases are invariably used in large A.C. power systems are that the use of three phases gives similarly greater effectiveness in transmission circuits and the three phases ensure that motors always run in the same direction, provided the sequence of connection of the phases is maintained.

The three windings of Figure 2.1(a) give voltages displaced in time or phase by 120°, as indicated in Figure 2.1(b). Because the voltage in the (a) phase reaches its peak 120° before the (b) phase and 240° before the (c) phase, the order of phase voltages reaching their maxima or phase sequence is a-b-c. Most countries use a, b, and c to denote the phases; however, R (Red), Y (Yellow), and B (Blue) has often been used. It is seen that the algebraic sum of the winding or phase voltages (and currents if the winding currents are equal) at every instant in time is zero. Hence, if one end of each winding is connected, then the electrical situation is unchanged and the three return lines can be dispensed with, yielding a three-phase, three-wire system, as shown in Figure 2.2(a). If the currents from the windings are not equal, then it is usual to connect a fourth wire (neutral) to the common connection or neutral point, as shown in Figure 2.2(b).

---

[1] Throughout the book, symbols in **bold** type represent complex (phasor) quantities requiring complex arithmetic. *Italic* type is used for magnitude (scalar) quantities within the text.

Electric Power Systems, Fifth Edition. B.M. Weedy, B.J. Cory, N. Jenkins, J.B. Ekanayake and G. Strbac.
© 2012 John Wiley & Sons, Ltd. Published 2012 by John Wiley & Sons, Ltd.

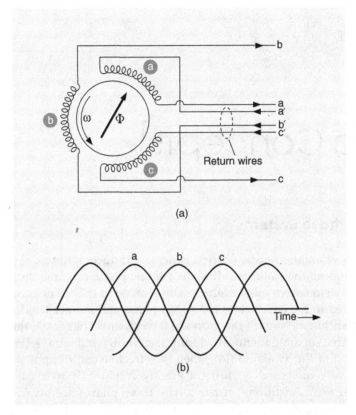

Figure 2.1 (a) Synchronous machine with three separate stator windings a, b and c displaced physically by 120°. (b) Variation of e.m.f.s developed in the windings with time

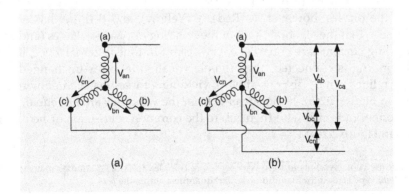

Figure 2.2 (a) Wye or star connection of windings, (b) Wye connection with neutral line

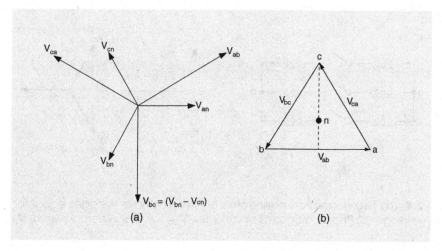

Figure 2.3 (a) Phasor diagram for wye connection, (b) Alternative arrangement of line-to-line voltages. Neutral voltage is at n, geometric centre of equilateral triangle

This type of winding connection is called 'wye' or 'star' and two sets of voltages exist:

1. the winding, phase, or line-to-neutral voltage, that is, $\mathbf{V}_{an}$, $\mathbf{V}_{bn}$, $\mathbf{V}_{cn}$; and
2. the line-to-line voltages, $\mathbf{V}_{ab}$, $\mathbf{V}_{bc}$, $\mathbf{V}_{ca}$. The subscripts here are important, $\mathbf{V}_{ab}$, means the voltage of line or terminal a with respect to b and $\mathbf{V}_{ba} = -\mathbf{V}_{ab}$.

The corresponding phasor diagram is shown in Figure 2.3(a) and it can be shown that[2]

$$|V_{ab}| = |V_{bc}| = |V_{ca}| = \sqrt{3}|V_{an}| = \sqrt{3}|V_{bn}| = \sqrt{3}|V_{cn}|.$$

The phase rotation of a system is very important. Consider the connection through a switch of two voltage sources of equal magnitude and both of rotation a-b-c. When the switch is closed no current flows. If, however, one source is of reversed rotation (easily obtained by reversing two wires), as shown in Figure 2.4, that is, a-c-b, a large voltage ($\sqrt{3}$ × phase voltage) exists across the switch contacts cb′ and bc′, resulting in very large currents if the switch is closed. Also, with reversed phase rotation the rotating magnetic field set up by a three-phase winding is reversed in direction and a motor will rotate in the opposite direction, often with disastrous results to its mechanical load, for example, a pump.

A three-phase load is connected in the same way as the machine windings. The load is balanced when each phase takes equal currents, that is, has equal impedance. With the wye connection the phase currents are equal to the current in the lines. The four-wire system is of particular use for low-voltage distribution networks in which

---

[2] Elsewhere, throughout the book, magnitude (scalar) quantities are represented by simple italics.

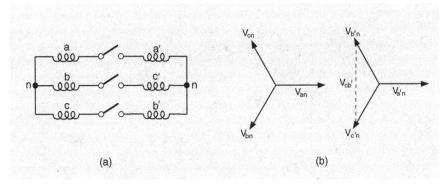

(a)   (b)

Figure 2.4 (a) Two generators connected by switch; phase voltages equal for both sets of windings, (b) Phasor diagrams of voltages. $\mathbf{V}_{cb'}$ = voltage across switch; $\mathbf{V}_{a'a}=0$

consumers are supplied with a single-phase supply taken between a line and neutral. This supply is often 230 V and the line-to-line voltage is 400 V. Distribution practice in the USA is rather different and the 220 V supply often comes into a house from a centre-tapped transformer, as shown in Figure 2.5, which in effect gives a choice of 220 V (for large domestic appliances) or 110 V (for lights, etc.).

The system planner will endeavour to connect the single-phase loads such as to provide balanced (or equal) currents in the three-phase supply lines. At any instant in time it is highly unlikely that consumers will take equal loads, and at the lower distribution voltages considerable unbalance occurs, resulting in currents in the neutral line. If the neutral line has zero impedance, this unbalance does not affect the load voltages. Lower currents flow in the neutral than in the phases and it is usual to install a neutral conductor of smaller cross-sectional area than the main line conductors. The combined or statistical effect of the large number of loads on the low-voltage network is such that when the next higher distribution voltage is considered, say 11 kV (line to line), which supplies the lower voltage network, the degree of unbalance is small. This and the fact that at this higher voltage, large three-phase, balanced motor loads are supplied, allows the three-wire system to be

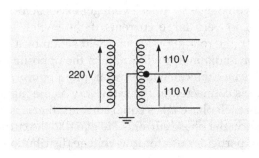

Figure 2.5 Tapped single-phase supply to give 220/110 V (centre-tap grounded), US practice

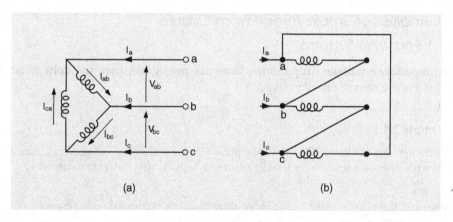

Figure 2.6 (a) Mesh or delta-connected load-current relationships, (b) Practical connections

used. The three-wire system is used exclusively at the higher distribution and transmission voltages, resulting in much reduced line costs and environmental impact.

In a balanced three-wire system a hypothetical neutral line may be considered and the conditions in only one phase determined. This is illustrated by the phasor diagram of line-to-line voltages shown in Figure 2.3(b). As the system is balanced the magnitudes so derived will apply to the other two phases but the relative phase angles must be adjusted by 120° and 240°. This single-phase approach is very convenient and widely used in power system analysis.

An alternative method of connection is shown in Figure 2.6. The individual phases are connected (taking due cognizance of winding polarity in machines and transformers) to form a closed loop. This is known as the mesh or delta connection. Here the line-to-line voltages are identical to the phase voltages, that is.

$$\mathbf{V}_{\text{line}} = \mathbf{V}_{\text{phase}}$$

The line currents are as follows:

$$\mathbf{I}_a = \mathbf{I}_{ab} - \mathbf{I}_{ca} \quad \mathbf{I}_b = \mathbf{I}_{bc} - \mathbf{I}_{ab} \quad \mathbf{I}_c = \mathbf{I}_{ca} - \mathbf{I}_{bc}$$

For balanced currents in each phase it is readily shown from a phasor diagram that $I_{\text{line}} = \sqrt{3} I_{\text{phase}}$. Obviously a fourth or neutral line is not possible with the mesh connection. The mesh or delta connection is seldom used for rotating-machine stator windings, but is frequently used for the windings of one side of transformers. A line-to-line voltage transformation ratio of $1{:}\sqrt{3}$ is obtained when going from a primary mesh to a secondary wye connection with the same number of turns per phase. Under balanced conditions the idea of the hypothetical neutral and single-phase solution may still be used (the mesh can be converted to a wye using the $\Delta \rightarrow Y$ transformation).

It should be noted that three-phase systems are usually described by their line-to-line voltage (e.g. 11, 132, 400 kV etc.).

2.1.1 Analysis of Simple Three-Phase Circuits

2.1.1.1 Four-Wire Systems

If the impedance voltage drops in the lines are negligible, then the voltage across each load is the source phase voltage.

Example 2.1

For the network of Figure 2.7(a), draw the phasor diagram showing voltages and currents and write down expressions for total line currents I_a, I_b, I_c, and the neutral current I_n.

Solution
Note that the power factor of the three-phase load is expressed with respect to the phase current and voltage.

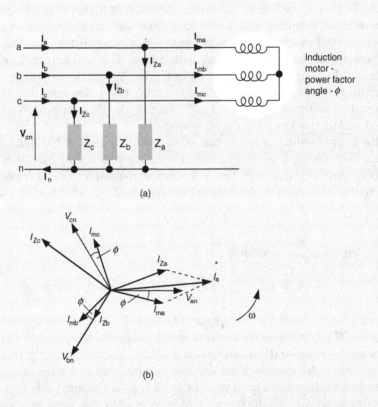

(a)

(b)

Figure 2.7 (a) Four-wire system with single-phase unbalanced loads and three phase balanced motor load, (b) Phasor diagram: only the phasors I_{za} and I_{ma} have been added to show I_a, I_b, and I_c can be found in a similar way. Note: V_{an} is the reference direction. Rotation is anticlockwise

The phasor diagram is shown in Figure 2.7(b). The line currents are as follows:

$$\mathbf{I_a} = \frac{V}{\sqrt{3}} \cdot \frac{1}{\mathbf{Z_a}} + I_m(\cos\phi - j\sin\phi)$$

$$\mathbf{I_b} = \frac{V}{\sqrt{3}}(-0.5 - j0.866) \cdot \frac{1}{\mathbf{Z_b}} + I_m\left[\cos\left(\frac{2\pi}{3} - \phi\right) - j\sin\left(\frac{2\pi}{3} - \phi\right)\right]$$

$$\mathbf{I_c} = \frac{V}{\sqrt{3}}(-0.5 + j0.866) \cdot \frac{1}{\mathbf{Z_c}} + I_m\left[\cos\left(\frac{4\pi}{3} - \phi\right) - j\sin\left(\frac{4\pi}{3} - \phi\right)\right]$$

The neutral current (as the induction motor currents do not contribute to $\mathbf{I_n}$)

$$\mathbf{I_n} = \mathbf{I_a} + \mathbf{I_b} + \mathbf{I_c} = \frac{V}{\sqrt{3}}[\mathbf{Y_a} + \mathbf{Y_b}(-0.5 - j0.866) + \mathbf{Y_c}(-0.5 + j0.866)]$$

where $\mathbf{Y_a} = \dfrac{1}{\mathbf{Z_a}}$, $\mathbf{Y_b} = \dfrac{1}{\mathbf{Z_b}}$ and $\mathbf{Y_c} = \dfrac{1}{\mathbf{Z_c}}$.

2.1.1.2 Three-Wire Balanced Systems

The system may be treated as a single-phase system using phase voltages. It must be remembered, however, that the total three-phase active power and reactive power are three times the values delivered by a single phase.

2.1.1.3 Three-Wire Unbalanced Systems

In more complex networks the method of symmetrical components is used (see Chapter 7), but in simple situations conventional network theory can be applied. Consider the source load arrangement shown in Figure 2.8, in which $\mathbf{Z_a} \neq \mathbf{Z_b} \neq \mathbf{Z_c}$. From the mesh method of analysis, the following equation is obtained.

$$\begin{bmatrix} \mathbf{Z_a} + \mathbf{Z_b} & -\mathbf{Z_b} \\ -\mathbf{Z_b} & \mathbf{Z_b} + \mathbf{Z_c} \end{bmatrix} \cdot \begin{bmatrix} i_1 \\ i_2 \end{bmatrix} = \begin{bmatrix} \mathbf{V_{ab}} \\ \mathbf{V_{bc}} \end{bmatrix} = \begin{bmatrix} V \\ V(-0.5 - j0.866) \end{bmatrix}$$

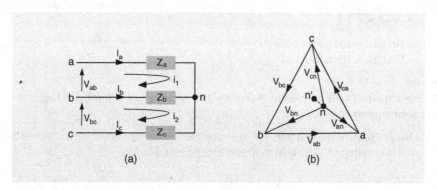

(a)   (b)

Figure 2.8 (a) Three-wire system with unbalanced load, i_1 and i_2 are loop currents, (b) Phasor diagram-voltage difference of neutral connection n from ground n′ → n; n′ = neutral voltage when load is balanced

Hence i_1 and i_2 are determined and from these $\mathbf{I_a}$, $\mathbf{I_b}$, and $\mathbf{I_c}$ are found together with the voltage of the neutral point (n) from ground (see Figure 2.8(b)). Depending on the severity of the imbalance, the voltage difference from the neutral point n to ground can attain values exceeding the phase voltage, that is, the point n can lie outside the triangle of line voltages. Such conditions produce damage to connected equipment and show the importance of earthing (grounding) the neutral.

The power and reactive power consumed by a balanced three-phase load for both wye or mesh connections are given by

$$P = \sqrt{3}VI \cos\phi \text{ and } Q = \sqrt{3}VI \sin\phi$$

where

$V =$ line-to-line voltage;
$I \ =$ line current;
$\phi \ =$ angle between load phase current and load phase voltage.

Alternatively

$$P = 3V_{ph}I_{ph} \cos\phi \text{ and } Q = 3V_{ph}I_{ph} \sin\phi$$

As the powers in each phase of a balanced load are equal, a single wattmeter may be used to measure total power. For unbalanced loads, two wattmeters are sufficient with three-wire circuits while for unbalanced four wire circuits, three wattmeters are needed.

Example 2.2

A three-phase wye-connected load is shown in Figure 2.9. It is supplied from a 200 V, three-phase, four-wire supply of phase sequence a-b-c. The neutral line has a resistance of 5 Ω. Two wattmeters are connected as shown. Calculate the power recorded on each wattmeter when:

a. $R_a = 10\,\Omega$, $R_b = 10\,\Omega$, $R_c = 10\,\Omega$
b. $R_a = 10\,\Omega$, $R_b = 10\,\Omega$, $R_c = 2\,\Omega$

Solution
The mesh method will be used to determine the currents in the lines and hence in the wattmeter current coils.

a. As the phase currents and voltages are balanced, no current flows in the neutral line. The line voltage $\mathbf{V_{ab}}$ is used as a reference phasor. Hence,

$$\mathbf{V_{ab}} = 200$$
$$\mathbf{V_{bc}} = (-0.5 - j0.866) \times 200$$

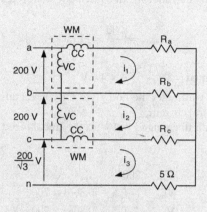

Figure 2.9 Use of two wattmeters to measure three-phase power, Example 2.2.
cc – current coil, vc – voltage coil

From Figure 2.9, with $i_3 = 0$

$$\begin{bmatrix} 20 & -10 \\ -10 & 20 \end{bmatrix} \cdot \begin{bmatrix} i_1 \\ i_2 \end{bmatrix} = \begin{bmatrix} 200 \\ -100 - j173.2 \end{bmatrix}$$

From which

$$i_1 = 10 - j5.75$$
$$i_2 = -j11.5$$

Power in wattmeter (1) = real part of $\mathbf{V}_{ab} \times \mathbf{I}_a^{*3} = 200(10 + j\,5.75)$

$$= 2\,\text{kW}$$

Power in wattmeter (2) = real part of $\mathbf{V}_{cb} \times \mathbf{I}_c^* = -\mathbf{V}_{bc} \times \mathbf{I}_c^*$

$$= (100 + j173.2)(i_2)^*$$
$$= (100 + j173.2)(j11.5)$$
$$= 2\,\text{kW}$$

The total measured power is therefore 4 kW.

---

[3] See Section 2.3 for use of the complex conjugate $\mathbf{I}^*$.

This can be checked by considering an equivalent balanced Y system.

$$V_{phase} = \frac{200}{\sqrt{3}} = 115.5 \, V$$

Actual power consumed $= 3 \times \dfrac{115.5^2}{10} = 4 \, kW$

As expected the two wattmeter method gives the correct answer for a three-phase connection with no neutral current.

b. The line voltages remain unchanged but with the unbalanced load it is now necessary to calculate i_3

$$\mathbf{V_{ab}} = 200$$
$$\mathbf{V_{bc}} = (-0.5 - j0.866) \times 200$$
$$\mathbf{V_{ca}} = (-0.5 + j0.866) \times 200$$

and

$$\mathbf{V_{cn}} = j\frac{200}{\sqrt{3}}$$

$$\begin{bmatrix} 20 & -10 & 0 \\ -10 & 12 & -2 \\ 0 & -2 & 7 \end{bmatrix} \cdot \begin{bmatrix} i_1 \\ i_2 \\ i_3 \end{bmatrix} = \begin{bmatrix} 200 \\ -100 - j173.2 \\ j\dfrac{200}{\sqrt{3}} \end{bmatrix}$$

Eliminate i_1,

$$\begin{bmatrix} 7 & -2 \\ -2 & 7 \end{bmatrix} \cdot \begin{bmatrix} i_2 \\ i_3 \end{bmatrix} = \begin{bmatrix} -j173.2 \\ j\dfrac{200}{\sqrt{3}} \end{bmatrix}$$

From which

$$i_3 = j10.26 \, A$$
$$i_2 = -j21.8 \, A$$

Therefore,

$$i_1 = 10 - j10.9 \, A$$

Now,

$$\mathbf{I_a} = i_1 = 10 - j10.9 \, A$$
$$\mathbf{I_b} = i_2 - i_1 = -j21.8 - 10 + j10.9 = -10 - j10.9 \, A$$
$$\mathbf{I_c} = i_3 - i_2 = j10.26 + j21.8 = j32.06$$

Power in wattmeter (1) $=$ real part of $\mathbf{V_{ab}} \times \mathbf{I_a^*} = \mathrm{Re}[200(10 + j10.9)] = 2 \, kW$

Power in wattmeter (2) = real part of $\mathbf{V_{cb}} \times \mathbf{I_c^*} = -\mathbf{V_{bc}} \times \mathbf{I_c^*}$

$$= \text{Re}[(100 + j173.2) \times j32.06]$$
$$= 5.55 \text{ kW}$$

Actual power consumed $= I_a^2 \times 10 + I_b^2 \times 10 + I_c^2 \times 2 + I_n^2 \times 5 = 6.96 \text{ kW}$

As the loads are not balanced and there is a return current in the neutral, the actual power consumed by the load is not the sum of the two wattmeter readings.

2.2 Three-Phase Transformers

The usual form of the three-phase transformer, that is, the core type, is shown in Figure 2.10. If the magnetic reluctances of the three limbs are equal, then the sum of the fluxes set up by the three-phase magnetizing currents is zero. In fact, the core is the magnetic equivalent of the wye-connected winding. It is apparent from the shape of Figure 2.10 that the magnetic reluctances are not exactly equal, but in an introductory treatment may be so assumed. An alternative to the three-limbed core is the use of three separate single-phase transformers. Although more expensive (about 20% extra), this has the advantage of lower weights for transportation, and

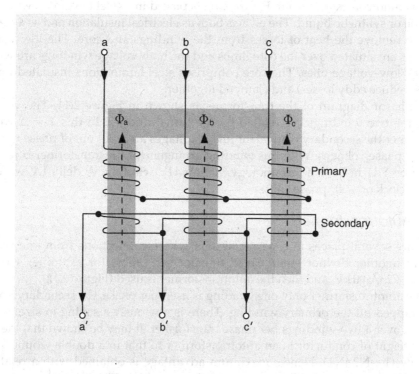

Figure 2.10 Three-phase core-type transformer. Primary connected in wye (star), secondary connected in mesh (delta)

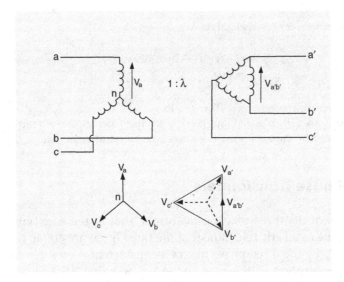

Figure 2.11 Windings and phasor diagram of a YΔ transformer

this aspect is crucial for large sizes. Also, with the installation of four single-phase units, a spare is available at reasonable cost.

The wound core, as shown in Figure 2.10, is placed in a steel tank filled with insulating oil or synthetic liquid. The oil acts both as electrical insulation and as a cooling agent to remove the heat of losses from the windings and core. The low voltage windings are situated over the core limbs and the high-voltage windings are wound over the low-voltage ones. The core comprises steel laminations insulated on one side (to reduce eddy losses) and clamped together.

The phasor diagram of the transformer is shown in Figure 2.11. The voltages across the two windings are related by the turns ratio (λ). In this $Y \rightarrow \Delta$ wound transformer the secondary equivalent phase voltages are $-30°$ out of phase with the primary phase voltages. With this winding arrangement the transformer is referred to as type Yd1 in British terminology. Y: star HV winding, d: delta LV winding: 1: one o'clock or $-30°$ phase shift.

2.2.1 Autotransformers

There are several places in a power system where connections from one voltage level to another do not entail large transformer ratios, for example, 400/275, 500/345, 725/500 kV, and then the autotransformer is used (Figure 2.12).

In the autotransformer only one winding is used per phase, the secondary voltage being tapped off the primary winding. There is obviously a saving in size, weight and cost over a two-windings per phase transformer. It may be shown that the ratio of the weight of conductor in an autotransformer to that in a double-wound one is given by ($1 - N2/N1$). Hence, maximum advantage is obtained with a relatively small difference between the voltages on the two sides. The effective reactance is reduced compared with the equivalent two winding transformer and this can give

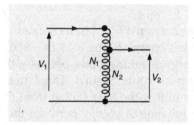

Figure 2.12 One phase of an autotransformer, $\dfrac{V_2}{V_1} = \dfrac{N_2}{N_1}$

rise to high short-circuit currents. The general constructional features of the core and tank are similar to those of double-wound transformers, but the primary and secondary voltages are now in-phase.

2.3 Active and Reactive Power

In the circuit shown in Figure 2.13, let the instantaneous values of voltage and current be

$$v = \sqrt{2}E \sin(\omega t) \text{ and } i = \sqrt{2}I \sin(\omega t + \phi)$$

The instantaneous power

$$p = vi = EI \cos\phi - EI \cos(2\omega t + \phi)$$

also

$$-EI \cos(2\omega t + \phi) = -EI(\cos 2\omega t \cos\phi - \sin 2\omega t \sin\phi)$$

$$\therefore p = vi = [EI \cos\phi - EI \cos(2\omega t)\cos\phi] + [EI \sin(2\omega t)\sin\phi]$$

$$= [\text{instantaneous real power}] + [\text{instantaneous reactive power}]$$

Figure 2.14 shows the instantaneous real and reactive power.

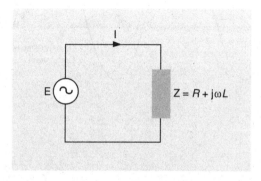

Figure 2.13 Voltage source and load

The mean active power $= EI \cos \phi$.

The mean value of $EI \sin (2\omega t) \sin \phi = 0$, but its maximum value $= EI \sin \phi$.

The voltage source supplies energy to the load in one direction only. At the same time an interchange of energy is taking place between the source and the load of average value zero, but of peak value $EI \sin \phi$. This latter quantity is known as the reactive power (Q) and the unit is the VAr (taken from the alternative name, Volt-Ampere reactive). The interchange of energy between the source and the inductive and capacitive elements (that is, the magnetic and electric fields) takes place at twice the supply frequency. Therefore, it is possible to think of an active or real power component P (watts) of magnitude $EI \cos \phi$ and a reactive power component Q (VAr) equal to $EI \sin \phi$ where ϕ is the power factor angle, that is the angle between **E** and **I**. It should be stressed, however, that the two quantities P and Q are physically quite different and only P can do real work.

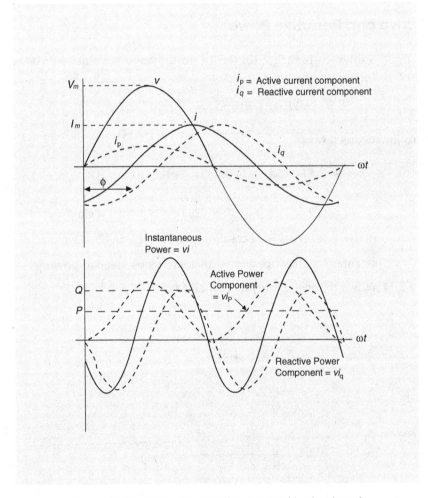

Figure 2.14 Real and reactive power (single phase)

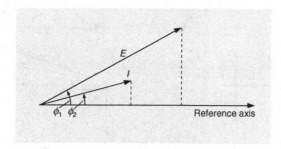

Figure 2.15 Phasor diagram

The quantity **S** (volt-amperes), known as the complex (or apparent) power, may be found by multiplying **E** by the conjugate of **I** or vice versa. Consider the case when **I** lags **E**, and assume $\mathbf{S} = \mathbf{E}^*\mathbf{I}$. Referring to Figure 2.15,

$$\mathbf{S} = Ee^{-j\phi_1} \times Ie^{j\phi_2} = EIe^{-j(\phi_1-\phi_2)}$$
$$= EIe^{-j\phi}$$
$$= P - jQ$$

Next assume

$$\mathbf{S} = \mathbf{EI}^*$$
$$= EIe^{j(\phi_1-\phi_2)}$$
$$= EIe^{j\phi}$$
$$= P + jQ$$

Obviously both the above methods give the correct magnitudes for P and Q but the sign of Q is different. The method used is arbitrarily decided and the usual convention to be adopted is that the volt-amperes reactive absorbed by an inductive load shall be considered positive, and by a capacitive load negative; hence $\mathbf{S} = \mathbf{EI}^*$. This convention is recommended by the International Electrotechnical Commission.

In a network the net stored energy is the sum of the various inductive and capacitive stored energies present. The net value of reactive power is the sum of the VArs absorbed by the various components present, taking due account of the sign. VArs can be considered as being either produced or absorbed in a circuit; a capacitive load can be thought of as generating VArs. Assuming that an inductive load is represented by $R + jX$ and that the $\mathbf{S} = \mathbf{EI}^*$ convention is used, then an inductive load absorbs positive VArs and a capacitive load produces VArs.

The various elements in a network are characterized by their ability to generate or absorb reactive power. Consider a synchronous generator which can be represented by the simple equivalent circuit shown in Figure 2.16. When the generator is over-excited, that is, its generated e.m.f. is high, it produces VArs and the complex power flow from the generator is $P + jQ$. When the machine is under-excited the generated current leads the busbar voltage and the apparent power from the generator is

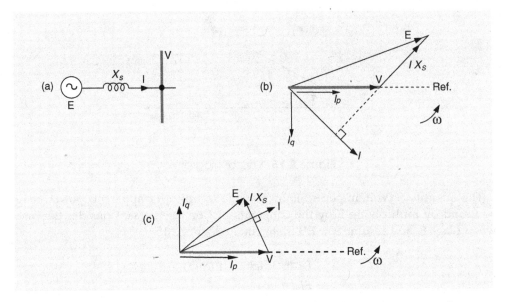

Figure 2.16 (a) Line diagram of system, (b) Overexcited generator-phasor diagram, (c) Underexcited generator-phasor diagram

$P - jQ$. It can also be thought of as absorbing VArs. The reactive power characteristics of various power-system components are summarized as follows: reactive power is generated by over-excited synchronous machines, capacitors, cables and lightly loaded overhead lines; and absorbed by under-excited synchronous machines, induction motors, inductors, transformers, and heavily loaded overhead lines, see Figure 2.17.

The reactive power absorbed by a reactance of X_L $[\Omega] = I^2 X_L$ [VAr] where I is the current. A capacitance with a voltage V applied produces V^2B [VAr], where $B = 1/Xc = \omega C$ [S] and $Xc = 1/\omega C$ [Ω]. At a node, $\Sigma P = 0$ and $\Sigma Q = 0$ in accordance with Kirchhoff's law.

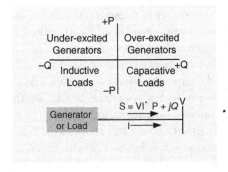

Figure 2.17 Power quadrant

2.4 The Per-Unit System

In the analysis of power networks, instead of using actual values of quantities it is usual to express them as fractions of reference quantities, such as rated or full-load values. These fractions are called per unit (denoted by p.u.) and the p.u. value of any quantity is defined as

$$\frac{\text{actual value (in any unit)}}{\text{base or reference value in the same unit}} \qquad (2.1)$$

Some authorities express the p.u. value as a percentage. Although the use of p.u. values may at first sight seem a rather indirect method of expression there are, in fact, great advantages; they are as follows:

1. The apparatus considered may vary widely in size; losses and volt drops will also vary considerably. For apparatus of the same general type the p.u. volt drops and losses are in the same order, regardless of size.
2. The use of $\sqrt{3}$'s in three-phase calculations is reduced.
3. By the choice of appropriate voltage bases the solution of networks containing several transformers is facilitated.
4. Per-unit values lend themselves more readily to digital computation.

2.4.1 Resistance and Impedance

Given that base value of voltage is V_b and base value of current is I_b, the base value of resistance is given by:

$$R_b = \frac{V_b}{I_b} \qquad (2.2)$$

Now from (2.1) and (2.2), the p.u. value of the resistance is given by:

$$R_{p.u.} = \frac{R(\Omega)}{R_b(\Omega)} = \frac{R(\Omega)}{V_b/I_b} = \frac{R(\Omega) \cdot I_b}{V_b} = \frac{\text{voltage drop across R at base current}}{\text{base voltage}}$$

Also

$$R_{p.u.} = \frac{R(\Omega)I_b^2}{V_b I_b}$$

$$= \frac{\text{power loss at base current}}{\text{base power or volt-amperes}}$$

$\therefore$ the power loss (p.u.) at base or rated current $= R_{p.u.}$
Power loss (p.u.) at $I_{p.u.}$ current $= R_{p.u.} I_{p.u.}^2$.

Similarly

$$\text{p.u. impedance} = \frac{\text{impedance in ohms}}{\left(\dfrac{\text{base voltage}}{\text{base current}}\right)} = \frac{Z(\Omega)I_b}{V_b}$$

Example 2.3

A d.c. series machine rated at 200 V, 100 A has an armature resistance of 0.1 Ω and field resistance of 0.15 Ω. The friction and windage loss is 1500 W. Calculate the efficiency when operating as a generator.

Solution
Total series resistance in p.u. is given by

$$R_{p.u.} = \frac{0.25}{200/100} = 0.125 \, \text{p.u.}$$

Where $V_{base} = 200 \, V$ and $I_{base} = 100 \, A$
Friction and windage loss

$$= \frac{1500}{200 \times 100} = 0.075 \, \text{p.u.}$$

At the rated load, the series-resistance loss $= 1^2 \times 0.125$
and the total loss $= 0.125 + 0.075 = 0.2 \, \text{p.u.}$
As the output $= 1 \, \text{p.u.}$, the efficiency

$$= \frac{1}{1 + 0.2} = 0.83 \, \text{p.u.}$$

2.4.2 Three-Phase Circuits

In three-phase circuits, a p.u. phase voltage has the same numerical value as the corresponding p.u. line voltage. With a measured line voltage of 100 kV and a rated line voltage of 132 kV, the p.u. value is 0.76. The equivalent phase voltages are $100/\sqrt{3}$ kV and $132/\sqrt{3}$ kV and hence the p.u. value is again 0.76. The actual values of R, X_L, and X_C for lines, cables, and other apparatus are phase values. When working with ohmic values it is less confusing to use the equivalent phase values of all quantities. In the p.u. system, three-phase values of voltage, current and power can be used without undue anxiety about the result being a factor of $\sqrt{3}$ incorrect.

It is convenient in a.c. circuit calculations to work in terms of base volt-amperes, (S_{base}). Thus,

$$S_{base} = V_{base} \times I_{base} \times \sqrt{3}$$

when V_{base} is the line voltage and I_{base} is the line current in a three-phase system. Hence

$$I_{base} = \frac{S_{base}}{\sqrt{3}V_{base}} \qquad (2.3)$$

Expression (2.3) shows that if S_{base} and V_{base} are specified, then I_{base} is determined. Only two base quantities can be chosen from which all other quantities in a three-phase system are calculated. Thus,

$$Z_{base} = \frac{V_{base}/\sqrt{3}}{I_{base}} = \frac{V_{base}/\sqrt{3}}{S_{base}/\sqrt{3}V_{base}} = \frac{V_{base}^2}{S_{base}} \qquad (2.4)$$

Hence

$$Z_{p.u.} = \frac{Z(\Omega)}{Z_{base}} = \frac{Z(\Omega) \times S_{base}}{V_{base}^2} \qquad (2.5)$$

Expression (2.5) shows that $Z_{p.u.}$ is directly proportional to the base VA and inversely proportional to the base voltage squared.

If we wish to calculate $Z_{p.u.}$ to a new base VA, then

$$Z_{p.u.}(\text{new base}) = Z_{p.u.}(\text{old base}) \times \frac{S_{\text{new base}}}{S_{\text{old base}}} \qquad (2.6)$$

If we wish to calculate $Z_{p.u.}$ to a new voltage base, then

$$Z_{p.u.}(\text{new base}) = Z_{p.u.}(\text{old base}) \times \frac{(V_{\text{old base}})^2}{(V_{\text{new base}})^2} \qquad (2.7)$$

2.4.3 Transformers

Consider a single-phase transformer in which the total series impedance of the two windings referred to the primary is Z_1 (Figure 2.18).

Then the p.u. impedance $Z_{p.u.} = \dfrac{Z_1}{V_1/I_1}$, where I_1, and V_1, are the base values of the primary circuit.

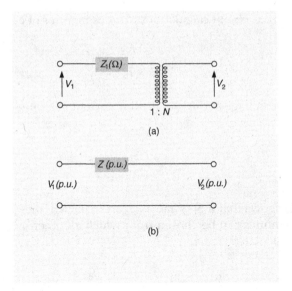

Figure 2.18 Equivalent circuit of single-phase transformer

The ohmic impedance referred to the secondary is

$$Z_2 = Z_1 N^2 \ [\Omega]$$

and this in p.u. notation is

$$Z_{p.u.} = Z_1 N^2 \left/ \dfrac{V_2}{I_2} \right.$$

V_2 and I_2 are base voltage and current of the secondary circuit. If they are related to the base voltage and current of the primary by the turns ratio of the transformer then.

$$Z_{p.u.} = Z_1 N^2 \dfrac{I_1}{N} \cdot \dfrac{1}{V_1 N} = \dfrac{Z_1 I_1}{V_1}$$

Hence provided the base voltages on each side of a transformer are related by the turns ratio, the p.u. impedance of a transformer is the same whether considered from the primary or the secondary side. The winding does not appear in the equivalent circuit (Figure 2.18b), the transformer impedance in per unit is only calculated once and Equation (2.7) is not required.

Example 2.4

In the network of Figure 2.19, two single-phase transformers supply a 10 kVA resistance load at 200 V. Show that the p.u. load is the same for each part of the circuit and calculate the voltage at point D.

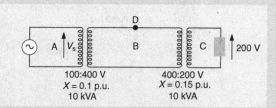

Figure 2.19 Network with two transformers–p.u. approach

Solution

The load resistance is $(200^2/10 \times 10^3)$, that is, $4\,\Omega$.

In each of the circuits A, B, and C a different voltage exists, so that each circuit will have its own base voltage, that is, 100 V in A, 400 V in B, and 200 V in C.

Although it is not essential for rated voltages to be used as bases, it is essential that the voltage bases used be related by the turns ratios of the transformers. If this is not so the simple p.u. framework breaks down. The same volt-ampere base is used for all the circuits as $V_1 I_1 = V_2 I_2$ on each side of a transformer and is taken in this case as 10 kVA. The per unit impedances of the transformers are already on their individual equipment bases of 10 kVA and so remain unchanged.

The base impedance in C

$$Z_{base} = \frac{V_{base}^2}{\text{base VA}} = \frac{200^2}{10000} = 4\,\Omega$$

The load resistance (p.u.) in C

$$= \frac{4}{4} = 1\,p.u.$$

In B the base impedance is

$$Z_{base} = \frac{V_{base}^2}{\text{base VA}} = \frac{400^2}{10000} = 16\,\Omega$$

and the load resistance (in ohms) referred to B is

$$= 4 \times N^2 = 4 \times 2^2 = 16\,\Omega$$

Hence the p.u. load referred to B

$$\frac{16}{16} = 1\,p.u.$$

Similarly, the p.u. load resistance referred to A is also 1 p.u. Hence, if the voltage bases are related by the turns ratios the load p.u. value is the same for all circuits.

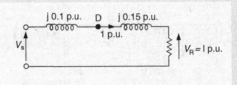

Figure 2.20 Equivalent circuit with p.u. values, of network in Figure 2.19

An equivalent circuit may be used as shown in Figure 2.20. Let the volt-ampere base be 10 kVA; the voltage across the load (V_R) is 1 p.u. (as the base voltage in C is 200 V).

The base current at voltage level C of this single phase circuit

$$I_{base} = \frac{\text{base VA}}{V_{base}} = \frac{10000}{200} = 50\,\text{A}$$

The corresponding base currents in the other circuits are 25 A in B, and 100 A in A.

The actual load current is 200 V/4 Ω = 50 A = 1 p.u.

Hence the supply voltage V_S

$$\mathbf{V_s} = 1(j0.1 + j0.15) + 1\,\text{p.u.}$$

$$V_S = \sqrt{1^2 + 0.25^2} = 1.03\,\text{p.u.}$$

$$= 1.03 \times 100 = 103\,\text{V}$$

The voltage at point D in Figure 2.17

$$\mathbf{V_D} = 1(j0.15) + 1\,\text{p.u.}$$

$$V_D = \sqrt{1^2 + 0.15^2} = 1.012\,\text{p.u.}$$

$$= 1.012 \times 400 = 404.8\,\text{V}$$

It is a useful exercise to repeat this example using ohms, volts and amperes.

A summary table of the transformation of the circuit of Figure 2.19 into per unit is shown.

Section of network	S_{base} (common for network)	V_{base} (chosen as transformer turns ratio)	I_{base} (calculated from S_{base}/V_{base})	Z_{base} (calculated from V_{base}^2/S_{base})
A	10 kVA	100 V	100 A	1 Ω
B	10 kVA	400 V	25 A	16 Ω
C	10 kVA	200 V	50 A	4 Ω

Example 2.5

Figure 2.21 shows the schematic diagram of a radial transmission system. The ratings and reactances of the various components are shown. A load of 50 MW at 0.8 p.f. lagging is taken from the 33 kV substation which is to be maintained at 30 kV. It is required to calculate the terminal voltage of the synchronous machine. The line and transformers may be represented by series reactances. The system is three-phase.

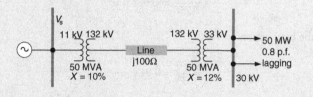

Figure 2.21 Line diagram of system for Example 2.5

Solution
It will be noted that the line reactance is given in ohms; this is usual practice. The voltage bases of the various circuits are decided by the nominal transformer voltages, that is, 11, 132, and 33 kV. A base of 100 MVA will be used for all circuits. The reactances (resistance is neglected) are expressed on the appropriate voltage and MVA bases.
  Base impedance for the line

$$= \frac{V_{base}^2}{S_{base}}$$

$$= \frac{(132 \times 10^3)^2}{100 \times 10^6}$$

$$= 174\,\Omega$$

Hence the p.u. reactance

$$= \frac{j100}{174} = j0.575 \text{ p.u.}$$

Per unit reactance of the sending-end transformer

$$j0.1 \times \frac{100}{50} = j0.2 \text{ p.u.}$$

Per unit reactance of the receiving-end transformer

$$j0.12 \times \frac{100}{50} = j0.24 \text{ p.u.}$$

$$\text{Load current} = \frac{50 \times 10^6}{\sqrt{3} \times 30 \times 10^3 \times 0.8} = 1203 \text{ A}$$

Base current for 33 kV, 100 MVA

$$= \frac{100 \times 10^6}{\sqrt{3} \times 33 \times 10^3} = 1750 \, \text{A}$$

Hence the p.u. load current

$$= \frac{1203}{1750} = 0.687 \, \text{p.u.}$$

Voltage of the load busbar

$$= \frac{30}{33} = 0.91 \, \text{p.u.}$$

The equivalent circuit is shown in Figure 2.22.
Also,

$$\mathbf{V_S} = 0.687 \times (0.8 - j0.6)(j0.2 + j0.575 + j0.24) + (0.91 + j0)$$
$$= 1.328 + j0.558 \, \text{p.u.}$$
$$V_S = 1.44 \, \text{p.u.}$$
$$= 1.44 \times 11 \, \text{kV} = 15.84 \, \text{kV}$$

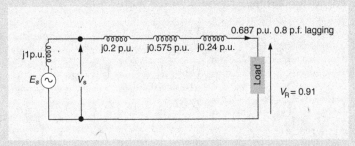

Figure 2.22 Equivalent circuit for Example 2.5

2.5 Power Transfer and Reactive Power

The circuit shown in Figure 2.23 represents the simplest electrical model for a source (with voltage $\mathbf{V_G}$) feeding into a power system represented by a load of $P + jQ$. It can also represent the power flows in a line connecting two busbars in an interconnected power system.

The voltage at the source end and load end are related by:

$$\mathbf{V_G} = \mathbf{V_L} + (R + jX)\mathbf{I} \tag{2.8}$$

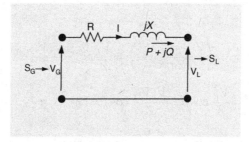

Figure 2.23 Power transfer between sources

If $R + jX = Z\angle\theta$ where $Z = \sqrt{R^2 + X^2}$ and $\theta = \tan^{-1}(X/R)$, then the current can be obtained as:

$$I = \frac{V_G - V_L}{Z\angle\theta}$$

Therefore the apparent power at the source end is given by:

$$S_G = V_G I^* = V_G \left(\frac{V_G - V_L}{Z\angle\theta}\right)^* \tag{2.9}$$

If the load end voltage is chosen as the reference and the phase angle between the load end and source end is δ:

$$V_L = V_L\angle 0° = V_L \text{ and } V_G = V_G\angle\delta \tag{2.10}$$

Substituting for V_L and V_G from equation (2.10) into (2.9) yields:

$$S_G = V_G e^{j\delta}\left(\frac{V_G e^{-j\delta} - V_L}{Z e^{-j\theta}}\right)$$

$$= \frac{V_G^2}{Z} e^{j\theta} - \frac{V_G V_L}{Z} e^{j(\theta+\delta)}$$

Therefore

$$P_G = \frac{V_G^2}{Z} \cos\theta - \frac{V_G V_L}{Z} \cos(\theta + \delta)$$

$$Q_G = \frac{V_G^2}{Z} \sin\theta - \frac{V_G V_L}{Z} \sin(\theta + \delta)$$

$\tag{2.11}$

Similarly,

$$\mathbf{S_L} = V_L\left(\frac{V_L - V_Ge^{-j\delta}}{Ze^{-j\theta}}\right) = \frac{V_L^2}{Z}e^{j\theta} - \frac{V_LV_G}{Z}e^{j(\theta-\delta)}$$

$$P_L = \frac{V_L^2}{Z}\cos\theta - \frac{V_LV_G}{Z}\cos(\theta - \delta)$$

$$Q_L = \frac{V_L^2}{Z}\sin\theta - \frac{V_LV_G}{Z}\sin(\theta - \delta)$$

(2.12)

The power output to the load is a maximum when $\cos(\theta - \delta) = 1$, that is, $\theta = \delta$

2.5.1 Calculation of Sending and Received Voltages in Terms of Power and Reactive Power

The determination of the voltages and currents in a network can obviously be achieved by means of complex notation, but in power systems usually power (P) and reactive power (Q) are specified and often the resistance of lines is negligible compared with reactance. For example, if $R = 0.1X$, the error in neglecting R is 0.49%, and even if $R = 0.4X$ the error is 7.7%.

From the transmission link shown in Figure 2.23:

For the load:

$$\mathbf{V_LI^*} = P + jQ$$

$$\mathbf{I} = \frac{P - jQ}{\mathbf{V_L^*}}$$

The voltage at the source and load are related by:

$$\mathbf{V_G} = \mathbf{V_L} + (R + jX)\mathbf{I}$$

$$= \mathbf{V_L} + (R + jX)\left[\frac{P - jQ}{\mathbf{V_L^*}}\right]$$

(2.13)

As $\mathbf{V_L} = \mathbf{V_L^*} = V_L$ (in this case):

$$\mathbf{V_G} = V_L + (R + jX)\left[\frac{P - jQ}{V_L}\right]$$

$$= \left[V_L + \frac{RP + XQ}{V_L}\right] + j\left[\frac{XP - RQ}{V_L}\right]$$

(2.14)

Equation (2.14) can be represented by the phasor diagram shown in Figure 2.24.

$$\Delta V_p = \frac{RP + XQ}{V}$$

(2.15)

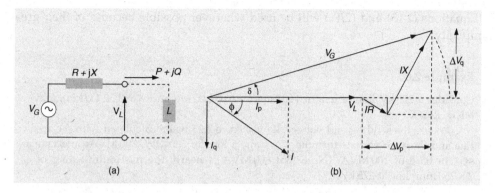

Figure 2.24 Phasor diagram for transmission of power through a line

and

$$\Delta V_q = \frac{XP - RQ}{V} \tag{2.16}$$

If δ is small (as is usually the case in Distribution circuits) then

$$\Delta V_q \ll V_L + \Delta V_p$$

then

$$V_G = V_L + \frac{RP + XQ}{V_L}$$

and

$$V_G - V_L = \frac{RP + XQ}{V_L}$$

Hence the arithmetic difference between the voltages is given approximately by

$$\frac{RP + XQ}{V_L}$$

In a transmission circuit, $R \simeq 0$ then

$$V_G - V_L = \frac{XQ}{V_L} \tag{2.17}$$

that is, the voltage magnitude depends only on Q.

The angle of transmission δ is obtained from $\sin^{-1}(\Delta V_q/V_G)$, and depends only on P.

Equations (2.15) and (2.16) will be used wherever possible because of their great simplicity.

Example 2.6

Consider a 275 kV line of length 160 km ($R = 0.034\,\Omega$/km and $X = 0.32\,\Omega$/km). Obviously, $R \ll X$.

Compare the sending end voltage V_G in Figure 2.25 when calculated with the accurate and the approximate formulae. Assume a load of 600 MW, 300 MVAr and take a system base of 100 MVA. (Note that 600 MVA is nearly the maximum rating of a $2 \times 258\,mm^2$ line at 275 kV.)

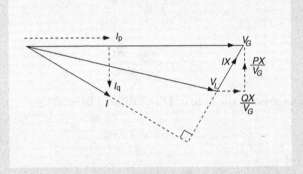

Figure 2.25 Phasor diagram when V_G Is specified

Solution
The base impedance is

$$\frac{\left(275 \times 10^3\right)^2}{100 \times 10^6} = 756\,\Omega$$

For line of length 160 km,

$$X(\Omega) = 160 \times 0.32 = 51.2\,\Omega$$

$$X = \frac{51.2}{756} = 0.0677\,\text{p.u.}$$

and let the received voltage be

$$V_L = 275\,\text{kV} = 1\,\text{p.u.}$$

then

$$XP = 0.0677 \times 6 = 0.406\,\text{p.u.}$$

and

$$V_L + QX = 1 + 3 \times 0.0677 = 1.203 \text{ p.u.}$$

Hence $\delta = \tan^{-1}\left(\dfrac{0.406}{1.203}\right) = 18°$

This is a significant angle between the voltages across a circuit and so the use of the approximate equations for this heavily loaded, long transmission circuit will involve some inaccuracy.

Using equation (2.14) with R neglected.

$$V_G^2 = (1 + 0.0677 \times 3)^2 + (0.0677 \times 6)^2$$
$$V_G = 1.27 \text{ p.u.}$$

An approximate value of V_G can be found from equation (2.17) as

$$V_L - V_G = \Delta V = \frac{0.0677 \times 3}{1} = 0.203$$
$$V_G = 1.203 \text{ p.u.}$$

that is, an error of 5.6%.

With a shorter, 80 km, length of this line at the same load (still neglecting R) gives

$$XP = 0.0338 \times 6 = 0.203 \text{ p.u.}$$

and

$$V_L + QX = 1 + 3 \times 0.0338 = 1.101 \text{ p.u.}$$

$$\delta = \tan^{-1}\left(\frac{.203}{1.101}\right) = 10°$$

The accurate formula gives

$$V_G^2 = \left((1 + 0.101)^2 + (0.203)^2\right)$$
$$V_G = 1.120 \text{ p.u.}$$

and the approximate one gives

$$V_G = 1 + .101 = 1.101 \text{ p.u.}$$

that is, an error of 1.7%.

In the solution above, it should be noted that p.u. values for power system calculations are conveniently expressed to three decimal places, implying a measured value to 0.1% error. In practice, most measurements will have an error of at least 0.2% and possibly 0.5%.

If V_G is specified and V_L is required, the phasor diagram in Figure 2.25 is used.

From this,

$$V_L = \sqrt{\left[\left(V_G - \frac{QX}{V_G}\right)^2 + \left(\frac{PX}{V_G}\right)^2\right]} \qquad (2.18)$$

If

$$\frac{PX}{V_G} \ll \frac{V_G^2 - QX}{V_G} \quad \text{then}$$

$$V_L = V_G - \frac{QX}{V_G} \qquad (2.19)$$

2.6 Harmonics in Three-Phase Systems

The sine waves of the currents and voltages in a power system may not be perfect and so may contain harmonics. Harmonics are created by domestic non-linear loads, particularly computer and TV power supplies but also by large industrial rectifiers (e.g. for aluminium smelters). Consider the instantaneous values of phase voltages in a balanced system containing harmonics up to the third:

$$v_a = V_1 \sin \omega t + V_2 \sin 2\omega t + V_3 \sin 3\omega t \qquad (2.20)$$

$$v_b = V_1 \sin\left(\omega t - \frac{2\pi}{3}\right) + V_2 \sin 2\left(\omega t - \frac{2\pi}{3}\right) + V_3 \sin 3\left(\omega t - \frac{2\pi}{3}\right)$$

$$= V_1 \sin\left(\omega t - \frac{2\pi}{3}\right) + V_2 \sin\left(2\omega t - \frac{4\pi}{3}\right) + V_3 \sin\left(3\omega t - 2\pi\right) \qquad (2.21)$$

$$= V_1 \sin\left(\omega t - \frac{2\pi}{3}\right) + V_2 \sin\left(2\omega t - \frac{4\pi}{3}\right) + V_3 \sin 3\omega t$$

$$v_c = V_1 \sin\left(\omega t - \frac{4\pi}{3}\right) + V_2 \sin 2\left(\omega t - \frac{4\pi}{3}\right) + V_3 \sin 3\left(\omega t - \frac{4\pi}{3}\right)$$

$$= V_1 \sin\left(\omega t - \frac{4\pi}{3}\right) + V_2 \sin\left(2\omega t - 2\pi - \frac{2\pi}{3}\right) + V_3 \sin\left(3\omega t - 2\pi\right) \qquad (2.22)$$

$$= V_1 \sin\left(\omega t - \frac{4\pi}{3}\right) + V_2 \sin\left(2\omega t - \frac{2\pi}{3}\right) + V_3 \sin 3\omega t$$

Here V_1, V_2 and V_3 are the magnitudes of the harmonic voltages.

From equations (2.20)–(2.22) the phasor diagram for fundamental, second and third harmonic components of v_a, v_b and v_c shown in Figure 2.26 can be obtained. As shown in Figure 2.26, the fundamental terms have the normal phase rotation (as do the fourth, seventh and tenth, etc., harmonics). However, the second harmonic terms possess a reversed phase rotation (also the fifth, eighth, eleventh, etc.), and the third harmonic terms are all in phase (also all harmonics of multiples three).

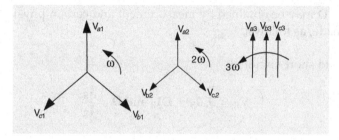

Figure 2.26 Phase rotation of harmonic voltage (fundamental, 2nd and 3rd harmonics)

When substantial harmonics are present the $\sqrt{3}$ relation between line and phase quantities no longer holds. As can be seen from Figure 2.26, harmonic voltages other than that of triple harmonics ($n = 3$, 6, 9, etc.) in successive phases are $2\pi/3$ out of phase with each other. In the resulting line voltages (as $\mathbf{V_{ab}}$ is given by ($\mathbf{V_a} - \mathbf{V_b}$)) no triple harmonics exist. The mesh connection forms a complete path for the triple harmonic currents which flow in phase around the loop.

When analysing the penetration of harmonics into the power network an initial approximation is to assume that the effective reactance for the nth harmonic is n times the fundamental value.

2.7 Useful Network Theory

2.7.1 Four-Terminal Networks

A lumped-constant circuit, provided that it is passive, linear and bilateral, can be represented by the four-terminal network shown in the diagram in Figure 2.27. The complex parameters $\mathbf{A}$, $\mathbf{B}$, $\mathbf{C}$ and $\mathbf{D}$ describe the network in terms of the sending- and receiving-end voltages and currents as follows:

$$\mathbf{V_S} = \mathbf{AV_R} + \mathbf{BI_R}$$
$$\mathbf{I_S} = \mathbf{CV_R} + \mathbf{DI_R}$$

and it can be readily shown that $\mathbf{AD} - \mathbf{BC} = 1$.

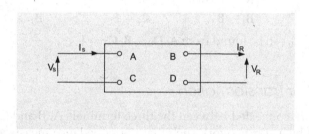

Figure 2.27 Representation of a four-terminal (two-port) network

A, B, C, and **D** may be obtained by measurement and certain physical interpretations can be made, as follows:

1. Receiving-end short-circuited:

$$\mathbf{V_R} = 0, \ \mathbf{I_S} = \mathbf{DI_R} \text{ and } \mathbf{D} = \frac{\mathbf{I_S}}{\mathbf{I_R}}$$

Also,

$$\mathbf{V_S} = \mathbf{BI_R} \text{ and } \mathbf{B} = \frac{\mathbf{V_S}}{\mathbf{I_R}} = \text{short circuit impedance}$$

2. Receiving-end open-circuited:
   Here

$$\mathbf{I_R} = 0, \ \mathbf{I_S} = \mathbf{CV_R} \ \text{ and } \mathbf{C} = \frac{\mathbf{I_S}}{\mathbf{V_R}}$$
$$\mathbf{V_S} = \mathbf{AV_R} \ \text{ and } \mathbf{A} = \frac{\mathbf{V_S}}{\mathbf{V_R}}$$

Expressions for the constants can be found by complex (that is, magnitude and angle) measurements carried out solely at the sending end with the receiving end open and short-circuited.

Often it is useful to have a single four-terminal network for two or more items in series or parallel, for example, a line and two transformers in series.

It is shown in most texts on circuit theory that the generalized constants for the combined network, $\mathbf{A_0}$, $\mathbf{B_0}$, $\mathbf{C_0}$, and $\mathbf{D_0}$ for the two networks (1) and (2) in cascade are as follows:

$$\mathbf{A_0} = \mathbf{A_1 A_2} + \mathbf{B_1 C_2} \quad \mathbf{B_0} = \mathbf{A_1 B_2} + \mathbf{B_1 D_2}$$
$$\mathbf{C_0} = \mathbf{A_2 C_1} + \mathbf{C_2 D_1} \quad \mathbf{D_0} = \mathbf{B_2 C_1} + \mathbf{D_1 D_2}$$

For two four-terminal networks in parallel it can be shown that the parameters of the equivalent single four-terminal network are:

$$\mathbf{A_0} = \frac{\mathbf{A_1 B_2} + \mathbf{A_2 B_1}}{\mathbf{B_1} + \mathbf{B_2}} \quad \mathbf{B_0} = \frac{\mathbf{B_1 B_2}}{\mathbf{B_1} + \mathbf{B_2}} \quad \mathbf{D_0} = \frac{\mathbf{B_2 D_1} + \mathbf{B_1 D_2}}{\mathbf{B_1} + \mathbf{D_2}}$$

$\mathbf{C_0}$ can be found from $\mathbf{A_0 D_0} - \mathbf{B_0 C_0} = 1$

2.7.2 Delta-Star Transformation

The delta network connected between the three terminals A, B and C of Figure 2.28 can be replaced by a star network such that the impedance measured between the terminals is unchanged. The equivalent impedances can be found as follows:

Impedance between terminals AB with C open-circuited is given by

$$Z_{OA} + Z_{OB} = Z_{AB}//(Z_{CA} + Z_{BC})$$

Similarly

$$Z_{OB} + Z_{OC} = Z_{BC}//(Z_{AB} + Z_{CA})$$

and

$$Z_{OC} + Z_{OA} = Z_{CA}//(Z_{BC} + Z_{AB})$$

From these three equations Z_{OA}, Z_{OB} and Z_{OC}, the three unknowns, can be determined as

$$Z_{OA} = \frac{Z_{AB}Z_{CA}}{Z_{AB} + Z_{BC} + Z_{CA}}$$

$$Z_{OB} = \frac{Z_{AB}Z_{BC}}{Z_{AB} + Z_{BC} + Z_{CA}} \tag{2.15}$$

$$Z_{OC} = \frac{Z_{BC}Z_{CA}}{Z_{AB} + Z_{BC} + Z_{CA}}$$

2.7.3 Star-Delta Transformation

A star-connected system can be replaced by an equivalent delta connection if the elements of the new network have the following values (Figure 2.28):

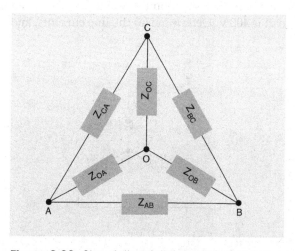

Figure 2.28 Star-delta, delta-star transformation

$$Z_{AB} = \frac{Z_{OA}Z_{OB} + Z_{OB}Z_{OC} + Z_{OC}Z_{OA}}{Z_{OC}}$$

$$Z_{BC} = \frac{Z_{OA}Z_{OB} + Z_{OB}Z_{OC} + Z_{OC}Z_{OA}}{Z_{OA}}$$

$$Z_{CA} = \frac{Z_{OA}Z_{OB} + Z_{OB}Z_{OC} + Z_{OC}Z_{OA}}{Z_{OB}}$$

Problems

2.1   The star-connected secondary winding of a three-phase transformer supplies 415 V (line to line) at a load point through a four-conductor cable. The neutral conductor is connected to the winding star point which is earthed. The load consists of the following components:

Between a and b conductors a 1 Ω resistor
Between a and neutral conductors a 1 Ω resistor
Between b and neutral conductors a 2 Ω resistor
Between c and neutral conductors a 2 Ω resistor

Connected to the a, b and c conductors is an induction motor taking a balanced current of 100 A at 0.866 power factor (p.f.) lagging.

Calculate the currents in the four conductors and the total power supplied.

Take the 'a' to neutral voltage as the reference phasor. The phase sequence is a-b-c.

(Answer: $I_A = 686 + j157$ A, $I_B = -506 + j361$ A, $I_C = -60 + j204$ A, 350 kW 350 kW)

2.2   The wye-connected load shown in Figure 2.29 is supplied from a transformer whose secondary-winding star point is solidly earthed. The line voltage supplied to the load is 400 V. Determine (a) the line currents, and (b) the voltage of

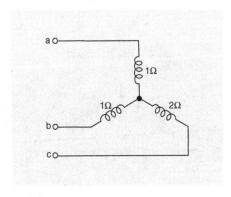

Figure 2.29 Circuit for Problem 2.2

the load star point with respect to ground. Take the 'a' to 'b' phase voltage as the reference phasor. The phase sequence is a-b-c.

(Answer: (a) $I_a = 200 - j69.3$ A; (b) $V_{ng} = -47$ V)

2.3 Two capacitors, each of 10 μF, and a resistor R, are connected to a 50 Hz three phase supply, as shown in Figure 2.30. The power drawn from the supply is the same whether the switch S is open or closed. Find the resistance of R.

(Answer: 142 Ω)

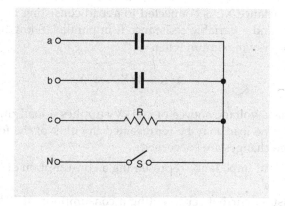

Figure 2.30 Circuit for Problem 2.3

2.4 The network of Figure 2.31 is connected to a 400 V three-phase supply, with phase sequence a-b-c. Calculate the reading of the wattmeter W.

(Answer: 2.33 kW)

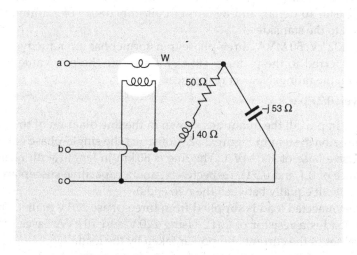

Figure 2.31 Circuit for Problem 2.4

2.5 A 400 V three-phase supply feeds a delta-connected load with the following branch impedances:

$$Z_{RY} = 100\Omega \quad Z_{YB} = j100\Omega \quad Z_{BR} = -j100\Omega$$

Calculate the line currents for phase sequences (a) RYB; (b) RBY.

(Answer: (a) 7.73, 7.73, 4 A: (b) 2.07, 2.07, 4 A)

2.6 A synchronous generator, represented by a voltage source in series with an inductive reactance X_1, is connected to a load consisting of a fixed inductive reactance X_2 and a variable resistance R in parallel. Show that the generator power output is a maximum when

$$1/R = 1/X_1 + 1/X_2$$

2.7 A single-phase voltage source of 100 kV supplies a load through an impedance $j100\,\Omega$. The load may be represented in either of the following ways as far as voltage changes are concerned:

a. by a constant impedance representing a consumption of 10 MW, 10 MVAr at 100 kV; or

b. by a constant current representing a consumption of 10 MW, 10 MVAr at 100 kV.

Calculate the voltage across the load using each of these representations.

(Answer: (a) $(90 - j8.2)$ kV; (b) $(90 - j10)$ kV)

2.8 Show that the p.u. impedance (obtained from a short-circuit test) of a star-delta three-phase transformer is the same whether computed from the star-side parameters or from the delta side. Assume a rating of G (volt-amperes), a line-to-line input voltage to the star-winding terminals of V volts, a turns ratio of 1 : N (star to delta), and a short circuit impedance of Z (ohms) per phase referred to the star side.

2.9 An 11/132 kV, 50 MVA, three-phase transformer has an inductive reactance of $j0.5\,\Omega$ referred to the primary (11 kV). Calculate the p.u. value of reactance based on the rating. Neglect resistance.

(Answer: 0.21 p.u.)

2.10 Express in p.u. all the quantities shown in the line diagram of the three-phase transmission system in Figure 2.32. Construct the single-phase equivalent circuit. Use a base of 100 MVA. The line is 80 km in length with resistance and reactance of 0.1 and 0.5 Ω, respectively, and a capacitive susceptance of 10 μS per km (split equally between the two ends).

2.11 A wye-connected load is supplied from three-phase 220 V mains. Each branch of the load is a resistor of 20 Ω. Using 220 V and 10 kVA bases calculate the p.u. values of the current and power taken by the load.

(Answer: 0.24 p.u.; 0.24 p.u.)

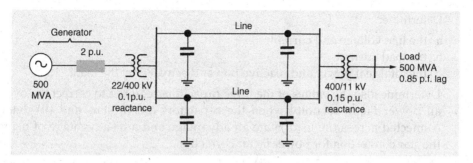

Figure 2.32 Circuit for Problem 2.10

2.12 A 440 V three-phase supply is connected to three star-connected loads in parallel, through a feeder of impedance $(0.1 + j0.5)\,\Omega$ per phase. The loads are as follows: 5 kW, 4 kVAr; 3 kW, 0 kVAr; 10 kW, 2 kVAr.
Determine:

a. line current;
b. power and reactive power losses in the feeder per phase;
c. power and reactive power from the supply and the supply power factor.

(Answer: (a) 24.9 A; (b) 62 W, 310 VAr; (c) 18.2 kW, 6.93 kVAr, 0.94)

2.13 Two transmission circuits are defined by the following **ABCD** constants: 1, 50, 0, 1, and $0.9\langle 2°$, $150\langle 79°$, $9 \times 10^{-4}\langle 91°$, $0.9\langle 2°$. Determine the **ABCD** constants of the circuit comprising these two circuits in series.

(Answer: $\mathbf{A} = 0.9\langle 4.85$; $\mathbf{B} = 165.1\langle 63.6$)

2.14 A 132 kV overhead line has a series resistance and inductive reactance per phase per kilometre of 0.156 and 0.4125 Ω, respectively. Calculate the magnitude of the sending-end voltage when transmitting the full line capability of 125 MVA when the power factor is 0.9 lagging and the received voltage is 132 kV, for 16 km and 80 km lengths of line. Use both accurate and approximate methods.

(Answer: 136.92 kV (accurate), 136.85 kV; 157.95 kV (accurate), 156.27 kV)

2.15 A synchronous generator may be represented by a voltage source of magnitude 1.7 p.u. in series with an impedance of 2 p.u. It is connected to a zero-impedance voltage source of 1 p.u. The ratio of X/R of the impedance is 10. Calculate the power generated and the power delivered to the voltage source if the angle between the voltage sources is 30°.

(Answer: 0.49 p.u.; 0.44 p.u.)

2.16 A three-phase star-connected 50 Hz generator generates 240 V per phase and supplies three delta-connected load coils each having a resistance of 10 Ω and an inductance of 47.75 mH.

Determine:

a. the line voltage and current;
b. the load current;
c. the total real power and reactive power dissipated by the load.

Determine also the values of the three capacitors required to correct the overall power factor to unity when the capacitors are (i) star- and (ii) delta-connected across the load. State an advantage and a disadvantage of using the star connection for power factor correction.

(Answer: (a) 416 V, 39.96 A; (b) 23.07 A; (c) 15.9 kW, 24.0 kVAr; (i) 441 μF; (ii) 147 μF)

(From Engineering Council Examination, 1996)

3

Components of a Power System

3.1 Introduction

Investigations of large interconnected electric power systems, either through manual calculations or computer simulation, use the simplest models of the various
components (for example, lines, cable, generators, transformers, and so on) that
show the phenomenon being studied without unnecessary detail. Representations
based on an equivalent circuit not only make the principles clearer but also these
simple models are used in practice. For more sophisticated treatments, especially of
synchronous machines and fast transients, the reader is referred to the more
advanced texts given in the Further Reading section at the end of the book.

Loads are considered as components even though their exact composition and
characteristics are not known with complete certainty and vary over time. When
designing a power supply system or extending an existing one a prediction of the
loads to be expected is required, statistical methods being used. This is the aspect of
power supply known with least precision.

Most of the equivalent circuits used are single phase and employ phase-to-neutral
values. This assumes that the loads are balanced three-phase which is reasonable for
normal steady-state operation. When unbalance exists between the phases, full
treatment of all phases is required, and special techniques for dealing with this are
described in Chapter 7.

3.2 Synchronous Machines

Large synchronous generators for power generation have one of two forms of rotor
construction. In the round or cylindrical rotor the field winding is placed in slots cut
axially along the rotor length as illustrated in Figure 3.1(a). The diameter is

Electric Power Systems, Fifth Edition. B.M. Weedy, B.J. Cory, N. Jenkins, J.B. Ekanayake and G. Strbac.
© 2012 John Wiley & Sons, Ltd. Published 2012 by John Wiley & Sons, Ltd.

relatively small (1–1.5 m) and the machine is suitable for operation at high speeds (3000 or 3600 r.p.m.) driven by a steam or gas turbine. Hence it is known as a turbo-generator. In a salient pole rotor the poles project, as shown in Figure 3.1(b), and the machine is driven by a hydro turbine or diesel engine at a lower rotational speed. The frequency of the generated voltage (*f*) and speed are related by

$$f = \frac{np}{60}$$

where *n* is speed in r.p.m. and *p* is the number of pairs of poles. A hydro turbine rotating at 150 r.p.m. thus needs 20 pole pairs to generate at 50 Hz and a large diameter rotor to accommodate the poles.

The three-phase currents in the stator winding or armature generate a magnetic field which rotates with the synchronously rotating rotor to which the d.c. field is fixed. The effect of the stator winding field on the rotor field is referred to as 'armature reaction'. Figure 3.1(b) indicates the paths of the main field fluxes in a salient pole machine from which it can be seen that these paths are mainly in iron, apart from crossing the air gap between stator and rotor. On the other hand, the

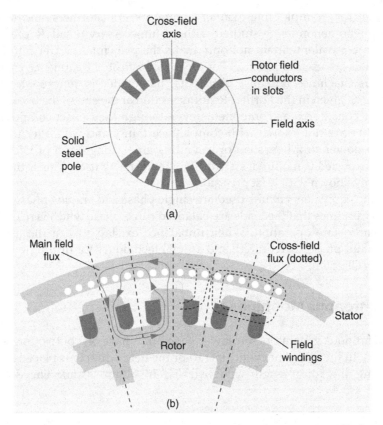

Figure 3.1 (a) Cylindrical rotor, (b) Salient pole rotor, and stator with flux paths

cross-field flux has a longer path in air and hence the reluctance of its path is greater than for the main field flux. Consequently, any e.m.f.s produced by the main field flux in the stator are larger than those produced by any cross-field fluxes.

Machine designers shape the poles in a salient machine or distribute the windings in a round-rotor machine to obtain maximum fundamental sinusoidal voltage on no-load in the stator windings. The stator winding is carefully designed and distributed such as to minimize the harmonic voltages induced in it and only the fundamental e.m.f. is significant for power system studies.

Whether a synchronous machine has a cylindrical or salient-pole rotor, its action is the same when generating power and can be best understood in terms of the 'primitive' representation shown in Figure 3.2(a). The rotor current I_f produces a magnetomotive force (m.m.f.) F_f rotating with the rotor and induces a sinusoidal voltage in the stator winding. When the stator windings are closed through an external circuit, current flows, thereby delivering power to the external load.

These stator currents produce an m.m.f., shown in Figure 3.2(a) for phase A as F_a, which interacts with the rotor m.m.f. F_f, thereby producing a retarding force on the rotor that requires torque (and hence power) to be provided from the prime mover. The vector m.m.f.s in Figure 3.2(b) are drawn for an instant at which the current I in

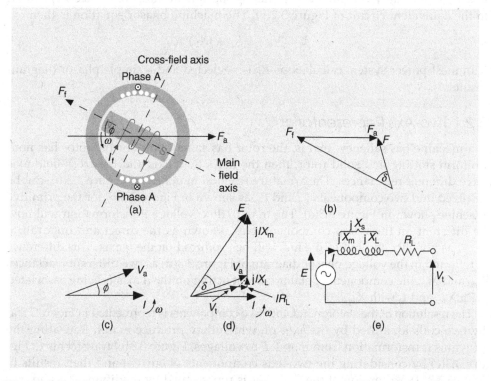

Figure 3.2 (a) Primitive machine at the instant of maximum current in phase A conductors and field m.m.f. F_f. (b) Corresponding space diagram of vector m.m.f.s with resultant F. (c) Terminal phasor diagram, (d) Phasor diagram with voltage representation of magnetic conditions, (e) Equivalent circuit for round-rotor machine

phase A is a maximum and the voltage is at an angle ϕ as shown in Figure 3.2(c). Currents in phases B and C of the stator (equal and opposite at this instant) also contribute along the axis of F_a. The resultant air-gap m.m.f. is shown as $F = F_a + F_f$, using vectorial addition as in Figure 3.2(b), and this gives rise to a displaced air-gap flux such that the phase of the stator-induced e.m.f. E is delayed from that produced on no-load for the same rotor position by the torque angle shown as δ in Figure 3.2(d).

Detailed consideration of generator action taking full account of magnetic path saturation requires a simulation of the fluxes, including all winding currents and eddy currents. However, for power system calculations the m.m.f. summation of Figure 3.2(b) can be transferred to the voltage-current phasor diagram of Figure 3.2(d) by rotation through 90° clockwise and scaling F to V_a. Then, F_f becomes E and F_a corresponds to IX_m where X_m is an equivalent reactance representing the effect of the magnetic conditions within the machine. Effects due to saturation can be allowed for by changing X_m and relating E to the field current I_f by the open circuit (no-load) saturation curve (see later). The machine terminal voltage V_t is obtained from V_a by recognizing that the stator phase windings have a small resistance R_L and a leakage reactance X_L (about 10% of X_m) resulting from flux produced by the stator but not crossing the air gap. The reactances X_L and X_m are usually considered together as the synchronous reactance X_S, shown in the equivalent circuit of Figure 3.2(e). The machine phasor equation is then

$$\mathbf{E} = \mathbf{V_t} + \mathbf{I}(R_L + jX_S)$$

In most power system calculations R_L is neglected and a simple phasor diagram results.

3.2.1 Two-Axis Representation

If a machine has saliency, that is, the rotor has salient poles or the rotor has non-uniform slotting in a solid rotor, then the main field axis and the cross-field axis have different reluctances. The armature reaction m.m.f., F_a in Figure 3.2(b), can be resolved into two components F_d and F_q, as shown in Figure 3.3(b) for the primitive machine shown in Figure 3.3(a). The m.m.f./flux/voltage transformation will now be different on these two component axes (known as the direct and quadrature axes) as less flux and induced e.m.f. will be produced on the q axis. This difference is reflected in the voltage phasor diagram of Figure 3.3(d) as two different reactances X_{ad} and X_{aq}, the component of stator current acting on the d axis I_d being associated with X_{ad}, and I_q with X_{aq}.

The resolution of the stator m.m.f. into two components is represented in Figure 3.3(a) by two coils identified by the axes on which they produce m.m.f. Repeating the previous transformation from m.m.f. F to voltages, Figure 3.3(b) transforms to Figure 3.3(c) by considering the two-axis components of current and then results in Figure 3.3(d). Saturation of the two axes is represented by modifying X_{ad} and X_{aq}. Because I has been resolved into two components it is no longer possible for a salient machine to be represented by an equivalent circuit. However, equations can be obtained for the V_d and F_q components of the terminal voltage V_t as:

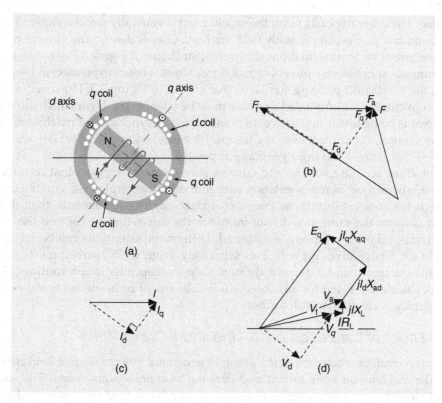

Figure 3.3 (a) Primitive machine – d and q axes, (b) The m.m.f. vectors resolved in d and q axes. F_d and F_q are components of F_a., (c) Current phasors resolved into I_d and I_q components, (d) Phasor diagram of voltages, E of Figure 3.2(d) now becomes E_q

$$V_q = E_q - jI_d(X_{ad} + X_L) - I_qR_L$$

$$V_d = -jI_q(X_{aq} + X_L) - I_dR_L$$

It will be seen that the resistive terms (if included) contain the currents from the other axis. The term E_q is the e.m.f. representing the action of the field winding and in many power system studies can be taken as constant if no automatic voltage regulator (AVR) action is assumed.

A single equation can be written in complex form as

$$\begin{aligned} E_q &= \mathbf{V} + \mathbf{I}R_L + j\mathbf{I}X_L + j(I_dX_{ad} + I_qX_{aq}) \\ &= \mathbf{V} + \mathbf{I}R_L + j(I_dX_d + I_qX_q) \end{aligned}$$

where $\qquad X_d = X_{ad} + X_L$

and $\qquad X_q = X_{aq} + X_L$

In a salient machine, $X_{ad} \neq X_{aq}$ and torque and power can be developed with no field current, the stator m.m.f. 'locking' in with the direct axis as in a reluctance machine. The effect depends upon the saliency but is generally small compared with the maximum power rating with field current. Consequently, for simple multi-machine power system calculations the equivalent circuit of Figure 3.2(e) can be used.

The transient behaviour of synchronous machines is often expressed in two axis terms, the basic concept being similar to that shown in Figure 3.3. However, a variety of conventions are employed and care must be taken to establish which direction of current is being taken as positive. In many texts, if a synchronous machine is running as a motor, the current entering the machine is taken as positive. Here we have assumed that current leaving a generator is positive.

To produce accurate voltage and current waveforms under transient conditions the two-axis method of representation with individually coupled coils on the d and q axes is necessary. Since these axes are at right-angles (orthogonal), then fluxes linking coils on the same axis do not influence the fluxes linking coils on the other axis, that is, no cross-coupling is assumed. Differential equations can be set up to account for flux changes on both axes separately, from which corresponding voltages and currents can be derived through step-by-step integration routines. This form of calculation is left for advanced study. The use of finite elements can extend the accuracy of calculation still further.

3.2.2 Effect of Saturation on X_S - the Short-Circuit Ratio

The open-circuit characteristic is the graph of generated voltage against field current with the machine on open circuit and running at synchronous speed. The short-circuit characteristic is the graph of stator current against field current with the terminals short-circuited. Both curves for a modern machine are shown in Figure 3.4. Here, X_S is equal to the open-circuit voltage produced by the same field current that produces rated current on short circuit, divided by this rated stator current. X_S is constant only over the linear part of the open-circuit characteristic (the air-gap line) when saturation is ignored. The actual value of X_S at full-load current will obviously be less than this value and several methods exist to allow for the effects of saturation.

The short-circuit ratio (SCR) of a generator is defined as the ratio between the field current required to give nominal open-circuit voltage and that required to circulate full-load current in the armature when short-circuited. In Figure 3.4 the short-circuit ratio is AH/AK, that is, 0.63. The SCR is also commonly calculated in terms of the air-gap line and the short-circuit curve, this giving an unsaturated value. To allow for saturation it is common practice to assume that the synchronous reactance is 1/SCR, which for this machine is 1.58 p.u. Economy demands the design of machines of low SCR and a value of 0.55 is common for modern machines. Unfortunately, transient stability margins are reduced as synchronous reactance increases (i.e. SCR reduces) so a conflict for design between economy and stability arises.

3.2.3 Turbogenerators

During the 1970s, the ratings of steam-driven turbine generators reached more than 1000 MW to obtain improvements in efficiency made possible by large turbine plant and their lower capital costs per megawatt.

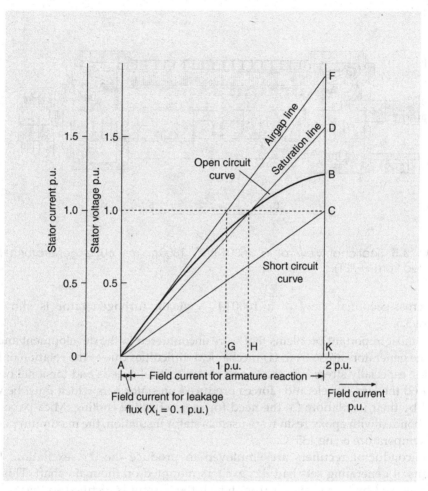

Figure 3.4 Open- and short-circuit characteristics of a synchronous machine. Unsaturated value of $X_S = FK/CK$. With operation near nominal voltage, the saturation line is used to give a linear characteristic with saturation $X_S = DK/CK$

Large generators are cooled by hydrogen circulating around the air-gap and de-ionized water pumped through hollow stator conductors, a typical statistic being:

500 MW, 588 MVA	winding cooling	rotor, hydrogen; stator, water;
	rotor diameter	1.12 m;
	rotor length	6.2 m;
	total weight	0.63 kg/kVA;
	rotor weight·	0.1045 kg/kVA.

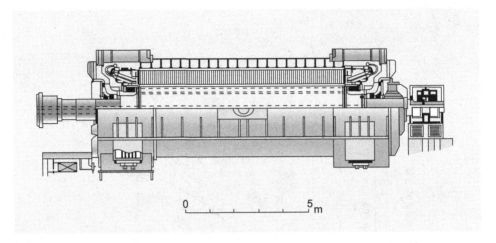

Figure 3.5 Sectional view of a 1000 MVA, 1800 r.p.m., 60 Hz generator (Figure adapted from IEE/IET)

A cross-sectional view of a 1000 MVA steam turbogenerator is shown in Figure 3.5.

The most important problems that were encountered in the development and use of large generators arose from (1) mechanical difficulties due to the rotation of large masses, especially stresses in shafts and rotors, critical speeds and torsional oscillations; (2) the large acceleration forces produced on stator bars which must be withstood by their insulation; (3) the need for more effective cooling. Mica paper and glass bonded with epoxy resin were used as stator insulation, the maximum permissible temperature being 135 °C.

Semiconductor rectifiers are employed to produce the d.c. excitation. Early designs of generating sets had d.c. exciters mounted on the main shaft. This type was followed by an a.c. exciter from which the current is rectified so that a.c. and d.c. slip-rings are required. In more modern machines, a.c. exciters with integral fused-diode or thyristor rectifiers are employed, thus avoiding any brush gear and consequent maintenance problems, the semiconductors rotating on the main shaft.

Reliability of turbogenerators is all-important, and the loss resulting from an outage of a large conventional generator for one day is roughly equivalent to 1% of the initial cost of the machine.

3.3 Equivalent Circuit Under Balanced Short-Circuit Conditions

A typical set of oscillograms of the currents in three stator phases when a synchronous generator is suddenly short-circuited is shown in Figure 3.6(a). In all three traces a direct-current component is evident and this is to be expected from a knowledge of transients in R-L circuits. The magnitude of the initial transient direct current in any phase present depends upon the instant at which the short circuit is applied and the speed of its decay depends on the power factor of the circuit. As

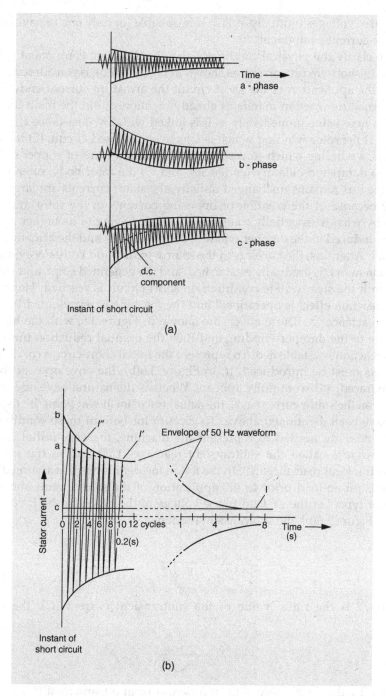

Figure 3.6 (a) Oscillograms of the currents in the three phases of a generator when a sudden short circuit is applied, (b) Trace of a short-circuit current when direct-current component is removed

there are three voltages mutually at 120° it is possible for only one to have the maximum direct-current component.

Often, to clarify the physical conditions, the direct current component is ignored and a trace of short-circuit current, as shown in Figure 3.6(b), is considered. Immediately after the application of the short circuit the armature current endeavours to create an armature reaction m.m.f., as already mentioned, but the main flux cannot change to a new value immediately as it is linked with low-resistance circuits consisting of: (1) the rotor winding which is effectively a closed circuit; (2) the damper bars, that is, a winding which consists of short-circuited turns of copper strip set in the poles to dampen oscillatory tendencies, and (3) the rotor body, often of forged steel. As the flux remains unchanged initially, the stator currents are large and can flow only because of the creation of opposing currents in the rotor and damper windings by what is essentially transformer action. Owing to its higher resistance, the current induced in the damper winding decays rapidly and the armature current starts to fall. After this, the currents in the rotor winding and body decay, the armature reaction m.m.f. is gradually established, and the generated e.m.f. and stator current fall until the steady-state condition on short circuit is reached. Here, the full armature reaction effect is operational and the machine is represented by the synchronous reactance X_S. These effects are shown in Figure 3.6, with the high initial current due to the damper winding and then the gradual reduction until the full armature reaction is established. To represent the initial short-circuit conditions two new models must be introduced. If, in Figure 3.6(b), the envelopes of the 50 Hz waves are traced, a discontinuity appears. Whereas the natural envelope continues to point 'a' on the stator current axis, the actual trace finishes at point 'b'; the reasons for this have been mentioned above. To account for both of these conditions, two new reactances are needed to represent the machine, the very initial conditions requiring what is called the subtransient reactance (X'') and in the subsequent period the transient reactance (X'). In the following definitions it is assumed that the generator is on no-load prior to the application of the short circuit and is of the round-rotor type. Let the no-load phase voltage of the generator be E volt (r.m.s.), then, from Figure 3.6(b), the subtransient reactance

$$X'' = \frac{E}{0b/\sqrt{2}}$$

where $0b/\sqrt{2}$ is the r.m.s. value of the subtransient current (I''); the transient reactance

$$X' = \frac{E}{0a/\sqrt{2}}$$

where $0a/\sqrt{2}$ is the r.m.s. value of the transient current (I'), and finally

$$X_S = \frac{E}{0c/\sqrt{2}}$$

Table 3.1 Constants of synchronous machines – 60 Hz (all values expressed as per unit on rating)

Type of machine	X_S (or X_d)	X_q	X'	X''	X_2	X_0	R_L
Turbo-alternator	1.2–2.0	1–1.5	0.2–0.35	0.17–0.25	0.17–0.25	0.04–0.14	0.003–0.008
Salient pole (hydroelectric)	0.16–1.45	0.4–1.0	0.2–0.5	0.13–0.35	0.13–0.35	0.02–0.2	0.003–0.0015
Synchronous compenstor	1.5–2.2	0.95–1.4	0.3–0.6	0.18–0.38	0.17–0.37	0.03–0.15	0.004–0.01

X_2 = negative sequence reactance.
X_0 = zero sequence reactance.
X' and X'' are the direct axis quantities.
R_L = a.c. resistance of the stator winding per phase.

Typical values of X'', X' and X_S for various types and sizes of machines are given in Table 3.1.

If the machine is previously on load, the voltage applied to the equivalent reactance, previously E, is now modified due to the initial load volt-drop. Consider Figure 3.7. Initially, the load current is I_L and the terminal voltage is V. The voltage behind the transient reactance X' is

$$\mathbf{E'} = \mathbf{I_L}(\mathbf{Z_L} + j X')$$

$$= \mathbf{V} + j\mathbf{I_L} X'$$

and hence the transient current on short circuit $= \dfrac{\mathbf{E'}}{jX'}$.

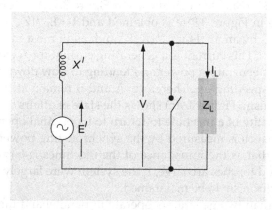

Figure 3.7 Modification of equivalent circuit to allow for initial load current

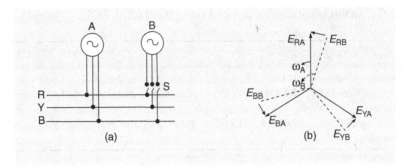

Figure 3.8 (a) Generators in parallel, (b) Corresponding phasor diagrams

3.4 Synchronous Generators in Parallel

Consider two machines A and B (as shown in Figure 3.8(a)), the voltages of which have been adjusted to be equal by their field regulators, and the speeds of which are slightly different. In Figure 3.8(b) the phase voltages are E_{RA}, and so on, and the speed of machine A is ω_A radians per second and of B, ω_B radians per second.

If the voltage phasors of A are considered stationary, those of B rotate at a relative velocity $(\omega_B - \omega_A)$ and hence there are resultant voltages across the switch S of $(E_{RA} - E_{RB})$, and so on, which reduce to zero once during each relative revolution. If the switch is closed at an instant of zero voltage, the machines are connected (synchronized) without the flow of large currents due to the resultant voltages across the armatures. When the two machines are in synchronism they have a common terminal-voltage, speed and frequency. Any tendency for one machine to accelerate relative to the other immediately results in a retarding or synchronizing torque being set up due to the current circulated.

Two machines operating in parallel, with $E_A = E_B$ and on no external load, are represented by the equivalent circuit shown in Figure 3.9(a). If A tries to gain speed the phasor diagram in Figure 3.9(b) is obtained and $I = E_R/(Z_A + Z_B)$. The circulating current I lags E_R by an angle $\tan^{-1}(X/R)$ and, as in most machines $X \gg R$, this angle approaches 90°. This current is a generating current for A and a motoring current for B; hence A is generating power and tending to slow down and B is receiving power from A and speeding up. Therefore, A and B remain at the same speed, 'in step', or in synchronism. Figure 3.9(c) shows the state of affairs when B tries to gain speed on A. The quality of a machine to return to its original operating state after a momentary disturbance is measured by the synchronizing power and torque. It is interesting to note that as the impedance of the machines is largely inductive, the restoring powers and torques are large; if the system were largely resistive it would be difficult for synchronism to be maintained.

Normally, more than two generators operate in parallel and the operation of one machine connected in parallel with many others is of great interest. If the remaining machines in parallel are of such capacity that the presence of the generator under

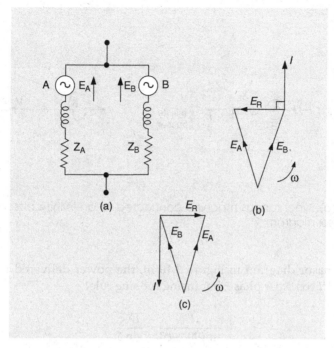

Figure 3.9 (a) Two generators in parallel–equivalent circuit, (b) Machine A in phase advance of machine B. (c) Machine B in phase advance of machine A

study causes no difference to the voltage and frequency of the other, they are said to comprise an infinite busbar system, that is, an infinite system of generation. In practice, a perfect infinite busbar is never fully realized but if, for example, a 600 MW generator is removed from a 30 000 MW system, the difference in voltage and frequency caused will be very slight.

3.5 The Operation of a Generator on an Infinite Busbar

In this section, in order to simplify the ideas as much as possible, the resistance of the generator will be neglected; in practice, this assumption is usually reasonable. Figure 3.10(a) shows the schematic diagram of a machine connected to an infinite busbar along with the corresponding phasor diagram. If losses are neglected the power output from the turbine is equal to the power output from the generator. The angle δ between the **E** and **V** phasors is known as the load angle and depends on the power input from the turbine shaft. With an isolated machine supplying its own load the load dictates the power required. When connected to an infinite busbar the load delivered by the machine is no longer directly dependent on the connected load. By changing the turbine output, and hence δ, the generator can be made to take on any load that the operator desires subject to economic and technical limits.

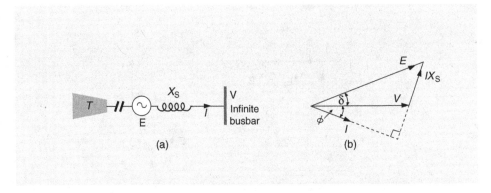

Figure 3.10 (a) Synchronous machine connected to an infinite busbar, (b) Corresponding phasor diagram

From the phasor diagram in Figure 3.10(b), the power delivered to the infinite busbar $= P = VI \cos \phi$ per phase, but (using the sine rule)

$$\frac{E}{\sin(90 + \phi)} = \frac{IX_S}{\sin \delta}$$

Hence

$$I \cos \phi = \frac{E}{X_S} \sin \delta$$

$$\text{Power delivered} = \frac{VE}{X_S} \sin \delta \tag{3.1}$$

This expression is of extreme importance as it governs, to a large extent, the operation of a power system. It could have been obtained by neglecting resistance and hence setting θ in Equation (2.11) to 90°.

Equation (3.1) is shown plotted in Figure 3.11. The maximum power is obtained at $\delta = 90°$. If δ becomes larger than 90° due to an attempt to obtain more than P_{max}, increase in δ results in less power output and the machine becomes unstable and loses synchronism. Loss of synchronism results in the interchange of current surges between the generator and network as the poles of the machine pull into synchronism and then out again; that is, the generator 'pole slips'.

If the power output of the generator is increased by small increments with the no-load voltage kept constant, the limit of stability occurs at $\delta = 90°$ and is known as the steady-state stability limit. There is another limit of stability due to a sudden large change in conditions, such as caused by a fault, and this is known as the transient stability limit; it is possible for the rotor to oscillate beyond 90° a number of times. If these oscillations diminish, the machine is stable. The load angle δ has a physical significance; it is the angle between like radial marks on the end of the rotor shaft of

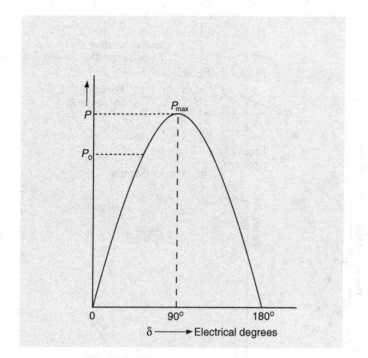

Figure 3.11 Power-angle curve of a synchronous machine. Resistance and saliency are neglected

the machine and on an imaginary rotor representing the system. The marks are in identical physical positions when the machine is on no-load. The synchronizing power coefficient

$$= \frac{dP}{d\delta} \quad \text{[watts per radian]}$$

and the synchronizing torque coefficient

$$= \frac{1}{\omega_S} \frac{dP}{d\delta}$$

3.5.1 The Performance Chart of a Synchronous Generator

It is convenient to summarize the operation of a synchronous generator connected to an infinite busbar in a single diagram or chart which allows an operator to see immediately whether the machine is operating within the limits of stability and rating.

Consider Figure 3.12(a), the phasor diagram for a round-rotor machine connected to a constant voltage busbar, ignoring resistance. The locus of constant IX_S, hence,

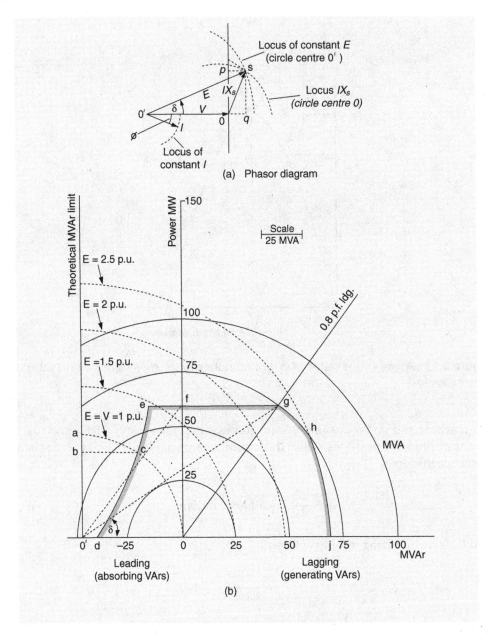

Figure 3.12 Performance chart of a synchronous generator

and constant MVA, is a circle, and the locus of constant excitation E is another circle. Therefore, provided V is constant, then

0s is proportional to VI or MVA
ps is proportional to $VI \sin \phi$ or MVAr
sq is proportional to $VI \cos \phi$ or MW.

To obtain the scaling factor for MVA, MVAr, and MW, the fact that at zero excitation, $E = 0$ and $IX_S = V$, is used, from which I is V/Xs at 90° leading to 00' corresponding to VAr/phase.

The construction of a chart for a 60 MW machine follows (Figure 3.12(b)).

Machine Data

60 MW, 0.8 p.f., 75 MVA
11.8 kV, SCR 0.63, 3000 r.p.m.

$$Z_{base} = \frac{11.8^2}{75} = 1.856\Omega$$

$$X_S = \frac{1}{0.63}\text{p.u.} = 1.59\text{pu} = 2.94\Omega/\text{phase}$$

The chart will refer to complete three-phase values of MW and MVAr. When the excitation and hence E are reduced to zero, the current leads V by 90° and is equal to (V/X_S), that is, $11\,800/(\sqrt{3} \times 2.94)$. The leading VArs correspond to this, given by $3V^2/X_S = 47.4$ MVAr.

With centre 0 a number of semicircles are drawn of radii equal to various MVA loadings, the most important being the 75 MVA circle. Also, arcs with 0' as centre are drawn with various multiples of 00' (or V) as radii to give the loci for constant excitations. Lines may also be drawn from 0 corresponding to various power factors, but for clarity only 0.8 p.f. lagging is shown, that is, the machine is generating VArs. The operational limits are fixed as follows. The rated turbine output gives a 60 MW limit which is drawn as shown, that is, a line for example, which meets the 75 MVA line at g. The MVA arc governs the thermal loading of the machine, that is, the stator temperature rise, so that over the portion gh the output is decided by the MVA rating. At point h the rotor heating becomes more decisive and the arc hj is decided by the maximum excitation current allowable, in this case assumed to be 2.5 p.u. The remaining limit is that governed by loss of synchronism at leading power factors. The theoretical limit is the line perpendicular to 00' at 0' (i.e. $\delta = 90°$), but, in practice, a safety margin is introduced to allow a further increase in load of either 10 or 20% before instability. In Figure 3.12(b) a 10% margin is used and is represented by ecd: it is constructed in the following manner. Considering point 'a' on the theoretical limit on the $E = 1$ p.u. arc, the power 0'a is reduced by 10% of the rated power (i.e. by 6 MW) to 0'b; the operating point must, however, still be on the same E arc and b is projected to c, which is a point on the new limiting curve. This is repeated for several excitations, finally giving the curve ecd.

The complete operating limit is shown shaded and the operator should normally work within the area bounded by this line, provided the generator is running at terminal rated voltage. If the voltage is different from rated, for example, at −5% of rated, the whole operating chart will shrink by a pro-rata amount, except for the excitation and turbine maximum power line. Note, therefore, that for the generator to produce rated power at reduced voltage, the stator requires to operate on over-current. This may be possible if an overcurrent rating is given.

Example 3.1

Use the chart of Figure 3.12 to determine the excitation and the load angle at full load (60 MW, 0.8 p.f. lag) and at unity power factor, rated output. Check by calculation.

Solution
From the diagram, the following values are obtained:

	Excitation E	Angle δ
0.8 p.f. lag	2.3 p.u.	33°
u.p.f.	1.7 p.u.	50°

Check

$$Power = \frac{VE}{X_S} \sin \delta$$

$$\sin \delta = \frac{60 \times 10^6 \times 2.94}{11800 \times (11800 \times 2.3)}$$

$$\delta = 0.55 \text{rad} = 33.4 \text{deg}$$

Repeating check in per unit (on 75 MVA, 11.8 kV base)

$$\sin \delta = \frac{0.8 \times 1.58}{2.3}$$

$$\delta = 0{,}55 \text{rad}$$

3.6 Automatic Voltage Regulators (AVRs)

Excitation of a synchronous generator is derived from a d.c. supply with variable voltage and requires considerable power to produce the required operating flux. Hence an exciter may require 1 MW or more to excite a large generator's field winding on the rotor. Nowadays, such a power amplifier is best arranged through an a.c. generator driven or overhung on the rotor shaft feeding the rotor through diodes, also mounted on the shaft and rotating with it (see Figure 3.13). The diodes can be replaced by thyristor controllers and the power amplifier can be eliminated.

In general the deviation of the terminal voltage from a prescribed value is passed to control circuits and thus the field current is varied. A general block diagram of a typical control system is given in Figure 3.14.

The most important aspect of the excitation system is its speed of response between a change in the reference signal V_r and the change in the excitation current I_f.

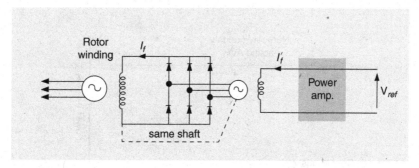

Figure 3.13 Excitation arrangements for a synchronous generator using rotating diodes

Control system theory using Bode diagrams and phase margins is required to design appropriate responses and stability boundaries of a generator excitation system.

3.6.1 Automatic Voltage Regulators and Generator Characteristics

The equivalent circuit used to represent the synchronous generator can be modified to account for the action of a regulator. Basically there are three conditions to consider:

1. Operation with fixed excitation and constant no-load voltage (E), that is, no regulator action. This requires the usual equivalent circuit of E in series with X_S.

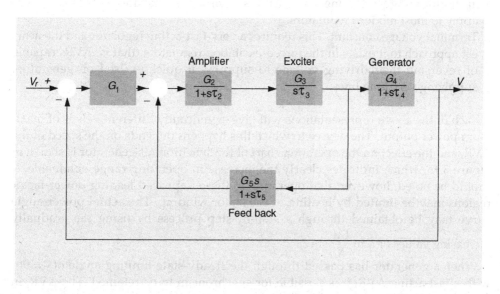

Figure 3.14 Block diagram of a continuously acting closed-loop automatic voltage regulator

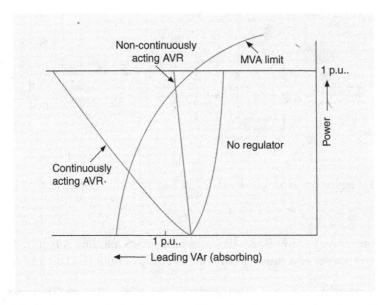

Figure 3.15 Performance chart as modified by the use of automatic voltage regulators

2. Operation with a regulator which is not continuously acting, that is, the terminal voltage varies with load changes. This can be simulated by E' and a reactance smaller than the synchronous value. It has been suggested by experience, in practice, that a reasonable value would be the transient reactance, although some authorities suggest taking half of the synchronous reactance. This mode will apply to most modern regulators.
3. Terminal voltage constant. This requires a very-fast-acting regulator and the nearest approach to it exists in the forced-excitation regulators (that is, AVRs capable of reversing their driving voltage to suppress it quickly) used on generators supplying very long lines.

Each of the above representations will give significantly different values of maximum power output. The degree to which this happens depends on the speed of the AVR, and the effect on the operation chart of the synchronous generator is shown in Figure 3.15, which indicates clearly the increase in operating range obtainable. It should be noted, however, that operation in these improved leading power-factor regions may be limited by heating of the stator winding. The actual power-angle curve may be obtained through a step-by-step process by using the gradually increasing values of E in $\dfrac{EV}{X} \sin \delta$.

When a generator has passed through the steady-state limiting angle of $\delta = 90°$ with a fast-acting AVR, it is possible for synchronism to be retained. The AVR, in forcing up the voltage, increases the power output of the machine so that instead of the power falling after $\delta = 90°$, it is maintained and $dP/d\delta$ is still positive. This is

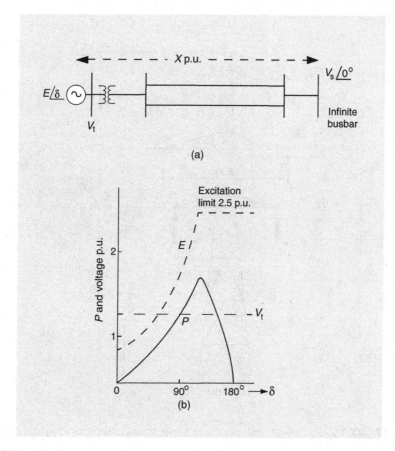

Figure 3.16 (a) Generator feeding into an infinite busbar (b) variation of output power P, generator voltage E, and terminal voltage V_t with load (transmission angle δ). Perfect AVR

illustrated in Figure 3.16(b), where the $P - \delta$ relation for the system in Figure 3.16(a) is shown. Without the AVR the terminal voltage of the machine V_t will fall with increased δ, the generated voltage E being constant and the power reaching a maximum at 90°. With a perfect AVR, V_t is maintained constant, E being increased with increase in δ. As $P = (EV_S/X) \sin \delta$, it is evident that power will increase beyond 90° until the excitation limit is reached, as shown in Figure 3.16(b).

3.7 Lines, Cables and Transformers

3.7.1 Overhead Lines –Types and Parameters

Overhead lines are suspended from insulators which are themselves supported by towers or poles. The insulation of the conductors is air. The span between two lowers depends on the allowable sag in the line, and for steel towers with very

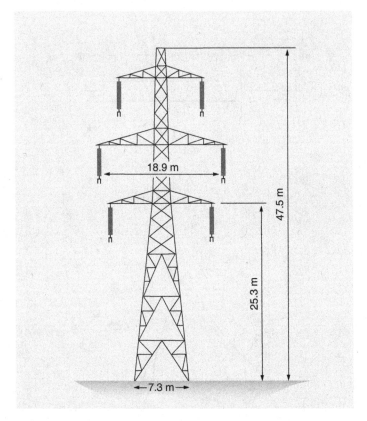

Figure 3.17 400 kV double-circuit overhead-line tower. Two conductors per phase (bundle conductors)

high-voltage lines the span is normally 370–460 m (1200–1500 ft). Typical supporting structures are shown in Figures 3.17 and 3.18. There are two main types of tower:

1. Those for straight runs in which the stress due to the weight of the line alone has to be withstood.
2. Those for changes in route, called deviation towers; these withstand the resultant forces set up when the line changes direction.

When specifying towers and lines, ice and wind loadings are taken into account, as well as extra forces due to a break in the conductors on one side of a tower. For lower voltages and distribution circuits, wood or reinforced concrete poles are used with conductors supported in horizontal formation.

The live conductors are insulated from the towers by insulators which take two basic forms: the pin-type and the suspension type. The pin-type insulator is shown in Figure 3.19 and it is seen that the conductor is supported on the top of the insulator. This type is used for lines up to 33 kV. The two or three porcelain 'sheds' or 'petticoats' provide an adequate leakage path from the conductor to earth and are

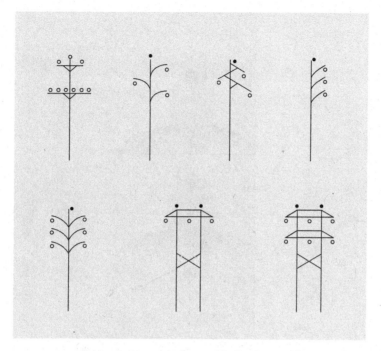

Figure 3.18 Typical pole-type structures

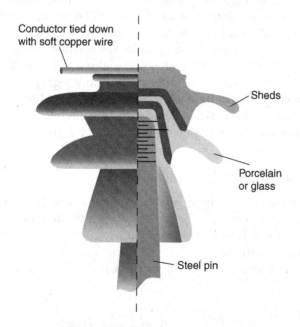

Figure 3.19 Pin-type insulators

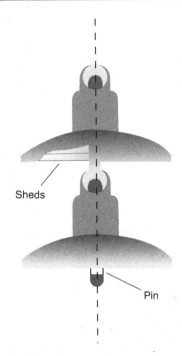

Figure 3.20 Suspension-type insulators

shaped to follow the equi-potentials of the electric field set up by the conductor-tower system. Suspension insulators (Figure 3.20) consist of a string of interlinking separate discs made of glass or porcelain. A string may consist of many discs depending upon the line voltage; for 400 kV lines, 19 discs of overall length 3.84 m (12 ft 7 in) are used. The conductor is held at the bottom of the string which is suspended from the tower. Owing to the capacitances existing between the discs, conductor and tower, the distribution of voltage along the insulator string is not uniform, the discs nearer the conductor being the more highly stressed. Methods of calculating this voltage distribution are available, but are of limited value owing to the shunting effect of the leakage resistance (see Figure 3.21). This resistance depends on the presence of pollution on the insulator surfaces and is considerably modified by rain and fog.

3.7.1.1 Inductance and Capacitance

The parameters of interest for circuit analysis are inductance, capacitance and resistance. The derivation of formulae for the calculation of these quantities is given in many reference and text books. It is intended here merely to quote these formulae and discuss their application.

The inductance of a single-phase two-wire line is

$$L = \frac{\mu_0}{4\pi}\left[1 + 4\ln\left(\frac{d-r}{r}\right)\right]H/m$$

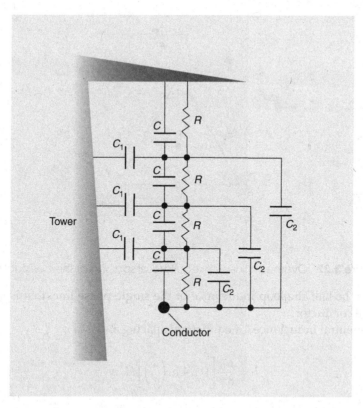

Figure 3.21 Equivalent circuit of a string of four suspension insulators. C = self capacitance of disc; C_1 = capacitance disc to earth; C_2 = capacitance disc to line; R = leakage resistance

Since $d \gg r$

$$L = \frac{\mu_0}{4\pi}\left[1 + 4\ln\left(\frac{d}{r}\right)\right]$$

$$= \frac{\mu_0}{4\pi}\left[4\ln\left(e^{1/4}\right) + 4\ln\left(\frac{d}{r}\right)\right] = \frac{\mu_0}{\pi}\ln\left(\frac{d}{re^{-1/4}}\right)$$

Defining $r' = re^{-1/4}$ and substituting $\mu_0 = 4\pi \times 10^{-7}$; $L = 4\ln\left(\frac{d}{r'}\right) \times 10^{-7}$

where d is the distance between the centres, and r is the radius, of the conductors.

When performing load flow and balanced-fault analysis on three-phase systems it is usual to consider one phase only, with the appropriate angular adjustments made for the other two phases. Therefore, phase voltages are used and the inductances and capacitances are the equivalent phase or line-to-neutral values. For a three-phase line with equilateral spacing (Figure 3.22) the inductance and capacitance with respect to the hypothetical neutral conductor are used, and this inductance can

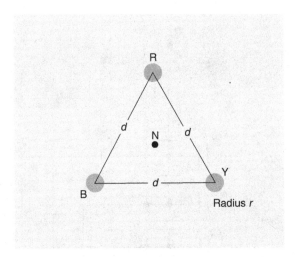

Figure 3.22 Overhead line with equilateral spacing of the conductors

be shown to be half the loop inductance of the single-phase line, that is, the inductance of one conductor.

The line-neutral inductance for equilateral spacing and $d \gg r$

$$L = \frac{\mu_0}{8\pi} \left[1 + 4\ln\left(\frac{d}{r}\right)\right] H/m$$

The capacitance of a single-phase line (again $d \gg r$)

$$C = \frac{\pi\varepsilon_0}{\ln[d/r]} \quad F/m$$

With three-phase conductors spaced equilaterally, the capacitance of each line to the hypothetical neutral is double that for the two-wire circuit, that is.

$$C = \frac{2\pi\varepsilon_0}{\ln[d/r]} \quad F/m$$

Although the conductors are rarely spaced in the equilateral formation, it can be shown that the average value of inductance or capacitance for any formation of conductors can be obtained by the representation of the system by one of equivalent equilateral spacing (Figure 3.23). The equivalent spacing d_{eq} between conductors is given by

$$d_{eq} = \sqrt[3]{d_{12} \times d_{23} \times d_{31}}$$

The use of bundle conductors, that is, more than one conductor per insulator, reduces the reactance; it also reduces conductor to ground surface voltage gradients

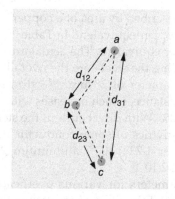

Figure 3.23 Geometric spacing of conductors

and hence corona loss and radio interference. When two circuits are situated on the same towers the magnetic interaction between them should be taken into account.

Unsymmetrical conductor spacing results in different inductances for each phase which causes an unbalanced voltage drop, even when the load currents are balanced. The residual or resultant voltage or current induces unwanted voltages into neighbouring communication lines. This can be overcome by the interchange of conductor positions at regular intervals along the route, a practice known as transposition (see Figure 3.24). In practice, lines are rarely transposed at regular intervals and transposition is carried out where physically convenient, for example at substations. In short lines the degree of unbalance existing without transposition is small and may be neglected in calculations.

3.7.1.2 Resistance

Overhead-line conductors usually comprise a stranded steel core (for mechanical strength) surrounded by aluminium wires which form the conductor. It should be noted that aluminium and ACSR (aluminium conductor steel reinforced)

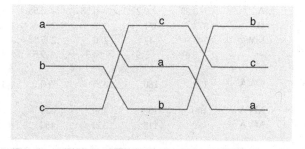

Figure 3.24 Transposition of conductors

conductors are sometimes described by area of a copper conductor having the same d.c. resistance, that is their copper equivalent. In Table 3.2a, 258 mm^2 (0.4 in^2) ACSR conductor implies the copper equivalent. The actual area of aluminium is approximately 430 mm^2, and including the steel core the overall cross-section of the line is about 620 mm^2, that is a diameter of 28 mm. Electromagnetic losses in the steel increase the effective a.c. resistance, which increases with current magnitude, giving an increase of up to about 10%. With direct current the steel carries 2–3% of the total current. The electrical resistivities of some conductor materials are as follows in ohm-metres at 20 °C: copper 1.72×10^{-8}; aluminium 2.83×10^{-8} (3.52×10^{-8} at 80 °C); aluminium alloy 3.22×10^{-8}.

Table 3.2a gives the parameters for various overhead-line circuits for the line voltages operative in Britain, Table 3.2b gives values for international lines, and Table 3.2c relates to the USA.

3.7.2 Representation of Lines

The manner in which lines and cables are represented depends on their length and the accuracy required. There are three broad classifications of length: short, medium and long. The actual line or cable is a distributed constant circuit, that is it has resistance, inductance and capacitance distributed evenly along its length, as shown in Figure 3.25.

Except for long lines the total resistance, inductance and capacitance of the line are concentrated to give a lumped-constant circuit. The distances quoted are only a rough guide.

Table 3.2a Overhead-line constants at 50 Hz (British) (per phase, per km)

Voltage: No. and area of conductors (mm)^2:		132 kV		275 kV		400 kV	
		1×113	1×258	2×113	2×258	2×258	4×258
Resistance (R)	Ω	0.155	0.068	0.022	0.034	0.034	0.017
Reactance (X_L)	Ω	0.41	0.404	0.335	0.323	0.323	0.27
Susceptance ($1/X_c$)	$S \times 10^{-6}$	7.59	7.59	9.52	9.52	9.52	10.58
Charging current (I_c)	A	0.22	0.22	0.58	0.58	0.845	0.945
Surge impedance	Ω	373	371	302	296	296	258
Natural load	MW	47	47	250	255	540	620
X_L/R ratio	—	2.6	5.9	4.3	9.5	9.5	15.8
Thermal rating:							
Cold weather (below 5 °C)	MVA	125	180	525	760	1100	2200
Normal (5–18 °C)	MVA	100	150	430	620	900	1800
Hot weather (above 18 °C)	MVA	80	115	330	480	790	1580

Table 3.2b Typical characteristics of bundled-conductor E.H.V. lines

Number of subconductors in bundle	Country	Line voltage and number of circuits (in parentheses)	Diameter of subconductors	Radius of circle on which subconductors are arranged	Resistance of bundle	Inductive reactance at 50 Hz	Susceptance at 50 Hz
		kV	mm	mm	Ω/km	Ω/km	μS/km
1	Japan	275 (2)	27.9	—	0.0744	0.511	3.01
	Canada	300 (2)	35.0	—	0.0451	0.492	2.33
	Australia	330 (1)	45.0	—	0.0367	0.422	2.78
	Russia	330 (1)	30.2	—	0.065	0.404	2.82
	Italy	380 (1)	50.0	—	0.0294	0.398	2.84
Average (one conductor):					0.048	0.442	2.75
2	Japan	275 (2)	25.3	200	0.0444	0.374	4.35
	Canada	360 (1)	28.1	299	0.04	0.314	3.57
	Australia	330 (1)	31.8	190.5	0.0451	0.341	3.34
	U.S.A.	345 (1)	30.4	228.6	0.0315	0.325	3.55
	Italy	380 (1)	31.5	200	0.0285	0.315	3.57
	Russia	330 (1)	28.0	200	0.04	0.321	3.43
Average (two conductors):					0.040	0.323	3.58
3	Sweden	380 (1)	31.68	260	0.018	0.29	3.85
	Russia	525 (1)	30.2	230	0.0212	0.294	3.88
4	Germany	380 (2)	21.7	202	0.0316	0.260	4.3
	Germany	500 (1)	22.86	323	0.0285	0.272	4.0
	Canada	735 (1)	35.04	323	0.0121	0.279	3.9

Table 3.2c Overhead-line parameters – USA

Line voltage (kV)	345	345	500	500	735	735
Conductors per phase at 18 in spacing	1	2	2	4	3	4
Conductor code	Expanded	Curlew	Chuker	Parakeet	Expanded	Pheasant
Conductor diameter (inches)	1.750	1.246	1.602	0.914	1.750	1.382
Phase spacing (ft)	28	28	38	38	56	56
GMD (ft)	35.3	35.3	47.9	47.9	70.6	70.6
60 Hz inductive reactance Ω/mile X_A	0.3336	0.1677	0.1529	0.0584	0.0784	0.0456
X_D	0.4325	0.4325	0.4694	0.4694	0.5166	0.5166
$X_A + X_D$	0.7661	0.6002	0.6223	0.5278	0.5950	0.5622
60 Hz capacitive reactance MΩ-miles X'	0.0777	0.0379	0.0341	0.0126	0.0179	0.0096
X'_D	0.1057	0.1057	0.1147	0.1147	0.1263	0.1263
$X'_A + X'_D$	0.1834	0.1436	0.1488	0.1273	0.1442	0.1359
$Z_0(\Omega)$	374.8	293.6	304.3	259.2	276.4	
Natural loading MVA	318	405	822	965	1844	1955
Conductor d.c. resistance at 25°C (Ω/mile)	0.0644	0.0871	0.0510	0.162	0.0644	0.0709
Conductor a.c. resistance (60 Hz) at 50°C (Ω/mile)	0.0728	0.0979	0.0599	0.179	0.0728	0.0805

X_A = component of inductive reactance due to flux within a 1 ft radius.
X_D = component due to other phases.
Total reactance per phase = $X_A + X_D$.

$$X_A = 0.2794 \log_{10} \left(\frac{1}{[N(GMR)(A)^{N-1}]^{1/N}} \right) \Omega/\text{mile}.$$

$X_D = 0.2794 \log_{10} (GMD) \, \Omega/\text{mile}.$
X'_A and X'_D are similarly defined for capacitive reactance and $X'_A = 0.0683 \log_{10} \left(\dfrac{1}{[Nr(A)^{N-1}]^{1/N}} \right) \text{M}\Omega\text{-miles}; \ X'_D = 0.0683 \log_{10}(GMD) \text{M}\Omega\text{-miles}.$
GMR = geometric mean radius in feet;
GMD = geometric mean diameter in feet;
N = number of conductors per phase;
$A = S/2 \sin (\pi/N); N > 1$
$A = 0; N = 1;$
S = bundle spacing in feet;
r = conductor radius in feet.
Data adapted from Edison Electric Institute.

Table 3.3 Underground cable constants at 50 Hz (per km)

Size	mm²	132 kV		275 kV		400 kV
		355	645	970	1130	1935
Rating (soil $g = 120\,°C$ cm/W)	A	550	870	1100	1100	1600
	MVA	125	200	525	525	1100
Resistance (R) at 85 °C	Ω	0.065	0.035	0.025	0.02	0.013
Reactance (X_L)	Ω	0.128	0.138	0.22	0.134	0.22
Charging current (I_c)	A	7.90	10.69	15.70	17.77	23.86
X_L/R Ratio		2.0	4.0	8.8	6.6	16.8

3.7.2.1 The Short Line (up to 80 km, 50 miles)

The equivalent circuit is shown in Figure 3.26 and it will be noticed that both shunt capacitance and leakage resistance have been neglected. The four-terminal network constants are (see Section 2.7):

$$A = 1 \quad B = Z \quad C = 0 \quad D = 1$$

The *regulation* of a circuit is defined as:

$$\frac{\text{received voltage on no load } (V_S) - \text{received voltage on load } (V_R)}{\text{received voltage on load } (V_R)}$$

It should be noted that if I leads V_R in phase, that is a capacitive load, then $V_R > V_S$, as shown in Figure 3.27.

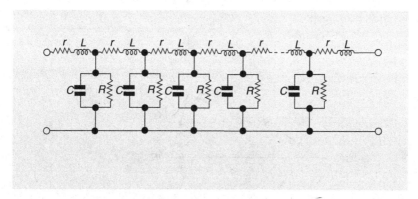

Figure 3.25 Distributed constant representation of a line: $L =$ inductance of line to neutral per unit length; $r =$ a.c. resistance per unit length; $C =$ capacitance line to neutral per unit length; $R =$ leakage resistance per unit length

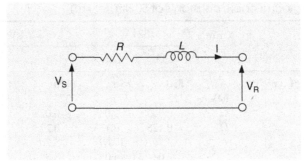

Figure 3.26 Equivalent circuit of a short line – representation under balanced three phase conditions

3.7.2.2 Medium-Length Lines (up to 240 km, 150 miles)

Owing to the increased length, the shunt capacitance is now included to form either a π or a T network. The circuits are shown in Figure 3.28. Of these two versions the π representation tends to be in more general use but there is little difference in accuracy between the two. For the π network:

$$\mathbf{V}_S = \mathbf{V}_R + \mathbf{IZ}$$
$$\mathbf{I} = \mathbf{I}_R + \mathbf{V}_R \frac{\mathbf{Y}}{2}$$
$$\mathbf{I}_S = \mathbf{I} + \mathbf{V}_S \frac{\mathbf{Y}}{2}$$

from which $\mathbf{V}_S$ and $\mathbf{I}_S$ are obtained in terms of $\mathbf{V}_R$ and $\mathbf{I}_R$ giving the following constants:

$$\mathbf{A} = \mathbf{D} = 1 + \frac{\mathbf{ZY}}{2}$$
$$\mathbf{B} = \mathbf{Z}$$
$$\mathbf{C} = \left(1 + \frac{\mathbf{ZY}}{4}\right)\mathbf{Y}$$

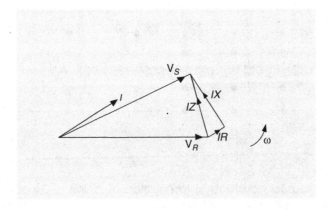

Figure 3.27 Phasor diagram for short line on leading power factor load

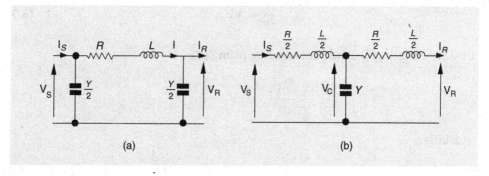

Figure 3.28 (a) Medium-length line – π equivalent circuit, (b) Medium-length line – T equivalent circuit

Similarly, for the T network:

$$V_S = V_C + \frac{I_S Z}{2}$$

$$V_C = V_R + \frac{I_R Z}{2}$$

$$I_S = I_R + V_C Y$$

giving

$$A = D = 1 + \frac{ZY}{2}$$

$$B = \left(1 + \frac{ZY}{4}\right) Y$$

$$C = Y$$

3.7.2.3 The Long Line (above 240 km, 150 miles)

Here the treatment assumes distributed parameters. The changes in voltage and current over an elemental length Δx of the line, x metres from the sending end, are determined, and given below:

$$\Delta V_x = z\Delta x \times I_x \text{ and } \Delta I_x = y\Delta x \times V_x$$

where

$z =$ impedance/unit length
$y =$ shunt admittance/unit length

If $\Delta x \to 0$, then

$$\frac{\partial V_x}{\partial x} = z I_x \qquad (3.2)$$

$$\frac{\partial I_x}{\partial x} = yV_x \tag{3.3}$$

By differentiating (3.2) and substituting from (3.3)

$$\frac{\partial^2 V_x}{\partial x^2} - zyV_x = 0 \tag{3.4}$$

Similarly,

$$\frac{\partial^2 I_x}{\partial x^2} - zyI_x = 0 \tag{3.5}$$

Solution to equations (3.4) and (3.5) takes the form $\mathbf{V}_x = A_1 \cosh \mathbf{P}x + A_2 \sinh \mathbf{P}x$ and $\mathbf{I}_x = B_1 \cosh \mathbf{P}x + B_2 \sinh \mathbf{P}x$. When $x = 0$, as $\mathbf{V}_x = \mathbf{V}_S$ and $\mathbf{I}_x = \mathbf{I}_S$, the voltage and current at x metres from the sending end are given by

$$\mathbf{V}_x = \mathbf{V}_S \cosh \mathbf{P}x - \mathbf{I}_S \mathbf{Z}_0 \sinh \mathbf{P}x$$
$$\mathbf{I}_x = \mathbf{I}_S \cosh \mathbf{P}x - \frac{\mathbf{V}_S}{\mathbf{Z}_0} \sinh \mathbf{P}x \tag{3.6}$$

where

$$\mathbf{P} = \text{propagation constant} = (\alpha + j\beta)$$
$$= \sqrt{\mathbf{zy}}$$
$$= \sqrt{(R + j\omega L)(G + j\omega C)}$$

and

$$\mathbf{Z}_0 = \text{characteristic impedance}$$
$$= \sqrt{\frac{\mathbf{z}}{\mathbf{y}}} = \sqrt{\frac{R + j\omega L}{G + j\omega C}} \tag{3.7}$$

where

$R =$ resistance/unit length
$L =$ inductance/unit length
$G =$ leakage/unit length
$C =$ capacitance/unit length

$\mathbf{Z}_0$ is the input impedance of an infinite length of the line; hence if any line is terminated in $\mathbf{Z}_0$ its input impedance is also $\mathbf{Z}_0$.

The propagation constant $\mathbf{P}$ represents the changes occurring in the transmitted wave as it progresses along the line; α measures the attenuation, and β the angular phase-shift. With a lossless line, where $R = G = 0$, $P = j\omega\sqrt{LC}$ and $\beta = \sqrt{LC}$. With a velocity of propagation of 3×10^5 km/s the wavelength of the transmitted voltage and current at 50 Hz is 6000 km. Thus, lines are much shorter than the wavelength of the transmitted energy.

Usually conditions at the load are required when $x = l$ in equations (3.6).

$$V_R = V_S \cosh Pl - I_S Z_0 \sinh Pl$$

$$I_R = I_S \cosh Pl - \frac{V_S}{Z_0} \sinh Pl$$

Alternatively,

$$V_S = V_R \cosh Pl + I_R Z_0 \sinh Pl$$

$$I_S = \frac{V_R}{Z_0} \cosh Pl + I_R \sinh Pl$$

The parameters of the equivalent four-terminal network are thus,

$$A = D = \cosh\sqrt{ZY}$$

$$B = \sqrt{\frac{Z}{Y}} \sinh\sqrt{ZY}$$

$$C = \sqrt{\frac{Y}{Z}} \sinh\sqrt{ZY}$$

$$(3.8)$$

where

Z = total series impedance of line
Y = total shunt admittance of line

The easiest way to handle the hyperbolic functions is to use the appropriate series.

$$A = D = \cosh\sqrt{ZY} = 1 + \frac{YZ}{2} + \frac{Y^2Z^2}{24} + \frac{Y^3Z^3}{720} + \cdots$$

$$B = Z\left(1 + \frac{YZ}{6} + \frac{Y^2Z^2}{120} + \frac{Y^3Z^3}{5040}\right)$$

$$C = Y\left(1 + \frac{YZ}{6} + \frac{Y^2Z^2}{120} + \frac{Y^3Z^3}{5040}\right)$$

Usually not more than three terms are required, and for (overhead) lines less than 500 km (312 miles) in length the following expressions for the constants hold

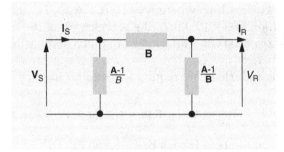

Figure 3.29 Equivalent circuit to represent accurately the terminal conditions of a long line

approximately:

$$A = D = 1 + \frac{ZY}{2}$$

$$B = Z\left(1 + \frac{ZY}{6}\right)$$

$$C = Y\left(1 + \frac{ZY}{6}\right)$$

An exact equivalent circuit for the long line can be expressed in the form of the π section shown in Figure 3.29. The application of simple circuit laws will show that this circuit yields the correct four-terminal network equations. Figure 3.29 is only for conditions at the ends of the line; if intermediate points are to be investigated, then the full equations must be used.

If only the first term of the expansions are used, then

$$B = Z$$

$$\frac{A - 1}{B} = \frac{Y}{2}$$

that is, the medium-length π representation.

Example 3.2

The conductors of a 1.6 km (1 mile) long, 3.3 kV, overhead line are in horizontal formation with 762 mm (30 in) between centres. The effective diameter of the conductors is 3.5 mm. The resistance per kilometre of the conductors is 0.41 Ω. Calculate the line-to-neutral inductance of the line. If the sending-end voltage is 3.3 kV (50 Hz) and the load is 1 MW at a lagging p.f. of 0.8, estimate the voltage at the load busbar and the power loss in the line.

Solution

The equivalent equilateral spacing is given by $d_e = \sqrt[3]{d_{12} \times d_{23} \times d_{31}}$

In this case $d_e = \sqrt[3]{762 \times 762 \times 1524}$

$$d_e = 762.\sqrt[3]{2} = 960\,mm$$

The inductance (line to neutral)

$$
\begin{aligned}
L &= \frac{\mu_0}{8\pi}\left[1 + 4\ln\left(\frac{d-r}{r}\right)\right]H/m \\
&= \frac{4\pi \times 10^{-7}}{8\pi}\left[1 + 4\ln\left(\frac{960 - 1.75}{1.75}\right)\right]H/m \\
&= 1.3105 \times 10^{-6} H/m
\end{aligned}
$$

The total inductance of 1.6 km

$$L_{total} = 2.11\,mH$$

The inductive reactance

$$X_L = 2\pi f L_{total} = 2\pi \times 50 \times 2.11 \times 10^{-3} = 0.66\,\Omega$$

Resistance of line

$$R_{total} = 1.6 \times 0.41 = 0.66\,\Omega$$

Impedance of the line

$$\mathbf{Z} = 0.66 + j0.66\,\Omega$$

An estimate of the voltage drop is required and so the distribution approximation will be used.

Choosing 1 MVA and 3.3 kV as the bases for the calculation

$$Z_{base} = \frac{V_{base}^2}{S_{base}} = \frac{3.3 \times 10^3}{10^6} = 10.89\,\Omega$$

$$\mathbf{Z} = 0.0602 + j0.0602 \text{ per unit}$$

$$P = 1\,MW\,(1\text{ per unit}) \quad Q = 0.75\,MVAr\,(0.75\text{ per unit})$$

Using the distribution approximation in per unit

$$\Delta V = PR + XQ = 1 \times 0.0602 + 0.75 \times 0.0602 = 0.1053 = 10.53\%$$

The voltage drop is therefore 347 V (3.3 kV line-line) and 201 V (1.9 kV line-neutral)

The apparent power of the load is 1.25 MVA or 1.25 per unit. Assuming the load voltage to be 1 per unit, this is also the current in per unit. The line loss is $1.25^2 \times 0.0602 = 0.094$ per unit or 94 kW for all three phases.

As the sending end voltage and receiving end power are specified, a more precise iterative calculation may be undertaken with Equation (2.13).

Example 3.3

A 150 km long overhead line with the parameters given in Table 3.2a for 400 kV, quad conductors is to be used to transmit 1800 MW (normal weather loading) to a load with a power factor of 0.9 lagging. Calculate the required sending end voltage using three line representations and compare the results.

Solution
From Table 3.2 a

$$R = 0.017\Omega/km$$

$$X_L = 0.27\Omega/km$$

$$\frac{1}{Xc} = 10.58 \times 10^{-6} S/km$$

Choosing a base of 2000 MVA and 400 kV.

$$Z_{base} = \frac{400^2}{2000} = 80\Omega$$

For a 150 km line

$$Z_{pu} = \frac{150}{80}(0.017 + j0.27) = 0.032 + j0.506 \text{ per unit}$$

Short-line representation:

Load power, $\mathbf{S} = 1800 + j870$ MVA
$\mathbf{S} = \mathbf{VI^*}$ and as the receiving end voltage is at 400 kV,

$$\mathbf{I} = \frac{1800}{2000} - j\frac{870}{2000} = 0.9 - j0.435 \text{ per unit}$$

Hence

$$\mathbf{V}s = \mathbf{V}_R + \mathbf{I}\mathbf{Z}$$

$$\mathbf{V}_s = 1.249 + j0.442 \text{ per unit}$$

$$|V_S| = 530 \text{ kV}$$

Medium-line representation:

$$V_s = AV_R + BI_R$$

$$A = 1 + \frac{ZY}{2}$$

$$B = Z$$

$$B = 0.032 + j0.506 \text{ per unit}$$

$$Y = 10.58 \times 10^{-6} \times 150 \times 80 = 0.127 \text{ per unit}$$

$$A = 1 + \frac{ZY}{2} = 0.968 + j0.002 \text{ per unit}$$

Hence

$$\begin{aligned}
V_s &= AV_R + BI_R \\
&= (0.968 + j0.002) + (0.032 + j0.506)(0.9 - j0.435) \\
&= 1.217 + j0.444 \\
|Vs| &= 518 \text{ kV}
\end{aligned}$$

Long-line representation:

$$A = \cosh\sqrt{ZY} = 1 + \frac{YZ}{2} + \frac{Y^2Z^2}{24} + \cdots$$

$$B = \sqrt{\frac{Z}{Y}} \sinh\sqrt{ZY} = Z(1 + \frac{YZ}{6} + \frac{Y^2Z^2}{120} + \cdots\cdots$$

$$YZ = -0.064 + j4.047 \times 10^{-3}$$

$$Y^2Z^2 = 4.115 \times 10^{-3} - j5.202 \times 10^{-4}$$

$$A = 0.968 + j2.002 \times 10^{-3}$$

$$B = 0.031 + j0.501$$

$$\begin{aligned}
V_S &= AV_R + BI_R \\
&= 1.214 + j0.439 \text{ per unit} \\
|Vs| &= 518 \text{ kV}
\end{aligned}$$

(Note: This is an extremely high voltage for normal operation and would not be tolerated. A 400 kV system would be designed for only about 10% steady-state over-voltage, that is $400 + 40 = 440$ kV. In practice, either the reactance of the line would be reduced by series capacitors and/or the power factor at the receiving end would be raised to at least 0.95 lag by the use of shunt capacitors or synchronous compensators).

3.7.2.4 The Natural Load

The characteristic impedance Z_0 (equation 3.7) is known as the surge impedance if the line is considered to be lossless and all resistances are neglected. When a line is terminated in its characteristic impedance the power delivered is known as the natural load. For a loss-free line supplying its natural load the reactive power absorbed by the line is equal to the reactive power generated, that is.

$$\frac{V^2}{X_C} = I^2 X_L$$

and

$$\frac{V}{I} = Z_0 = \sqrt{X_L X_C} = \sqrt{\frac{L}{C}}$$

At this load, V and I are in phase all along the line and optimum transmission conditions are obtained. In practice, however, the load impedances are seldom in the order of Z_0. Values of Z_0 for various line voltages are as follows (corresponding natural loads are shown in parentheses): 132 kV, 150 Ω (50 MW); 275 kV, 315 Ω (240 MW); 380 kV, 295 Ω (490 MW). The angle of the impedance varies between 0 and $-15°$. For underground cables Z_0 is roughly one-tenth of the overhead-line value. Lines are operated above the natural loading, whereas cables operate below this loading.

3.7.3 Parameters of Underground Cables

In traditional paper insulated cables with three conductors contained within a lead or aluminium sheath, the electric field set up has components tangential to the layers of impregnated paper insulation in which direction the dielectric strength is poor. Therefore, at voltages over 11 kV, each conductor is separately screened to ensure only radial stress through the paper. Cables using cross-linked polyethylene (XPLE) insulation are now commonly used and again the electric stresses are arranged to be radial. The capacitance of single-conductor and individually screened three-conductor cables is readily calculated. For three-conductor unscreened cables one must resort to empirical design data. A typical high-voltage XLPE insulated cable is shown in Figure 3.30. The capacitance (C) of single-core cables may be calculated from design data by the use of the formula

$$C = \frac{2\pi\varepsilon_0\varepsilon_r}{\ln\dfrac{R}{r}}$$

where r and R are the inner and outer radii of the dielectric and ε_r is the relative permittivity of the dielectric. This expression also holds for three core cables with each conductor separately screened.

Owing to the symmetry of the cable the positive phase-sequence values of C and L are the same as the negative phase-sequence values (i.e. for reversed phase rotation). The series resistance and inductance are complicated by the magnetic interaction

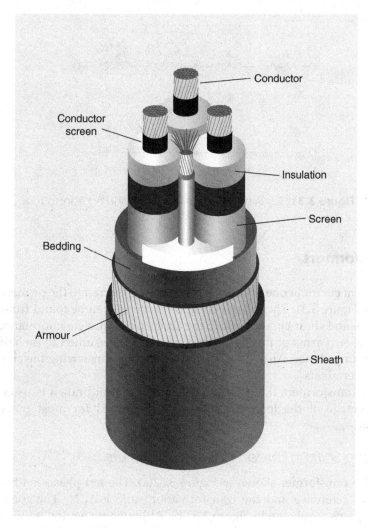

Figure 3.30 150 kV XLPE insulated 3 core cable with fibre optic communications

between the conductor and sheath. The effective resistance of the conductor is the direct current resistance modified by the following factors: the skin effect in the conductor; the eddy currents induced by adjacent conductors (the proximity effect); and the equivalent resistance to account for the I^2R losses in the sheath. The determination of these effects is complicated and is left for advanced study.

The parameters of the cable having been determined, the same equivalent circuits are used as for overhead lines, paying due regard to the selection of the correct model for the appropriate length of cable. Owing to the high capacitance of cables the charging current, especially at high voltages, is an important factor in deciding the permissible length to be used.

Table 3.3 gives the charging currents for self-contained low-pressure oil-filled (LPOF) cables.

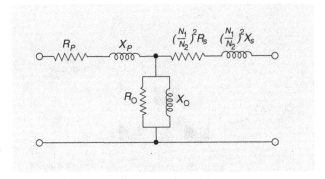

Figure 3.31 Equivalent circuit of a two-winding transformer

3.8 Transformers

The equivalent circuit of one phase of a transformer referred to the primary winding is shown in Figure 3.31. The resistances and reactances can be found from the well-known open- and short-circuit tests. In the absence of complete information for each winding, the two arms of the T network can each be assumed to be half the total transformer impedance. Also, little accuracy is lost in transferring the shunt branch to the input terminals.

In power transformers the current taken by the shunt branch is usually a very small percentage of the load current and is neglected for most power system calculations.

3.8.1 Phase Shifts in Three-Phase Transformers

Consider the transformer shown in Figure 3.32(a). The red phases on both circuits are taken as reference and the transformation ratio is 1: N. The corresponding phasor diagrams are shown in Figure 3.32(b). Although no neutral point is available in the delta side, the effective voltages from line to earth are denoted by $E_{R'n}$, $E_{Y'n}$, and $E_{B'n}$. Comparing the two phasor diagrams, the following relationships are seen:

$$E_{R'n} = \text{line-earth voltage on the delta-side}$$
$$= NE_{Rn}\angle 30°$$

that is, the positive sequence or normal balanced voltage of each phase is advanced through 30°. Similarly, it can be shown that the positive sequence currents are advanced through 30°.

By consideration of the negative phase-sequence phasor diagrams (these are phasors with reversed rotation, that is R-B-Y) it will readily be seen that the phase currents and voltages are shifted through −30°. When using the per unit system the transformer ratio does not directly appear in calculations and the phase shifts are often neglected.

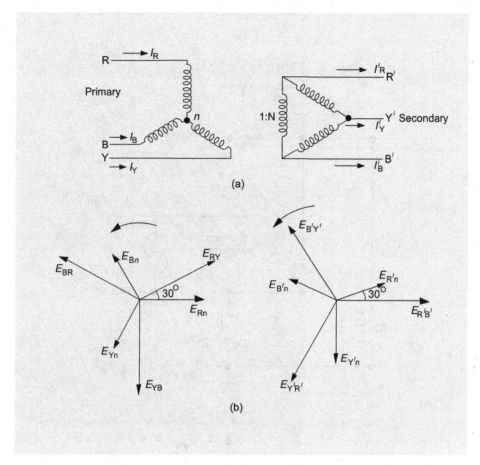

Figure 3.32 (a) Star-delta transformer with turns ratio 1: N. (b) Corresponding phasor diagrams (N taken as 1 in diagrams)

In star-star and delta-delta connected transformers there are no phase shifts. Hence transformers having these connections and those with star-delta connections should not be connected in parallel. To do so introduces a resultant voltage acting in the local circuit formed by the usually low transformer impedances. Figure 3.33 shows the general practice on the British network with regard to transformers with phase shifts. It is seen that the reference phasor direction is different at different voltage levels. The larger than 30° phase shifts are obtained by suitable rearrangement of the winding connections.

3.8.2 Three-Winding Transformers

Many transformers used in power systems have three windings per phase, the third winding being known as the tertiary. This winding is provided to enable compensation equipment to be connected at an economic voltage (e.g. 13 kV) and to provide a

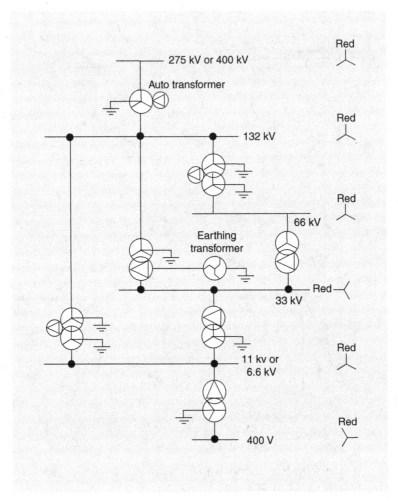

Figure 3.33 Typical phase shifts in a power system–British

circulating current path for third harmonics so that these currents do not appear outside the transformer.

The three-winding transformer can be represented under balanced three phase conditions by a single-phase equivalent circuit of three impedances star-connected as shown in Figure 3.34. The values of the equivalent impedances Z_p, Z_s and Z_t may be obtained by test. It is assumed that the no-load currents are negligible.

Let

Z_{ps} = impedance of the primary when the secondary is short-circuited and the tertiary open

Z_{pt} = impedance of the primary when the tertiary is short-circuited and the secondary open

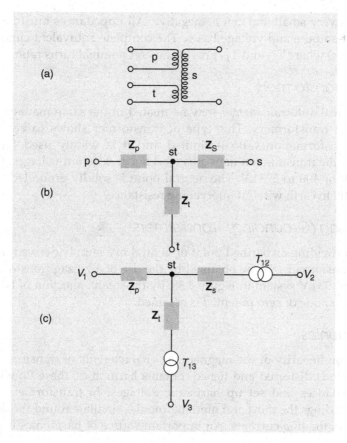

Figure 3.34 (a) Three-winding transformer (b) and (c) equivalent circuits

Z_{st} = impedance of the secondary when the tertiary is short-circuited and the primary open

The above impedances are in ohms referred to the same voltage base. Hence,

$$Z_{ps} = Z_p + Z_s$$
$$Z_{pt} = Z_p + Z_t$$
$$Z_{st} = Z_s + Z_t$$
$$Z_p = \frac{1}{2}\left(Z_{ps} + Z_{pt} - Z_{st}\right)$$
$$Z_s = \frac{1}{2}\left(Z_{ps} + Z_{st} - Z_{pt}\right)$$
$$Z_t = \frac{1}{2}\left(Z_{pt} + Z_{st} - Z_{ps}\right)$$

It should be noted that the star point st in Figure 3.34b is purely fictitious and that the diagram is a single-phase equivalent circuit. In most large transformers the value of Zs is very small and can be negative. All impedances must be referred to common volt-ampere and voltage bases. The complete equivalent circuit is shown in Figure 3.34(c) where T_{12} and T_{13} provide any off nominal turns ratio.

3.8.3 Autotransformers

The symmetrical autotransformer may be treated in the same manner as two and three-winding transformers. This type of transformer shows to best advantage when the transformation ratio is limited and it is widely used for the interconnection of the transmission networks working at different voltages, for example 275 to 132 kV or 400 to 275 kV. The neutral point is solidly grounded, that is connected directly to earth without intervening resistance.

3.8.4 Earthing (Grounding) Transformers

A means of providing an earthed point or neutral in a supply derived from a delta-connected transformer may be obtained by the use of a zigzag transformer, shown grounding the 33 kV system in Figure 3.33. By the interconnection of two windings on each limb, a node of zero potential is obtained.

3.8.5 Harmonics

Due to the non-linearity of the magnetizing characteristic of transformers the current waveform is distorted and hence contains harmonics; these flow through the system impedances and set up harmonic voltages. In transformers with delta-connected windings the third and ninth harmonics circulate round the delta and are less evident in the line currents. An important source of harmonics is power electronic devices.

On occasions, the harmonic content can prove important due mainly to the possibility of resonance occurring in the system, for example, resonance has occurred with fifth harmonics. Also, the third-harmonic components are in phase in the three conductors of a three-phase line, and if a return path is present these currents add and cause interference in neighbouring communication circuits and increase the neutral return current.

When analysing systems with harmonics it is often sufficient to use the normal values for series inductance and shunt capacitance but one must remember to calculate reactance and susceptance at the frequency of the harmonic. The effect on resistance is more difficult to assess: however, it is usually only required to assess the presence of harmonics and the possibility of resonance.

3.8.6 Tap-Changing Transformers

A method of controlling the voltages in a network lies in the use of transformers, the turns ratio of which may be changed. In Figure 3.35(a) a schematic diagram of an off-load tap changer is shown; this, however, requires the disconnection of the transformer when the tap setting is to be changed. Many larger transformers have on-load changers, one basic form of which is shown in Figure 3.35(b). In the position

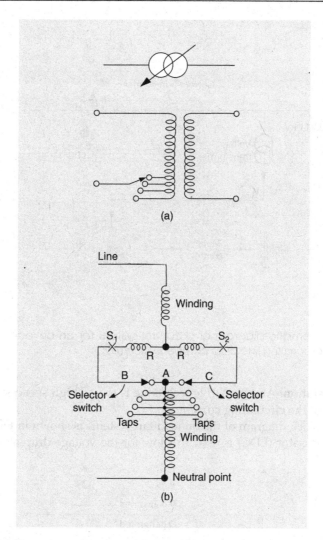

Figure 3.35 (a) Tap-changing transformer, (b) On-load tap-changing transformer (reactor type). S1, S2 transfer switches, R centre-tapped reactor

shown, the current divides equally in the two halves of reactor R, resulting in zero-resultant flux and minimum impedance. S_1 opens and the total current passes through the other half of the reactor. Selector switch B then moves to the next contact and S_1 closes. A circulating current now flows in the reactor R superimposed on the load current. Now S_2 opens and C moves to the next tapping; S_2 then closes and the operation is complete. Six switch operations are required for one change in tap position. The voltage change between taps is often 1–1.25% of the nominal voltage. This small change is necessary to avoid large voltage disturbances at consumer bus-bars. Figure 3.35(b) shows reactors used to limit the circulating current during the

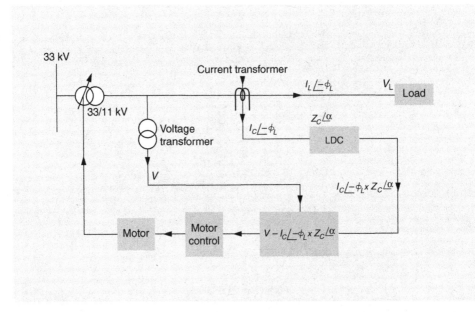

Figure 3.36 Schematic diagram of a control system for an on-load tap-changing transformer incorporating line drop compensation (LDC)

tap-change operation. An alternative technique is to use high speed switching with a resistor to limit the circulating current.

A schematic block diagram of the on-load tap systems is shown in Figure 3.36. The line drop compensator (LDC) is used to allow for the voltage drop along the feeder

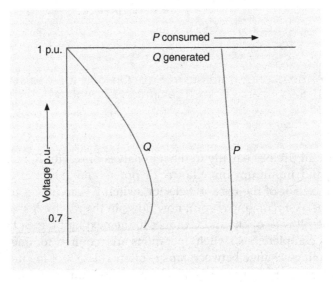

Figure 3.37 P- V and Q-V curves for a synchronous motor

to the load point, so that the actual load voltage is seen and corrected by the transformer. The total range of tapping varies with the transformer usage, a typical figure for generator transformers is +2 to −16% in 18 steps each of 1%.

3.8.7 Typical Parameters for Transformers

The leakage reactances of two-winding transformers increase slightly with their rating for a given voltage, that is from 3.2% at 20 kVA to 4.3% at 500 kVA at 11 kV. For larger sizes, that is 20 MVA upwards, 10–20% is a typical value at all voltages. For autotransformers the impedances are usually less than for double wound transformers. Parameters for large transformers are as follows:

1. 400/275 kV autotransformer, 500 MVA, 12% impedance, tap range +10 to −20%;
2. 400/132 kV double wound, 240 MVA, 20% impedance, tap range +5 to −15%.

It should be noted that a 10% reactance implies up to ten times rated current on short circuit. Winding forces depend upon current squared and so a transformer must be designed to withstand high forces caused by short circuits.

3.9 Voltage Characteristics of Loads

The variation of the power and reactive power taken by a load with various voltages is of importance when considering the manner in which such loads are to be represented in load flow and stability studies. Usually, in such studies, the load on a substation has to be represented and is a composite load consisting of industrial and domestic consumers. A typical composition of a substation load is as follows:

Induction motors	50–70% (air-conditioners, freezers, washers, fans, pumps, etc.)
Lighting and heating	20–25% (water heating, resistance heaters, etc.)
Synchronous motors	10%
(Transmission and distribution losses 10–12%)	

Increasingly electronic equipment draws a significant fraction of the load.

3.9.1 Lighting

Incandescent (filament) lights are independent of frequency and consume no reactive power. The power consumed does not vary as the $(voltage)^2$, but approximately as $(voltage)^{1.6}$. However, in many countries the use of incandescent lights is being reduced due to their low efficiency. Fluorescent (both traditional and compact) and sodium/mercury lamps, can take distorted currents and so contribute to network harmonics.

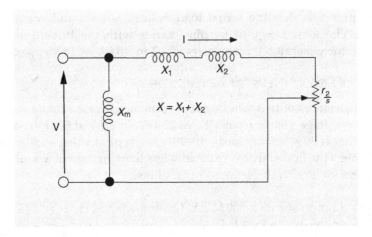

Figure 3.38 Equivalent circuit of an induction motor: X_1 = stator leakage reactance; X_2 = rotor leakage reactance referred to the stator; X_m = magnetizing reactance; r_2 = rotor resistance; s = slip p.u. Magnetizing losses have been ignored and the stator losses are lumped in with the line losses

3.9.2 Heating

This maintains constant resistance with voltage change and hence the power varies with (voltage)^2.

The above loads may be described as static.

3.9.3 Synchronous Motors

The power consumed is approximately constant with the applied voltage. For a given excitation the VArs change in a leading direction (i.e. reactive power is generated) with network voltage reduction. The P-V, Q-V, generalized characteristics are shown in Figure 3.37.

3.9.4 Induction Motors

The P-V, Q-V characteristics may be determined by the use of the simplified circuit shown in Figure 3.38. It is assumed that the mechanical load on the shaft is constant.

The electrical power

$$P_{electrical} = P_{mechanical} = P$$

$$P = \frac{3I^2 r_2}{s}$$

The reactive power consumed

$$= \frac{3V^2}{X_m} + 3I^2(X_1 + X_2)$$

Also, from Figure 3.38

$$X = X_1 + X_2$$

$$P = 3I^2 \frac{r_2}{s} = \frac{3V^2}{\left[\dfrac{r^2}{s^2} + X^2\right]} \frac{r_2}{s} \tag{3.9}$$

$$= \frac{3V^2 r_2 s}{r_2^2 + (sX)^2}$$

The well-known power-slip curves for an induction motor are shown in Figure 3.39. It is seen that for a given mechanical torque there is a critical voltage and a corresponding critical slip scr. If the voltage is reduced further the motor becomes unstable and stalls. This critical point occurs when

$$\frac{dP}{ds} = 0$$

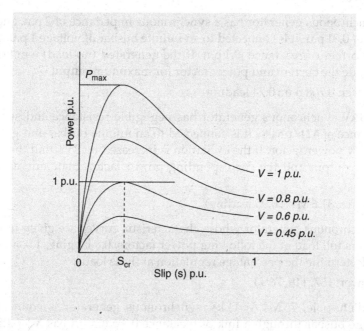

Figure 3.39 Power-slip curves for an induction motor. If voltage falls to 0.6 p.u. at full load $P = 1$, the condition is critical (slip s_{cr})

that is when

$$V^2 r_2 \frac{r_2^2 - (sX)^2}{[r_2^2 + (sX)^2]^2} = 0$$

so that

$$s = \frac{r_2}{X}$$

$$P_{max} = \frac{3V^2}{2X}$$

Alternatively, for a given output power, P

$$V_{critical} = \sqrt{\frac{2}{3}PX}$$

Problems

(Note: All machines are three-phase unless stated otherwise.)

3.1 When two four pole, 50 Hz synchronous generators are paralleled their phase displacement is 2° mechanical. The synchronous reactance of each machine is 10 Ω/phase and the common busbar voltage is 6.6 kV. Calculate the synchronizing torque.

(Answer: 968 Nm)

3.2 A synchronous generator has a synchronous impedance of 2 p.u. and a resistance of 0.01 p.u. It is connected to an infinite busbar of voltage 1 p.u. through a transformer of reactance j0.1 p.u. If the generated (no-load) e.m.f. is 1.1 p.u. calculate the current and power factor for maximum output.

(Answer: 0.708 p.u.; 0.74 leading)

3.3 A 6.6 kV synchronous generator has negligible resistance and synchronous reactance of 4 Ω/phase. It is connected to an infinite busbar and gives 2000 A at unity power factor. If the excitation is increased by 25% find the maximum power output and the corresponding power factor. State any assumptions made.

(Answer: 31.6 MW; 0.95 leading)

3.4 A synchronous generator whose characteristic curves are given in Figure 3.4 delivers full load at the following power factors: 0.8 lagging, 1.0, and 0.9 leading. Determine the percentage regulation at these loads.

(Answer: 167, 119, 76%)

3.5 A salient-pole, 75 MVA, 11 kV synchronous generator is connected to an infinite busbar through a link of reactance 0.3 p.u. and has $X_d = 1.5$ p.u. and $X_q = 1$ p.u., and negligible resistance. Determine the power output for a load angle 30° if the excitation e.m.f. is 1.4 times the rated terminal voltage. Calculate the synchronizing coefficient in this operating condition. All p.u. values are on a 75 MVA base.

(Answer: $P = 0.48$ p.u.; $dP/d\delta = 0.78$ p.u.)

3.6 A synchronous generator of open-circuit terminal voltage 1 p.u. is on no-load and then suddenly short-circuited; the trace of current against time is shown in Figure 3.6(b). In Figure 3.6(b) the current $0c = 1.8$ p.u., $0a = 5.7$ p.u., and $0b = 8$ p.u. Calculate the values of Xs, X' and X". Resistance may be neglected. If the machine is delivering 1 p.u. current at 0.8 power factor lagging at the rated terminal voltage before the short circuit occurs, sketch the new envelope of the 50 Hz current waveform.

(Answer: Xs $= 0.8$ p.u.; X' $= 0.25$ p.u.; X" $= 0.18$ p.u.)

3.7 Construct a performance chart for a 22 kV, 500 MVA, 0.9 p.f. generator having a short-circuit ratio of 0.55.

3.8 A 275 kV three-phase transmission line of length 96 km is rated at 800 A. The values of resistance, inductance and capacitance per phase per kilometre are $0.078 \, \Omega$, $1.056 \, \text{mH}$ and $0.029 \, \mu\text{F}$, respectively. The receiving-end voltage is 275 kV when full load is transmitted at 0.9 power factor lagging. Calculate the sending-end voltage and current, and the transmission efficiency, and compare with the answer obtained by the short-line representation. Use the nominal π and T methods of solution. The frequency is 60 Hz.

(Answer: Vs $= 179$ kV per phase)

3.9 A 220 kV, 60 Hz three-phase transmission line is 320 km long and has the following constants per phase per km:

Inductance 0.81 mH
Capacitance 12.8 μF
Resistance 0.038 Ω

Ignore leakage conductance.
If the line delivers a load of 300 A, 0.8 power factor lagging, at a voltage of 220 kV, calculate the sending-end voltage. Determine the π circuit which will represent the line.

(Answer: $V_s = 241$ kV)

3.10 Calculate the A B C D constants for a 275 kV overhead line of length 83 km. The parameters per kilometre are as follows:

Resistance 0.078 Ω
Reactance 0.33 Ω
Admittance (shunt capacitative) 9.53×10^{-6} S

The shunt conductance is zero.

(Answer: [A $= 0.98917 + j \, 0.00256$; B $= 6.474 + j \, 27.39$; C $= (-1.0126 \times 10^{-6} + j \, 7.8671 \times 10^{-4})$])

3.11 A 132 kV, 60 Hz transmission line has the following generalized constants:

$A = 0.9696 \angle 0.49°$

$B = 52.88 \angle 74.79° \, \Omega$

$C = 0.001177 \angle 90.15° S$

If the receiving-end voltage is to be 132 kV when supplying a load of 125 MVA 0.9 p.f. lagging, calculate the sending-end voltage and current.

(Answer: 165 kV, 498 A)

3.12 Two identical transformers each have a nominal or no-load ratio of 33/11 kV and a leakage reactance of 2 Ω referred to the 11 kV side; resistance may be neglected. The transformers operate in parallel and supply a load of 9 MVA, 0.8 p.f. lagging. Calculate the current taken by each transformer when they operate five tap steps apart (each step is 1.25% of the nominal voltage). Also calculate the kVAr absorbed by this tap setting.

(Answer: 307 A, 194 A for three-phase transformers, 118 kVAr)

3.13 An induction motor, the equivalent circuit of which is shown in Figure 3.40 is connected to supply busbars which may be considered as possessing a voltage and frequency which is independent of the load. Determine the reactive power consumed for various busbar voltages and construct the Q-V characteristic. Calculate the critical voltage at which the motor stalls and the critical slip, assuming that the mechanical load is constant.

(Answer: $s_{cr} = 0.2$, $V_c = 0.63$ p.u.)

3.14 A 100 MVA round-rotor generator of synchronous reactance 1.5 p.u. supplies a substation (L) through a step-up transformer of reactance 0.1 p.u., two lines each of reactance 0.3 p.u. in parallel and a step-down transformer of reactance 0.1 p.u. The load taken at L is 100 MW at 0.85 lagging. L is connected to a local generating station which is represented by an equivalent generator of 75 MVA and synchronous reactance of 2 p.u. All reactances are expressed on a base of 100 MVA. Draw the equivalent single-phase network. If the voltage at L is to

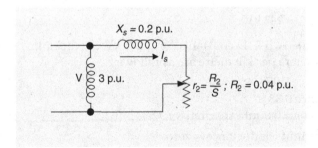

Figure 3.40 Equivalent circuit of 500 kW, 6.6 kV induction motor in Problem 3.13. All p.u. values refer to rated voltage and power ($P = 1$ p.u. and $V = 1$ p.u.)

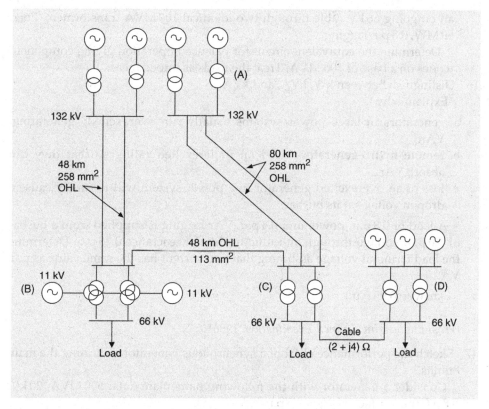

Figure 3.41 Line diagram of system in Problem 3.15

be 1 p.u. and the 75 MVA machine is to deliver 50 MW, 20 MVAr, calculate the internal voltages of the machines.

(Answer: $E_1 = 2$ p.u.; $E_2 = 1.72$ p.u.; $\delta_{2L} = 35.45°$)

3.15 The following data applies to the power system shown in Figure 3.41.

Generating station A: Four identical turboalternators, each rated at 16 kV, 125 MVA, and of synchronous reactance 1.5 p.u. Each machine supplies a 125 MVA, 0.1 p.u. transformer connected to a busbar sectioned as shown.

Substation B: Two identical 150 MVW, three-winding transformers, each having the following reactances between windings: 132/66 kV windings 10%; 66/11 kV windings 20%; 132/11 kV windings 20%; all on a 150 MVA base.

The secondaries supply a common load of 200 MW at 0.9 p.f. lagging. To each tertiary winding is connected a 30 MVA synchronous compensator of synchronous reactance 1.5 p.u.

Substation C: Two identical 150 MVA transformers, each of 0.15 p.u. reactance, supply a common load of 300 MW at 0.85 p.f. lagging.

Generating station D: Three identical 11 kV, 75 MVA generators, each of 1 p.u. synchronous reactance, supply a common busbar which is connected to an outgoing 66 kV cable through two identical 100 MVA transformers. Load 50 MW, 0.8 p.f. lagging.

Determine the equivalent circuit for balanced operation giving component values on a base of 100 MVA. Treat the loads as impedances.

3.16 Distinguish between kW, kVA, and kVAr.

Explain why

a. generators in large power systems usually run overexcited, 'generating' VAr.

b. remote hydro-generators need an underexcited rating so that they can absorb VAr.

c. loss of an overexcited generator in a power system will normally cause a drop in voltage at its busbar.

A load of 0.8 p.u. power and 0.4 p.u. VAr lagging is supplied from a busbar of 1.0 p.u. voltage through an inductive line of reactance 0.15 p.u. Determine the load terminal voltage assuming that p.u. current has the same value as p.u. VA.

(Answer: 0.95 p.u.)

(From Engineering Council Examination, 1996)

3.17 Sketch the performance chart of a synchronous generator indicating the main limits.

Consider a generator with the following nameplate data: 500 MVA, 20 kV, 0.8 p.f. (power factor), $X_s = 1.5$ p.u.

a. Calculate the internal voltage and power angle of the generator operating at 400 MW with $\cos\varphi = 0.8$ (lagging) with a 1 p.u. terminal voltage.

b. What is the maximum reactive power this generator can absorb from the system?

c. What is the maximum reactive power this generator can deliver to the system, assuming a maximum internal voltage of 2.25 p.u.

d. Place the numerical values calculated on the performance chart.

A graphical solution is acceptable.

(Answer: (a) 2.25 p.u., 32°; (b) 333 MVAr; (c) 417 MVAr)

(From Engineering Council Examination, 1997)

4

Control of Power and Frequency

4.1 Introduction

In a large power system, the control of power and frequency is related only weakly to the control of reactive power and voltage. For many purposes the operation of the governors controlling the power of the prime movers of generating units can be considered independently to the AVRs that control the excitation and hence the reactive power and voltage of the generators. By dealing with power and frequency separately from voltage and reactive power control, a better appreciation of the operation of power systems can be obtained. This separation is followed in Chapter 4 (Control of Power and Frequency) and Chapter 5 (Control of Voltage and Reactive Power).

In a large interconnected system, many synchronous generators, big and small, are directly connected and hence all have the same frequency. In many power systems (e.g. in Great Britain) the control of power is carried out by the decisions and actions of control engineers, as opposed to systems in which the control and allocation of load to machines is effected completely automatically. Fully automatic control systems are sometimes based on a continuous load-flow calculation by computers.

The allocation of the required power amongst the generators has to be decided before the load appears. Therefore the load must be predicted in advance. An analysis is made of the loads experienced over the same period in previous years; account is also taken of the value of the load immediately previous to the period under study and of the weather forecast. The probable load to be expected, having been decided, is then allocated to the various turbine-generators.

Load cycles of three power systems are shown in Figure 1.1. The PJM control area in the USA has peaks of 140 GW in the summer, a lower demand in the winter and a rate of increase of around 15 GW/h. For the isolated Great Britain system the rate of

Electric Power Systems, Fifth Edition. B.M. Weedy, B.J. Cory, N. Jenkins, J.B. Ekanayake and G. Strbac.
© 2012 John Wiley & Sons, Ltd. Published 2012 by John Wiley & Sons, Ltd.

increase during the week shown was around 4 GW/h. For the smaller Sri Lanka power system the rate of increase was 250 MW/h, provided mainly by hydro generators. The ability of machines to increase their output quickly from zero to full load, and subsequently reduce their output is important.

It is extremely unlikely that the output of the machines at any instant will exactly equal the load on the system. If the output is higher than the demand the machines will tend to increase in speed and the frequency will rise, and vice versa. Hence the frequency is not a constant quantity but continuously varies; these variations are normally small and have no noticeable effect on most consumers. The frequency is continuously monitored against standard time-sources and when long-term tendencies to rise or fall are noticed, the control engineers take appropriate action by regulating the generator outputs.

Should the total generation available be insufficient to meet the demand, the frequency will fall. If the frequency falls by more than around 1 Hz the reduced speed of power station pumps and fans may limit the output of the power stations and a serious situation arises. When there is insufficient generation, although the lower frequency will cause some reduction in power demand, the system voltage must be reduced (which leads to a further reduction in load), and if this is not sufficient then loads will have to be disconnected and continue to be disconnected until the frequency rises to a reasonable level. All utilities have a scheme of planned load shedding based on under-frequency relays set to reduce loads in blocks to prevent complete shut-down of the power system in extreme emergencies.

Figure 4.1 shows the frequency of the Great Britain power system when two large generating units tripped in rapid succession, for unrelated reasons. The frequency dropped quickly and the fall was only finally arrested by load shedding when it reached 48.8 Hz. Figure 4.1 also shows the morning increase in system load of ∼ 18 GW from around 05:30–10:30 and a similar reduction in load in the evening. Apart from during the incident around 11:35 when two generating units tripped, the frequency was maintained close to 50 Hz by the governors of generators.

When an increase in load occurs on the system, the speed and frequency of all the interconnected generators fall, since the increased energy requirement is met from the kinetic energy of the machines. This causes an increase in steam or water admitted to the turbines due to the operation of the governors and hence a new load balance is obtained. Initially, the boilers have a thermal reserve of steam, in their boiler drums, by means of which sudden changes can be supplied until a new firing rate has been established. Modern gas turbines have an overload capability for a few minutes which can usefully be exploited in emergency situations.

The stored kinetic energy of all the generators (and spinning loads) on the system is given by

$$KE = \frac{1}{2}I\omega^2 \quad [J]$$

$I = moment\ of\ inertia\ of\ all\ generators\,(kgm^2)$

$\omega = rotational\ speed\ of\ all\ generators(rad/s)$

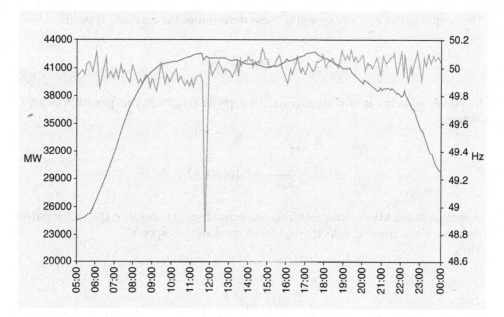

Figure 4.1 Variation of load (LH axis) and frequency (RH axis) of GB system. Two generators tripped, for unrelated reasons, at around 11:35 (Figure adapted from National Grid)

When $P_m = P_e$ (Figure 4.2) the rotational speed of the generators is maintained and the frequency is constant at 50 (or 60) Hz. When $P_m < P_e$ the rotational speed of the generators, and hence system frequency reduces. When $P_m > P_e$ the rotational speed of the generators, and hence system frequency increases.

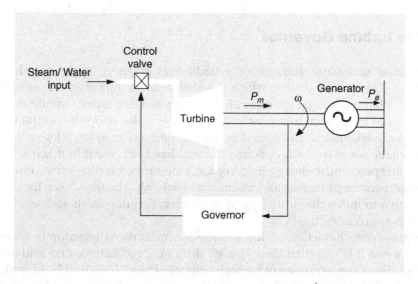

Figure 4.2 Control of frequency. Angular speed is measured and controls an inlet valve of the turbine fluid

The torque balance of any spinning mass determines the rotational speed.

$$T_m - T_e = I\frac{d\omega}{dt}$$

In power systems it is conventional to express the inertia in per unit as an H constant

$$H = \frac{\frac{1}{2}I\omega_S^2}{S_{rated}} \quad [\text{Ws/VA}]$$

where S_{rated} is the MVA rating of either an individual generator or the entire power system, ω_S is the angular velocity (rad/s) at synchronous speed.
Thus

$$\frac{d\omega}{dt} = \frac{\omega_s^2}{2HS_{rated}}(T_m - T_e)$$

which in per unit may be written as

$$\frac{d\omega}{dt} = \frac{1}{2H}(P_m - P_e)$$

Thus the rate of change of rotational speed and hence frequency depends on the power imbalance and the inertia of the spinning masses.

4.2 The Turbine Governor

A simplified schematic diagram of a traditional governor system is shown in Figure 4.3. The sensing device, which is sensitive to change of speed, is the time-honoured Watt centrifugal governor. In this, two weights move radially outwards as their speed of rotation increases and thus move a sleeve on the central spindle. This sleeve movement is transmitted by a lever mechanism to the pilot-valve piston and hence the servo-motor is operated. A dead band is present in this mechanism, that is, the speed must change by a certain amount before the valve commences to operate, because of friction and mechanical backlash. The time taken for the main steam valve to move due to delays in the hydraulic pilot-valve and servo-motor systems is appreciable, 0.2–0.3 s.

The governor characteristic for a large steam turbo alternator is shown in Figure 4.4 and it is seen that there is a 4% drop in speed between no load and full load. Because of the requirement for high response speed, low dead band, and accuracy in speed and load control, the mechanical governor has been replaced in large modern turbo generators by electro hydraulic governing. The method normally

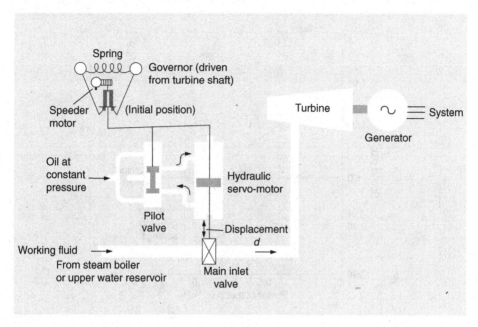

Figure 4.3 Governor control system employing the Watt governor as sensing device and a hydraulic servo-system to operate main supply valve. Speeder-motor gear determines the initial setting of the governor position

used to measure the speed is based on a toothed wheel on the generator shaft and magnetic-probe pickup. The use of electronic controls requires an electro hydraulic conversion stage, using conventional servo-valves.

An important feature of the governor system is the mechanism by means of which the governor sleeve and hence the main-valve positions can be changed and adjusted quite apart from when actuated by the speed changes. This is accomplished by the speed changer, or 'speeder motor', as it is sometimes termed. The

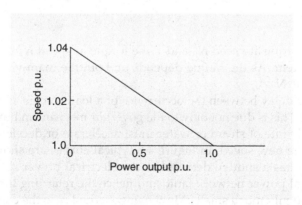

Figure 4.4 Idealized governor characteristic of a turboalternator with 4% droop from zero to full load

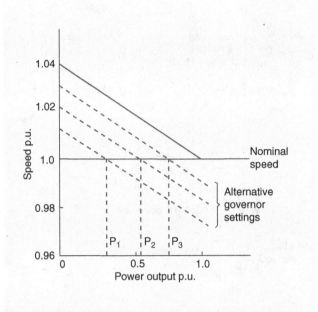

Figure 4.5 Effect of speeder gear on governor characteristics. P_1, P_2, and P_3 are power outputs at various settings but at the same speed

effect of this adjustment is the production of a family of parallel characteristics, as shown in Figure 4.5. Hence the power output of the generator at a given speed may be adjusted independently of system frequency and this is of extreme importance when operating at optimum economy.

The torque of the turbine may be considered to be approximately proportional to the displacement d of the main inlet valve. The expression for the change in torque with speed may be expressed approximately by the equation

$$T = T_0(1 - kN) \tag{4.1}$$

where T_0 is the torque at speed N_0 and T the torque at speed N; k is a constant for the governor system. As the torque depends on both the main-valve position and the speed, $T = f(d, N)$.

There is a time delay between the occurrence of a load change and the new operating conditions. This is due not only to the governor mechanism but also to the fact that the new flow rate of steam or water must accelerate or decelerate the rotor in order to attain the new speed. In Figure 4.6 typical curves are shown for a turbo-generator which has a sudden decrease in the electrical power required, perhaps due to an external power network fault, and hence the retarding torque on the turbine shaft is suddenly much smaller. In the ungoverned case the considerable time-lag between the load change and the attainment of the new steady speed is obvious. With the regulated or governed machine, due to the dead band in the governor

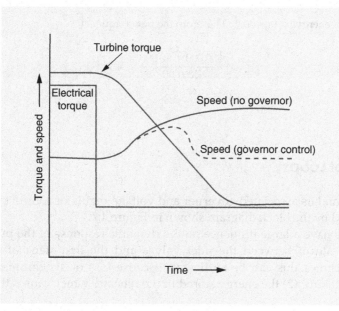

Figure 4.6 Graphs of turbine torque, electrical torque, and speed against time when the load on a generator suddenly falls

mechanism, the speed-time curve starts to rise, the valve then operates, and the fluid supply is adjusted. It is possible for damped oscillations to be set up after the load change.

An important factor regarding turbines is the possibility of overspeeding, when the load on the shaft is lost completely, with possible drastic mechanical breakdown. To avoid this, special valves are incorporated to automatically cut off the energy supply to the turbine. In a turbogenerator normally running at 3000 r.p.m. this over-speed protection operates at about 3300 r.p.m.

Example 4.1

An isolated 75 MVA synchronous generator feeds its own load and operates initially at no-load at 3000 r.p.m., 50 Hz. A 20 MW load is suddenly applied and the steam valves to the turbine start to open after 0.5 s due to the time-lag in the governor system. Calculate the frequency to which the generated voltage drops before the steam flow meets the new load. The stored energy for the machine is 4 kWs per kVA of generator capacity.

Solution
For this machine the stored energy at 3000 r.p.m.

$$= 4 \times 75\,000 = 300\,000 \text{ kWs}$$

Before the steam valves start to open the machine loses $20\,000 \times 0.5 = 10\,000$ kWs of the stored energy in order to supply the load.

The stored energy $\propto$ (speed)^2. Therefore the new frequency

$$= \sqrt{\frac{300\,000 - 10\,000}{300\,000}} \times 50\,Hz$$

$$= 49.2\,Hz$$

4.3 Control Loops

The machine and its associated governor and voltage-regulator control systems may be represented by the block diagram shown in Figure 4.7.

Two factors have a large influence on the dynamic response of the prime mover: (1) entrained steam between the inlet valves and the first stage of the turbine (in large machines this can be sufficient to cause loss of synchronism after the valves have closed); (2) the energy stored in the reheater which causes the output of

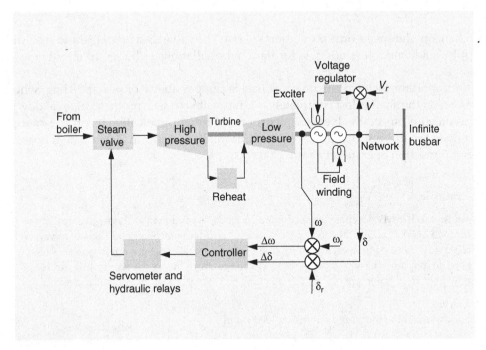

Figure 4.7 Block diagram of complete turboalternator control systems. The governor system is more complicated than that shown in Figure 4.3 owing to the inclusion of the load-angle δ in the control loop. Suffixes r refer to reference quantities and Δ to the error quantities. The controller modifies the error signal from the governor by taking into account the load angle

the low-pressure turbine to lag behind that of the high-pressure side. The transfer function

$$\frac{\text{prime mover torque}}{\text{valve opening}}$$

accounting for both these effects is

$$\frac{G_1 G_2}{(1 + \tau_t s)(1 + \tau_r s)} \tag{4.2}$$

where

G_1 = entrained steam constant;
G_2 = reheater gain constant;
τ_t = entrained steam time constant;
τ_r = reheater time constant.

The transfer function relating steam-valve opening d to changes in speed ω due to the governor feedback loop is

$$\frac{\Delta d}{\Delta \omega}(s) = \frac{G_3 G_4 G_5}{(1 + \tau_g s)(1 + \tau_1 s)(1 + \tau_2 s)} \tag{4.3}$$

where

τ_g = governor-relay time constant
τ_1 = primary-relay time constant
τ_2 = secondary-valve-relay time constant
$G_3 G_4 G_5$ = constants relating system-valve lift to speed change

By a consideration of the transfer function of the synchronous generator with the above expressions the dynamic response of the overall system may be obtained.

4.4 Division of Load between Generators

The use of the speed changer enables the steam input and electrical power output at a given frequency to be changed as required. The effect of this on two machines can be seen in Figure 4.8. The output of each machine is not therefore determined by the governor characteristics but can be varied by the operating personnel to meet economic and other considerations. The governor characteristics only completely decide the outputs of the machines when a sudden change in load occurs or when machines are allowed to vary their outputs according to speed within a prescribed

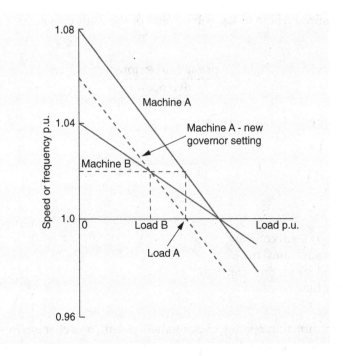

Figure 4.8 Two machines connected to an infinite busbar. The speeder gear of machine A is adjusted so that the machines share load equally

range in order to keep the frequency constant. This latter mode of operation is known as free-governor action.

It has been shown in Chapter 2 (Equation (2.15)) that when δ is small the voltage difference between the two ends of an interconnector of total impedance $R + jX$ is given by

$$\Delta V_p = V_G - V_L = \frac{RP + XQ}{V_L}$$

Also the angle between the voltage phasors (that is, the transmission angle) δ is given by

$$\sin^{-1}\left(\frac{\Delta V_q}{V_G}\right)$$

where

$$\Delta V_q = \frac{XP - RQ}{V_L}$$

When $X \gg R$, that is for most transmission networks,

$$\Delta V_q \propto P \quad \Delta V_p \propto Q$$

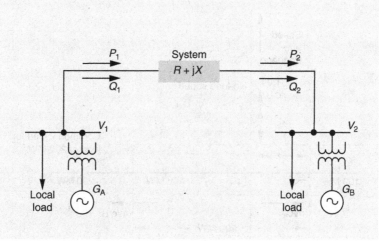

Figure 4.9 Two generating stations linked by an interconnector of impedance $(R+jX)$. The rotor of A is in phase advance of B

Hence, (1) the flow of power between two nodes is determined largely by the transmission angle; (2) the flow of reactive power is determined by the scalar voltage difference between the two nodes.

These two facts are of fundamental importance to the understanding of the operation of power systems.

Consider the two generator power system of Figure 4.9. The angular advance of G_A is due to a greater relative energy input to turbine A than to B. The provision of this extra steam (or water) to A is possible because of the action of the speeder gear without which the power outputs of A and B would be determined solely by the nominal governor characteristics. The following simple example illustrates these principles.

Example 4.2

Two synchronous generators operate in parallel and supply a total load of 200 MW. The capacities of the machines are 100 MW and 200 MW and both have governor droop characteristics of 4% from no load to full load. Calculate the load taken by each machine, assuming free governor action.

Solution
Let x megawatts be the power supplied from the 100 MW generator. Referring to Figure 4.10,

$$\frac{4}{100} = \frac{\alpha}{x}$$

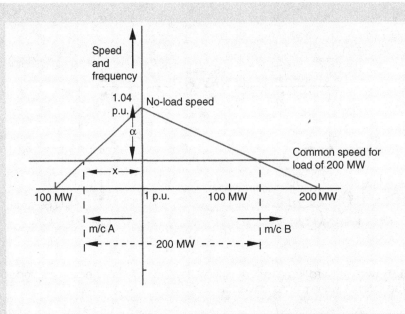

Figure 4.10 Speed-load diagram for Example 4.2

For the 200 MW machine,

$$\frac{4}{200} = \frac{\alpha}{200 - x}$$

$$\therefore \frac{4x}{100} = \frac{800 - 4x}{200}$$

and $x = 66.6\,\text{MW} = \text{load}$ on the 100 MW machine. The load on the 200 MW machine $= 133.3\,\text{MW}$.

It will be noticed that when the governor droops are the same the machines share the total load in proportion to their capacities or ratings. Hence it is often advantageous for the droops of all turbines to be equal.

Example 4.3

Two units of generation maintain 66 kV and 60 kV (line) at the ends of an interconnector of inductive reactance per phase of 40 Ω and with negligible resistance and shunt capacitance. A load of 10 MW is to be transferred from the 66 kV unit to the other end. Calculate the necessary conditions between the two ends, including the power factor of the current transmitted.

Solution
(using Equations (2.15) and (2.16))

As $R \simeq 0$

$$\Delta V_q = \frac{XP}{V_L} = \frac{40 \times 3.33 \times 10^6}{60\,000/\sqrt{3}} = 3840 \text{ V}$$

also

$$\frac{\Delta V_q}{66\,000/\sqrt{3}} = \sin \delta = 0.101$$

Hence the 66 kV busbars are 5° 44' in advance of the 60 kV busbars.

$$\Delta V_p = \frac{66\,000 - 60\,000}{\sqrt{3}} = \frac{XQ}{V_L} = \frac{40Q}{60\,000/\sqrt{3}}$$

hence

$Q = 3$ MVAr per phase (9 MVAr total)

The p.f. angle $\varphi = \tan^{-1}(Q/P) = 42°$ and hence the p.f. $= 0.74$.

4.5 The Power-Frequency Characteristic of an Interconnected System

The change in power for a given change in the frequency in an interconnected system is known as the stiffness of the system. The smaller the change in frequency for a given load change the stiffer the system. The power-frequency characteristic may be approximated by a straight line and $\Delta P/\Delta f = K$, where K is a constant (MW per Hz) depending on the governor and load characteristics.

Let ΔP_G be the change in generation with the governors operating 'free acting' resulting from a sudden increase in load ΔP_L. The resultant out-of balance power in the system

$$\Delta P = \Delta P_L - \Delta P_G \tag{4.4}$$

and therefore

$$K = \frac{\Delta P_L}{\Delta f} - \frac{\Delta P_G}{\Delta f} \tag{4.5}$$

$\dfrac{\Delta P_L}{\Delta f}$ measures the effect of the frequency characteristics of the load and $\Delta P_G \propto (P_T - P_G)$, where P_T is the turbine capacity connected to the network and P_G the output of the associated generators. When steady conditions are again reached, the load P_L is equal to the generated power P_G (neglecting losses): hence, $K = K_1 P_T - K_2 P_L$, where K_1 and K_2 are the power-frequency coefficients relevant to the turbines and load respectively.

K can be determined experimentally by connecting two large separate systems by a single link, breaking the connection and measuring the frequency change. For the British system, tests show that $K = 0.8P_T - 0.6P_L$ and lies between 2000 and

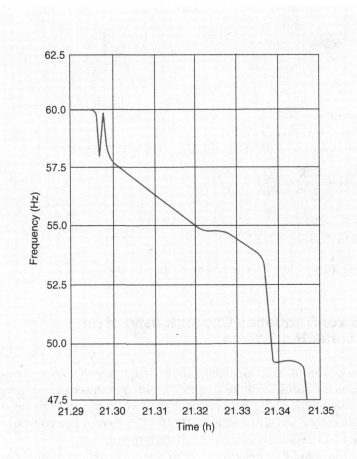

Figure 4.11 Decline of frequency with time of New York City system when isolated from external supplies (Reproduced with permission from IEEE © 1977)

5500 MW per Hz, that is, a change in frequency of 0.1 Hz requires a change in the range 200–550 MW, depending on the amount of plant connected. In smaller systems the change in frequency for a reasonable load change is relatively large and rapid-response electrical governors have been introduced to improve the power-frequency characteristic.

In 1977, owing to a series of events triggered off by lightning, New York City was cut off from external supplies and the internal generation available was much less than the city load. The resulting fall in frequency with time is shown in Figure 4.11, illustrating the time-frequency characteristics of an isolated power system.

4.6 System Connected by Lines of Relatively Small Capacity

Let K_A and K_B be the respective power-frequency constants of two separate power systems A and B, as shown in Figure 4.12. Let the change in the power transferred

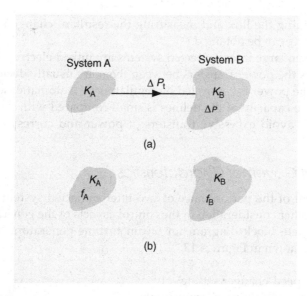

Figure 4.12 (a) Two interconnected power systems connected by a tie-line, (b) The two systems with the tie-line open

from A to B when a change resulting in an out-of-balance power ΔP occurs in system B, be ΔP_t, where ΔP_t is positive when power is transferred from A to B. The change in frequency in system B, due to an extra load ΔP and an extra input of ΔP_t, from A, is $-(\Delta P - \Delta P_t)/K_B$ (the negative sign indicates a fall in frequency). The drop in frequency in A due to the extra load ΔP_t is $-\Delta P_t/K_A$, but the changes in frequency in each system must be equal, as they are electrically connected. Hence,

$$-(\Delta P - \Delta P_t)/K_B = -\Delta P_t/K_A$$

$$\therefore \Delta P_t = +\left(\frac{K_A}{K_A + K_B}\right)\Delta P \qquad (4.6)$$

Next, consider the two systems operating at a common frequency f with A exporting ΔP_t to B. The connecting link is now opened and A is relieved of ΔP_t and assumes a freqency f_A, and B has ΔP_t more load and assumes f_B.
    Hence

$$f_A = f + \frac{\Delta P_t}{K_A} \quad \text{and} \quad f_B = f - \frac{\Delta P_t}{K_B}$$

from which

$$\frac{\Delta P_t}{f_A - f_B} = \frac{K_A K_B}{K_A + K_B} \qquad (4.7)$$

Hence, by opening the link and measuring the resultant change in f_A and f_B the values of K_A and K_B can be obtained.

In practice, when large interconnected systems are linked electrically to others by means of tie-lines the power transfers between them are usually decided by mutual agreement and the power is controlled by regulators or Automatic Generation Control (AGC). As the capacity of the tielines is small compared with the systems, care must be taken to avoid excessive transfers of power and corresponding cascade tripping.

4.6.1 Effect of Governor Characteristics

A fuller treatment of the performance of two interconnected systems in the steady state requires further consideration of the control aspects of the generators.

A more complete block diagram for steam turbine-generators connected to a power system is shown in Figure 4.13.

$\Delta P'=$ change in speed-changer setting;
$\Delta P =$ change in power output of prime movers;
$\Delta L =$ change load power;
$\Delta f =$ change in frequency;
$R =$ governor droop, that is drop in speed or frequency when machines of an area range from no load to full load

Figure 4.13 can be used to represent a number of coherent generators with similar characteristics.

For this system the following equation holds:

$$Ms\Delta f + D\Delta f = \Delta P - \Delta L$$

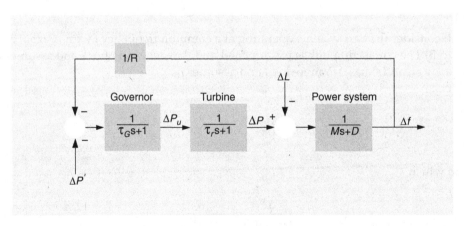

Figure 4.13 Block diagram for a turbine generator connected to a power system

where:

M is the angular momentum of the spinning generators and loads (see Chapter 8)
D is the damping coefficient of the load that is, change of power drawn by the load
with frequency

Therefore, the change from normal speed or frequency,

$$\Delta f = \frac{1}{Ms + D}(\Delta P - \Delta L)$$

This analysis holds for steam-turbine generation; for hydro-turbines, the large inertia of the water must be accounted for and the analysis is more complicated.

The representation of two systems connected by a tie-line is shown in Figure 4.14. The general analysis is as before except for the additional power terms due to the tie-line. The machines in each individual power system are considered to be closely coupled and to possess one equivalent rotor.

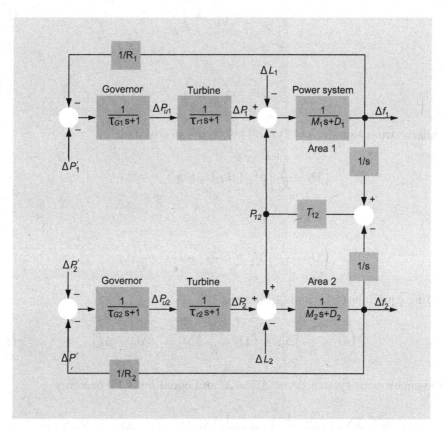

Figure 4.14 Block control diagram of two power systems connected by a tie-line

For system (1),

$$M_1 s\Delta f_1 + D_1 \Delta f_1 + T_{12}(\delta_1 - \delta_2) = \Delta P_1 - \Delta L_1 \tag{4.8}$$

where T_{12} is the synchronizing torque coefficient of the tie-line.
For system (2),

$$M_2 s\Delta f_2 + D_2 \Delta f_2 + T_{12}(\delta_2 - \delta_1) = \Delta P_2 - \Delta L_2 \tag{4.9}$$

The steady-state analysis of two interconnected systems may be obtained from the transfer functions given in the block diagram.
The response of the Governor is given by

$$\Delta P_U = \frac{-1}{\tau_G s + 1} \left(\frac{1}{R} \Delta f + \Delta P' \right) \tag{4.10}$$

In the steady state, from equation (4.10),

$$\Delta P_1 = \left(\frac{1}{R_1} \Delta f \right) \tag{4.11}$$

and

$$\Delta P_2 = \left(\frac{1}{R_2} \Delta f \right) \tag{4.12}$$

Similarly, from equations (4.8) and (4.9), in the steady state,

$$\left(D_1 + \frac{1}{R_1} \right) \Delta f_1 + T_{12}(\delta_1 - \delta_2) = -\Delta L_1 \tag{4.13}$$

and

$$\left(D_2 + \frac{1}{R_2} \right) \Delta f_2 + T_{12}(\delta_2 - \delta_1) = -\Delta L_2 \tag{4.14}$$

Adding equations (4.13) and (4.14) gives

$$\left(D_1 + \frac{1}{R_1} \right) \Delta f_1 + \left(D_2 + \frac{1}{R_2} \right) \Delta f_2 = -\Delta L_1 - \Delta L_2 \tag{4.15}$$

In a synchronous system, $\Delta f_1 = \Delta f_2 = \Delta f$ and equation (4.15) becomes

$$\left(D_1 + \frac{1}{R_1} + D_2 + \frac{1}{R_2} \right) \Delta f = -\Delta L_1 - \Delta L_2$$

and

$$\Delta f = \frac{-\Delta L_1 - \Delta L_2}{\left(D_1 + D_2 + \dfrac{1}{R_1} + \dfrac{1}{R_2}\right)} \tag{4.16}$$

From equations (4.13) and (4.14),

$$T_{12}(\delta_1 - \delta_2) = \frac{-\Delta L_1 (D_2 + 1/R_2) + \Delta L_2 (D_1 + 1/R_1)}{(D_2 + 1/R_2) + (D_1 + 1/R_1)} \tag{4.17}$$

Example 4.4

Two power systems, A and B, each have a regulation R of 0.1 p.u. and a damping factor D of 1(on their respective capacity bases). The capacity of system A is 1500 MW and of B 1000 MW. The two systems are interconnected through a tie-line and are initially at 60 Hz. If there is a 100 MW load increase in system A, calculate the change in the steady state values of frequency and power transfer.

Solution

$$D_A = 1500 \text{ MW/Hz}$$

$$D_B = 1000 \text{ MW/Hz}$$

$$R_A = \frac{0.1 \times 60}{1500} = \frac{6}{1500} \text{ Hz/MW}$$

$$R_B = \frac{0.1 \times 60}{1000} = \frac{6}{1000} \text{ Hz/MW}$$

From equation (4.16),

$$\Delta f = \frac{-\Delta L_A - \Delta L_B}{\left(D_A + D_B + \dfrac{1}{R_A} + \dfrac{1}{R_B}\right)} = \frac{-100}{1500 + 1000 + \dfrac{1500}{6} + \dfrac{1000}{6}} = -0.034 \, Hz$$

$$P_{AB} = T_{AB}(\delta_A - \delta_B) = \frac{-\Delta L_A (D_B + 1/R_B) + \Delta L_B (D_A + 1/R_A)}{(D_B + 1/R_B) + (D_A + 1/R_A)}$$

$$= \frac{-100\left(1000 + \dfrac{1000}{6}\right)}{1500 + 1000 + \dfrac{1500}{6} + \dfrac{1000}{6}} = \frac{-100\left(\dfrac{7000}{6}\right)}{\dfrac{17500}{6}} = -40 \, MW$$

4.6.2 Frequency-Bias-Tie-Line Control

Consider three interconnected power systems A, B and C, as shown in Figure 4.15, the systems being of similar size. Assume that initially A and B export to C, their previously agreed power transfers. If C has an increase in load the overall frequency tends to decrease and hence the generation in A, B and C increases. This results in increased power transfers from A and B to C. These transfers, however, are limited by the tie-line power controller to the previously agreed values and therefore instructions are given for A and B to reduce generation and hence C is not helped. This is a severe drawback of what is known as straight tie-line control, which can be overcome if the systems are controlled by using consideration of both load transfer and frequency, such that the following equation holds:

$$\sum \Delta P + B\Delta f = 0 \tag{4.18}$$

where $\sum \Delta P$ is the net transfer error and depends on the size of the system and the governor characteristic, and Δf is the frequency error and is positive for high frequency. B is known as the frequency bias factor and is derived from Equation (4.17).

In the case above, after the load change in C, the frequency error is negative (i.e. low frequency) for A and B and the sum of ΔP for the lines AC and BC is positive. For correct control,

$$\sum P_A + B_A \Delta f = \sum P_B + B_B \Delta f = 0$$

Systems A and B take no regulating action despite their fall in frequency.

In C, $\sum \Delta P_C$ is negative as it is importing from A and B and therefore the governor speeder motors in C operate to increase output and restore frequency. This system is known as frequency-bias-tie-line control and is often implemented automatically in interconnected systems.

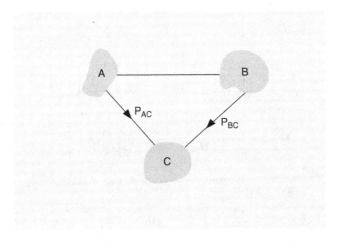

Figure 4.15 Three power systems connected by tie-lines

Problems

4.1 A 500 MVA, 2 pole, turbo-alternator delivers 400 MW to a 50 Hz system. If the generator circuit breaker is suddenly opened and the main steam valves take 400 ms to operate what will be the over-speed of the generator? The stored energy of the machine (generator and turbine) at synchronous speed is 4 kWs/kVA.

(Answer 3117 r.p.m or 52 Hz)

4.2 Two identical 60 MW synchronous generators operate in parallel. The governor settings on the machines are such that they have 4 and 3% droops (no-load to full-load percentage speed drop). Determine (a) the load taken by each machine for a total of 100 MW; (b) the percentage adjustment in the no-load speed to be made by the speeder motor if the machines are to share the load equally.

(Answer: (a) 42.8 and 57.2 MW; (b) 0.83% increase in no-load speed on the 4% droop machine)

4.3 a. Explain how the output power of a turbine-generator operating in a constant frequency system is controlled by adjusting the setting of the governor. Show the effect on the generator power-frequency curve.
  b. Generator A of rating 200 MW and generator B of rating 350 MW have governor droops of 5 and 8%, respectively, from no-load to full-load. They are the only supply to an isolated system whose nominal frequency is 50 Hz. The corresponding generator speed is 3000 r.p.m. Initially, generator A is at 0.5 p.u. load and generator B is at 0.65 p.u. load, both running at 50 Hz. Find the no load speed of each generator if it is disconnected from the system.
  c. Also determine the total output when the first generator reaches its rating.

(Answer: (b) Generator B 3156 r.p.m; generator A 3075 r.p.m; (c) 537 MW)

(From Engineering Council Examination, 1996)

4.4 Two power systems A and B are interconnected by a tie-line and have power frequency constants K_A and K_B MW/Hz. An increase in load of 500 MW on system A causes a power transfer of 300 MW from B to A. When the tie-line is open the frequency of system A is 49 Hz and of system B 50 Hz. Determine the values of K_A and K_B, deriving any formulae used.

(Answer: K_A 500 MW/Hz; K_B 750 MW/Hz)

4.5 Two power systems, A and B, having capacities of 3000 and 2000 MW, respectively, are interconnected through a tie-line and both operate with frequency-bias-tieline control. The frequency bias for each area is 1% of the system capacity per 0.1 Hz frequency deviation. If the tie-line interchange for A is set at 100 MW and for B is set (incorrectly) at 200 MW, calculate the steady-state change in frequency.

(Answer: 1 Hz; use $\Delta P_A + B_A f = \Delta P_B + B_B f$)

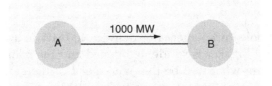

Figure 4.16 Interconnected systems of Problem 4.6 (b)

4.6 a. i. Why do power systems operate in an interconnected arrangement?
         ii. How is the frequency controlled in a power system?
         iii. What is meant by the stiffness of a power system?
     b. Two 50 Hz power systems are interconnected by a tie-line, which carries
        1000 MW from system A to system B, as shown in Figure 4.16. After the out-
        age of the line shown in the figure, the frequency in system A increases to
        50.5 Hz, while the frequency in system B decreases to 49 Hz.
         i. Calculate the stiffness of each system.
         ii. If the systems operate interconnected with 1000 MW being transferred
            from A to B, calculate the flow in the line after outage of a 600 MW gener-
            ator in system B.

(Answer: (b) (i) K_A 2000 MW/Hz, K_B 1000 MW/Hz; (ii) 1400 MW)

(From Engineering Council Examination, 1997)

5

Control of Voltage and Reactive Power

5.1 Introduction

The approximate relationship between the magnitude of the voltage difference of two nodes in a network and the flow of power was shown in Chapter 2 to be

$$\Delta V \approx \Delta V_p = \frac{RP + XQ}{V} \quad \text{(from 2.15)}$$

Also it was shown that the transmission angle δ is proportional to

$$\delta \propto \Delta V_q = \frac{XP - RQ}{V} \quad \text{(from 2.16)}$$

Hence it may be seen that for networks where $X \gg R$, that is, most high voltage power circuits, ΔV, the voltage difference, is determined mainly by Q while the angle δ is controlled by P.

Consider the simple system linking two generating stations A and B, as shown in Figure 5.1(a). Initially the system is considered to be only reactive and R is ignored. The machine at A is in phase advance of that at B and V_1 is greater than V_2; hence there is a flow of real power from A to B. This can be seen from the phasor diagram shown in Figure 5.1(b). It is seen that I_d and hence P is determined by $\angle\delta$ and the value of I_q and hence Q mainly, by $V_1 - V_2$. In this case $V_1 > V_2$ and reactive power is transferred from A to B. By varying the generator excitations such that $V_2 > V_1$, the direction of the reactive power is reversed, as shown in Figure 5.1(c).

Hence, real power can be sent from A to B or B to A by suitably adjusting the amount of steam (or water) admitted to the turbine, and reactive power can be sent

---

Electric Power Systems, Fifth Edition. B.M. Weedy, B.J. Cory, N. Jenkins, J.B. Ekanayake and G. Strbac.
© 2012 John Wiley & Sons, Ltd. Published 2012 by John Wiley & Sons, Ltd.

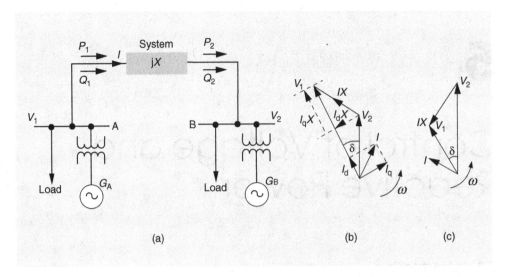

Figure 5.1 (a) System of two generators interconnected, (b) Phasor diagram when $V_1 > V_2$. I_d and I_q are components of I. (c) Phasor diagram when $V_2 > V_1$

in either direction by adjusting the voltage magnitudes. These two operations are approximately independent of each other if $X \gg R$, and the flow of reactive power can be studied almost independently of the real power flow.

The phasor diagrams show that if a scalar voltage difference exists across a largely reactive link, the reactive power flows towards the node of lower voltage. From another point of view, if, in a network, there is a deficiency of reactive power at a point, this has to be supplied from the connecting lines and hence the voltage at that point falls. Conversely if there is a surplus of reactive power generated (for example, lightly loaded cables generate positive VArs), then the voltage will rise. This is a convenient way of expressing the effect of the power factor of the transferred current, and although it may seem unfamiliar initially, the ability to think in terms of VAr flows, instead of exclusively with power factors and phasor diagrams, will make the study of power networks much easier.

If it can be arranged that Q_2 in the system in Figure 5.1(a) is zero, then there will be no voltage drop between A and B, a very satisfactory state of affairs.

Now assume that the interconnecting system shown in Figure 5.1(a) has some resistance and that V_1 is constant. Consider the effect of keeping V_2, and hence the voltage drop ΔV, constant. From equation (2.15)

$$Q_2 = \frac{V_2 \Delta V - RP_2}{X} = K - \frac{R}{X}P_2 \tag{5.1}$$

where K is a constant and R is the resistance of the system.

If this value of Q_2 does not exist naturally in the circuit then it will have to be obtained by artificial means, such as the connection at B of capacitors or inductors.

If the value of the power changes from P_2 to P'_2 and if V_2 remains constant, then the reactive power at B must change to Q'_2 such that

$$Q'_2 - Q_2 = \frac{R}{X}\left(P'_2 - P_2\right)$$

that is, an increase in real power causes an increase in the reactive power needed to maintain V_2. The change, however, is proportional to (R/X), which is normally small.

It is seen that voltage can be controlled by the injection into the network of reactive power of the correct sign. Other methods of a more obvious kind for controlling voltage are the use of tap-changing transformers.

5.2 The Generation and Absorption of Reactive Power

5.2.1 Synchronous Generators

Synchronous generators can be used to generate or absorb reactive power. An over-excited machine, that is, one with greater than nominal excitation, generates reactive power whilst an under-excited machine absorbs it. Synchronous generators are the main source of supply to the power system of both positive and negative VArs.

The ability to generate or absorb reactive power is shown by the performance chart of a synchronous generator. Reactive power generation (lagging power factor operation) is limited by the maximum excitation voltage allowable before the rotor currents lead to overheating. In Figure 3.12 this is 2.5 p.u.

The ability to absorb reactive power is determined by the short-circuit ratio (1/synchronous reactance) as the distance between the power axis and the theoretical stability-limit line in Figure 3.12 is proportional to the short-circuit ratio. In modern machines the value of the short-circuit ratio is made low for economic reasons, and hence the inherent ability to operate at leading power factors (absorbing VArs) is not large. For example, a 200 MW 0.85 p.f. machine with a 10% stability allowance has a capability of absorbing 45 MVAr at full power output. The VAr absorption capacity can, however, be increased by the use of continuously acting voltage regulators, as explained in Chapter 3.

5.2.2 Overhead Lines and Transformers

When fully loaded, overhead lines absorb reactive power. With a current I amperes flowing in a line of reactance per phase $X(\Omega)$ the VArs absorbed are I^2X per phase. On light loads the shunt capacitances of longer lines may become dominant and high voltage overhead lines then become VAr generators.

Transformers always absorb reactive power. A useful expression for the quantity may be obtained for a transformer of reactance X_T p.u. and a full load rating of $3V_\phi I_{rated}$

The ohmic reactance

$$X(\Omega) = X_T \frac{V_\phi}{I_{rated}}$$

Therefore the VArs absorbed

$$= 3I^2 X_T \frac{V_\phi}{I_{rated}}$$

$$= 3I^2 \frac{V_\phi^2}{V_\phi I_{rated}} X_T$$

$$= \frac{(3IV_\phi)^2}{3I_{rated} V_\phi} X_T$$

$$= \frac{(VA \text{ of load})^2}{\text{rated } VA \text{ of transformer}} \cdot X_T$$

5.2.3 Cables

Cables are generators of reactive power owing to their high shunt capacitance.
    A 275 kV, 240 MVA cable produces 6.25–7.5 MVAr per km; a 132 kV cable roughly 1.9 MVAr per km; and a 33 kV cable, 0.125 MVAr per km.

5.2.4 Loads

A load at 0.95 power factor implies a reactive power demand of 0.33 kVAr per kW of power, which is more appreciable than the mere quoting of the power factor would suggest. In planning a network it is desirable to assess the reactive power requirements to ascertain whether the generators are able to operate at the required power factors for the extremes of load to be expected. An example of this is shown in Figure 5.2, where the reactive losses are added for each item until the generator power factor is obtained.

Example 5.1

In the radial transmission system shown in Figure 5.2, all p.u. values are referred to the voltage bases shown and 100 MVA. Determine the power factor at which the generator must operate.

Solution
Voltage drops in the circuits will be neglected and the nominal voltages assumed.
    Starting with the consumer load, the VArs for each section of the circuit are added in turn to obtain the total
    Busbar A,

$$P = 0.5 \text{ p.u. } Q = 0$$

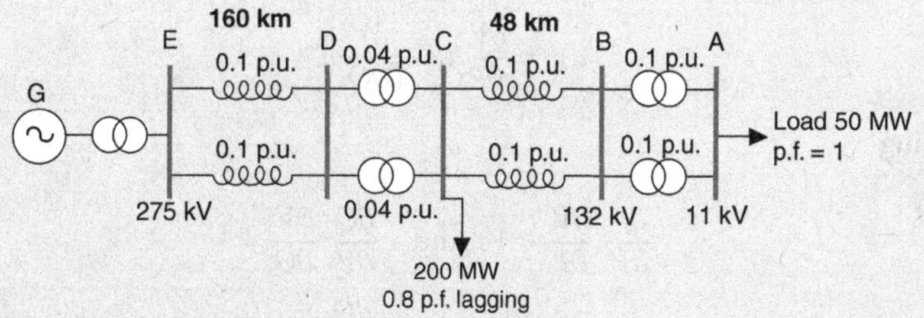

Figure 5.2 Radial transmission system with intermediate loads. Calculation of reactive-power requirement

$I^2 \times$ loss in 132 kV lines and transformers

$$= \frac{P^2 + Q^2}{V^2} X_{CA} = \frac{0.5^2}{1^2} 0.1$$

$$= 0.025 \, \text{p.u.}$$

Busbar C,

$$P = 2 + 0.5 \, \text{p.u.} = 2.5 \, \text{p.u.}$$

$$Q = 1.5 + 0.025 \, \text{p.u.} = 1.525 \, \text{p.u.}$$

I^2X loss in 275 kV lines and transformers

$$= \frac{2.5^2 + 1.525^2}{1^2} 0.07$$

$$= 0.6 \, \text{p.u.}$$

If the I^2X loss in the large generator-transformer is ignored, the generator must deliver $P = 2.5$ and $Q = 2.125$ p.u. and operate at a power factor of 0.76 lagging.

5.3 Relation between Voltage, Power, and Reactive Power at a Node

The voltage V at a node is a function of P and Q at that node, that is.

$$V = f(P, Q)$$

The voltage also depends on that of adjacent nodes and the present treatment assumes that these are infinite busbars.

The total differential of V,

$$dV = \frac{\partial V}{\partial P} dP + \frac{\partial V}{\partial Q} dQ$$

and using

$$\frac{\partial P}{\partial V} \cdot \frac{\partial V}{\partial P} = 1 \quad \text{and} \quad \frac{\partial Q}{\partial V} \cdot \frac{\partial V}{\partial Q} = 1$$

$$dV = \frac{dP}{(\partial P/\partial V)} + \frac{dQ}{(\partial Q/\partial V)}$$

(5.2)

It can be seen from equation (5.2) that the change in voltage at a node is defined by the two quantities

$$\left(\frac{\partial P}{\partial V}\right) \quad \text{and} \quad \left(\frac{\partial Q}{\partial V}\right)$$

As an example, consider a line with series impedance $(R + jX)$ and zero shunt admittance as shown in Figure 5.3. From equation (2.15),

$$(V_1 - V)V - PR - XQ = 0$$

(5.3)

where V_1, the sending-end voltage, is constant, and V, the receiving-end voltage, depends on P and Q.

From equation (5.3)

$$\frac{\partial P}{\partial V} = \frac{V_1 - 2V}{R}$$

(5.4)

Also,

$$\frac{\partial Q}{\partial V} = \frac{V_1 - 2V}{X}$$

(5.5)

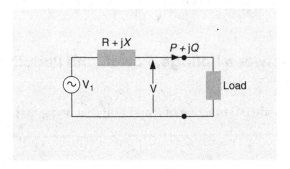

Figure 5.3 Single-phase equivalent circuit of a line supplying a load of $P + jQ$ from an infinite busbar of voltage V_1

Hence,

$$dV = \frac{dP}{\partial P/\partial V} + \frac{dQ}{\partial Q/\partial V} = \frac{RdP + XdQ}{V_1 - 2V} \tag{5.6}$$

For constant V and ΔV, $RdP + XdQ = 0$ and $dQ = -(R/X)dP$, which is obtainable directly from Equation (5.1).

Normally, $\partial Q/\partial V$ is the quantity of greater interest. It can be found experimentally using a load-flow calculation (see Chapter 6) by the injection of a known quantity of VArs at the node in question and calculating the difference in voltage produced. From the results obtained,

$$\frac{\Delta Q}{\Delta V} = \frac{Q_{after} - Q_{before}}{V_{after} - V_{before}}$$

ΔV should be small for this test, a few per cent of the normal voltage, thereby giving the sensitivity of the node to the VAr change.

From the expression,

$$\frac{\partial Q}{\partial V} = \frac{V_1 - 2V}{X}$$

proved for a single line, it is evident that the smaller the reactance associated with a node, the larger the value of $\partial Q/\partial V$ for a given voltage drop, that is, the voltage drop is inherently small. The greater the number of lines meeting at a node, the smaller the resultant reactance and the larger the value of $\partial Q/\partial V$. Obviously, $\partial Q/\partial V$ depends on the network configuration, but a high value would lie in the range 10–15 MVAr/kV. If the natural voltage drop at a point without the artificial injection of VArs is, say, 5 kV, and the value of $\partial Q/\partial V$ at this point is 10 MVAr/kV, then to maintain the voltage at its no-load level would require 50 MVAr. Obviously, the greater the value of $\partial Q/\partial V$, the more expensive it becomes to maintain voltage levels by injection of reactive power.

5.3.1 $\partial Q/\partial V$ and the Short-Circuit Current at a Node

It has been shown that for a connecting circuit of reactance X with a sending-end voltage V_1 and a received voltage V

$$\frac{\partial Q}{\partial V} = \frac{V_1 - 2V}{X}$$

If the three-phases of the connector are now short-circuited at the receiving end (i.e. a three-phase symmetrical short circuit applied), the current flowing in the lines

$$I = \frac{V_1}{X} \quad \text{assuming} \quad R \ll X$$

With the system on no-load

$$V = V_1 \quad \text{and} \quad \left|\frac{\partial Q}{\partial V}\right| = \left|\frac{V_1}{X}\right|$$

Hence the magnitude of $\partial Q/\partial V$ is equal to the short-circuit current. With normal operation, V is within a few per cent of V_1 and hence the value of $\partial Q/\partial V$ at $V = V_1$ gives useful information regarding reactive power/voltage characteristics for small excursions from the nominal voltage. This relationship is especially useful as the short-circuit current will normally be known at all substations.

Example 5.2

Three supply points A, B, and C are connected to a common busbar M. Supply point A is maintained at a nominal 275 kV and is connected to M through a 275/132 kV transformer (0.1 p.u. reactance) and a 132 kV line of reactance 50 Ω. Supply point B is nominally at 132 kV and is connected to M through a 132 kV line of 50 Ω reactance. Supply point C is nominally at 275 kV and is connected to M by a 275/132 kV transformer (0.1 p.u. reactance) and a 132 kV line of 50 Ω reactance.

If, at a particular system load, the line voltage of M falls below its nominal value by 5 kV, calculate the magnitude of the reactive volt-ampere injection required at M to re-establish the original voltage.

The p.u. values are expressed on a 500 MVA base and resistance may be neglected throughout.

Solution
The line diagram and equivalent single-phase circuit are shown in Figures 5.4 and 5.5.

It is necessary to determine the value of dQ/dV at the node or busbar M; hence the current flowing into a three-phase short-circuit at M is required.

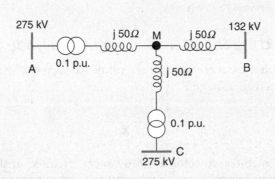

Figure 5.4 Schematic diagram of the system for Example 5.2

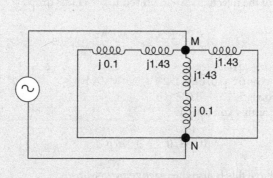

Figure 5.5 Equivalent single-phase network with the node M short-circuited to neutral)

The base value of reactance in the 132 kV circuit assuming a 500 MVA system base is

$$Z_{base} = \frac{132^2}{500} = 35\,\Omega$$

Therefore the line reactances

$$X_L = \frac{j50}{35} = j1.43\,\text{p.u.}$$

The equivalent reactance from M to N = $j0.5$ p.u.
  Hence the fault MVA at M

$$= \frac{500}{0.5} = 1000\,\text{MVA}$$

and the fault current

$$= \frac{1000 \times 10^6}{\sqrt{3} \times 132 \times 10^3} = 4380\,\text{A}$$

It has been shown that $\partial Q_M / \sqrt{3}\partial V_M$ = three-phase short-circuit current when Q_M and V_M are three-phase and line values

$$\frac{\partial Q_M}{\partial V_M} = 4380 \times \sqrt{3} = 7.6\,\text{MVAr/kV}$$

Assuming the natural voltage drop at M = 5 kV.

Therefore the value of the injected VArs required to offset this drop

$$= 7.6 \times 5 = 38 \, \text{MVAr}$$

An alternative approach is to consider:
The source impedance is 0.5 p.u. on a 132 kV, 500 MVA base
  A 5 kV voltage drop is $5/132 = 0.038$ p.u.
  The current flow for this volt drop is

$$0.038/0.5 = 0.076 \, \text{p.u.}$$

At close to 1 p.u. voltage this is also the reactive power flow.

$$Q = 0.076 \, \text{p.u.}$$

$$= 0.076 \times 500 = 38 \, \text{MVAr}$$

5.4 Methods of Voltage Control: (a) Injection of Reactive Power

In transmission systems with $X \gg R$, busbar voltages can be controlled by the injection or absorption of reactive power. However, controlling network voltage through reactive power flow is less effective in distribution networks where the higher circuit resistances lead to the reactive power flows having less effect on voltage and causing an increase in real power losses.

Although reactive power does no real work, it does lead to an increase in the magnitude of current in the networks and hence real power losses. Electricity suppliers often penalize loads with a poor power factor by applying charges based on kVAh (or even kVArh) in addition to kWh or even basing part of the charge on peak kVA drawn. The provision of static capacitors to improve the power factors of factory loads has been long established. The capacitance required for the power-factor improvement of loads for optimum economy is determined as follows.

Let the tariff of a consumer be based on both kVA and kWh

$$\text{charge} = \$A \times kVA + \$B \times kWh$$

A load of P kilowatts at power factor φ_1, lagging has a kVA of $P/\cos\varphi_1$. If this power factor is improved to $\cos \varphi_2$, the new kVA is $P/\cos \varphi_2$. The saving is therefore

$$\text{saving} = \$PA \left(\frac{1}{\cos \varphi_1} - \frac{1}{\cos \varphi_2} \right)$$

The reactive power required from the correcting capacitors

$$P(\tan \varphi_1 - \tan \varphi_2) \quad \text{kVAr}$$

Let the cost per annum in interest and depreciation on the capacitor installation be $C per kVAr or

$$\$CP(\tan \varphi_1 - \tan \varphi_2)$$

The net saving

$$= P\left[\$A \left(\frac{1}{\cos \varphi_1} - \frac{1}{\cos \varphi_2} \right) - \$C(\tan \varphi_1 - \tan \varphi_2) \right]$$

This saving is a maximum when

$$\frac{d(\text{saving})}{d\varphi_2} = P\left[\$A \left(-\frac{\sin \varphi_2}{\cos^2 \varphi_2} \right) + \$C \left(\frac{1}{\cos^2 \varphi_2} \right) \right] = 0$$

that is when $\sin \varphi_2 = \$C/\$A$.

It is interesting to note that the optimum power factor is independent of the original one. The improvement of load power factors in such a manner will help to alleviate the whole problem of VAr flow in the distribution system.

The main effect of transmitting power at non-unity power factors is to increase losses and reduce the ability of the circuits to transport active power. Thus both operating and capital costs are increased by low power factor. It is evident from equation (2.15) that, for circuits with a significant X/R ratio, the voltage drop is largely determined by the reactive power Q. At non-unity power factors the line currents are larger, giving increased I^2R losses and hence reduced thermal capability. One of the obvious places for the artificial injection of reactive power is at the loads themselves.

In general, four methods of injecting reactive power are available, involving the use of:

1. static shunt capacitors;
2. static series capacitors;
3. synchronous compensators;
4. static VAr compensators and STATCOMs.

5.4.1 Shunt Capacitors and Reactors

Shunt capacitors are used to compensate lagging power factor loads, whereas reactors are used on circuits that generate VArs such as lightly loaded cables. The effect of these shunt devices is to supply or absorb the requisite reactive power to maintain the magnitude of the voltage. Capacitors are connected either directly to a busbar or to the tertiary winding of a main transformer. In the USA they are often installed along the routes of distribution circuits to minimize the losses and voltage drops. Unfortunately, as the voltage reduces, the VArs produced by a shunt capacitor or absorbed by a reactor fall as the square of the voltage; thus, when needed most, their effectiveness drops. Also, with light network load when the voltage is high, the capacitor output is large and the voltage tends to rise to excessive levels, requiring some capacitors or cable circuits to be switched out by local overvoltage relays.

5.4.2 Series Capacitors

Capacitors can be connected in series with overhead lines and are then used to reduce the inductive reactance between the supply point and the load. One major drawback is the high overvoltage produced across the capacitor when a short-circuit current flows through the circuit, and special protective devices need to be incorporated (e.g. spark gaps) and non-linear resistors. The phasor diagram for a line with a series capacitor is shown in Figure 5.6(b).

The relative merits between shunt and series capacitors may be summarized as follows:

1. If the load VAr requirement is small, series capacitors are of little use.
2. With series capacitors the reduction in line current is small; hence if thermal considerations limit the current, little advantage is obtained and shunt compensation should be used.
3. If voltage drop is the limiting factor, series capacitors are effective; also, voltage fluctuations due to arc furnaces, and so on, are evened out.

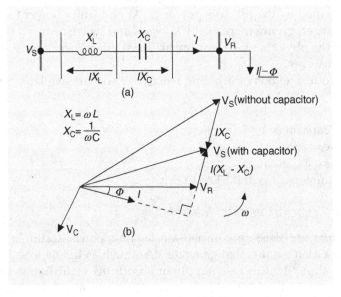

Figure 5.6 (a) Line with series capacitor, (b) Phasor diagram for fixed V_R

4. If the total line reactance is high, series capacitors are very effective in reducing voltage drops and stability is improved.

Both shunt and series capacitors need to be applied with care as they can both lead to resonance with the inductive reactance of the power system. Shunt capacitors are benign as long as their network is connected to the main power system and the voltage is controlled. However, if a section of network containing both shunt capacitors and induction generators is isolated then self-excitation of the induction generators can lead to very high resonant voltages. The use of series capacitors, although very effective in reducing voltage drop on heavily loaded circuits, can lead to sub-synchronous resonance with rotating machines. Capacitors are not commonly used in distribution systems in the UK, partly because of concerns over resonance.

5.4.3 Synchronous Compensators

A synchronous compensator is a synchronous motor running without a mechanical load and, depending on the value of excitation, it can absorb or generate reactive power. As the losses are considerable compared with static capacitors, the power factor is not zero. When used with a voltage regulator the compensator can automatically run overexcited at times of high load and underexcited at light load. A typical

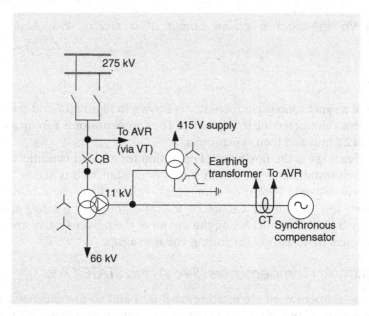

Figure 5.7 Typical installation with synchronous compensator connected to tertiary (delta) winding of main transformer. A neutral point is provided by the earthing transformer shown. The Automatic Voltage Regulator (AVR) on the compensator is controlled by a combination of the voltage on the 275 kV system and the current output; this gives a droop to the voltage-VAr output curve which may be varied as required

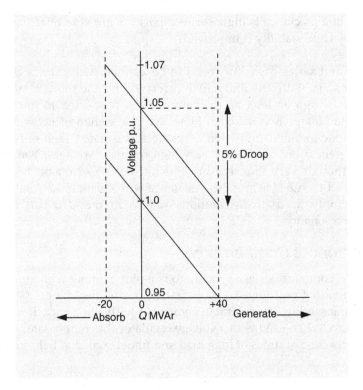

Figure 5.8 Voltage-reactive power output of a typical 40 MVAr synchronous compensator

connection of a synchronous compensator is shown in Figure 5.7 and the associated Volt-VAr output characteristic in Figure 5.8. The compensator is run up as an induction motor in 2.5 min and then synchronized.

A great advantage is the flexibility of operation for all load conditions. Although the cost of such installations is high, in some circumstances it is justified, for example at the receiving-end busbar of a long high-voltage line where transmission at power factors less than unity cannot be tolerated. Being a rotating machine, its stored energy is useful for increasing the inertia of the power system and for riding through transient disturbances, including voltage sags.

5.4.4 Static VAr Compensators (SVCs) and STATCOMs

Synchronous compensators are rotating machines and so are expensive and have mechanical losses. Hence they are being superseded increasingly by power electronic compensators: SVCs and STATCOMs.

SVCs use shunt connected reactors and capacitors controlled by thyristors. The reactive power is provided by the shunt elements (capacitors and inductors), as discussed in Section 5.4.1 but these are controlled by thyristors. The output of the

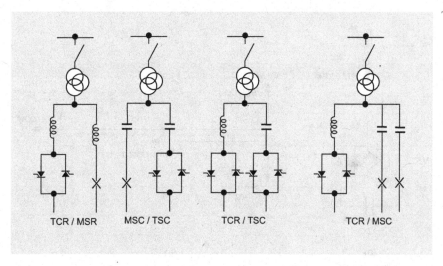

TCR / MSR MSC / TSC TCR / TSC TCR / MSC

Figure 5.9 Possible combinations of controlled reactors and capacitors forming an SVC. TCR: Thyristor Controlled Reactor, MSR: Mechanically Switched Reactor, MSC: Mechanically Switched Capacitor, TSC: Thyristor Switched Capacitor

Thyristor Controlled Reactor (TCR) is controlled by delaying the switching on of the thyristor within the 50/60 Hz cycle. The thyristor switches off when the current drops to zero. The firing angle of the thyristor can be varied within each cycle and hence the VAr absorption by the TCR controlled. As shown in Figure 5.9, TCRs may be used with Mechanically or Thyristor Switched Capacitors to create an SVC to export and import VArs.

When a capacitor is connected to a strong voltage source, very large currents can flow. Hence Thyristor Switched Capacitors are only operated in integral cycles and the operation of the thyristors is timed so that they switch when there is no instantaneous voltage across the capacitor.

A STATCOM (Static Compensator) is also a power electronic device to provide reactive power but it operates on a different principle (Figure 5.10). A STATCOM consists of a Voltage Source Converter (VSC) connected to the power system through a coupling reactance (L). The VSC uses very large transistors that can be turned on and off to synthesize a voltage sine wave of any magnitude and phase. $V_{STATCOM}$ is a 50/60 Hz sine wave kept in phase with $V_{terminal}$ (Figure 5.10). If the magnitude of $V_{STATCOM}$ is greater than that of $V_{terminal}$ then reactive power is generated by the STATCOM while if the magnitude of $V_{STATCOM}$ is less than that of $V_{terminal}$ then reactive power is absorbed by the STATCOM. A very small phase angle is introduced between $V_{STATCOM}$ and $V_{terminal}$ so that a small amount of real power flows into the STATCOM to charge the DC capacitor and provide for the losses of the converter. However, the principle of operation is that the reactive power is provided by the interaction of the two voltage magnitudes across the reactor. The DC capacitor is only used to operate the power electronics and control

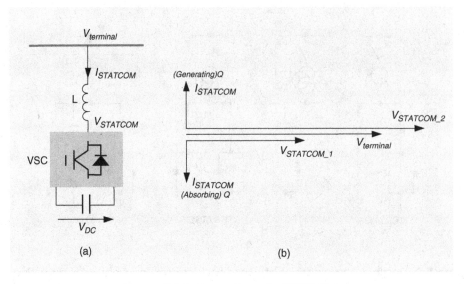

Figure 5.10 Operation of a STATCOM

the ripple current. STATCOMs can be controlled very fast and have a smaller physical equipment footprint than SVCs.

5.5 Methods of Voltage Control: (b) Tap-Changing Transformers

The basic operation of the tap-changing transformer has been discussed in Chapter 3. By changing the transformation ratio, the voltage in the secondary circuit is varied. Hence voltage and reactive power control is obtained.

   In distribution circuits, tap-changing transformers are the primary method of voltage control. In a distribution transformer, the tap-changer compensates for the voltage drop across the reactance of the transformer but also for the variations in the voltage applied to the primary winding caused by changes of load within the high voltage network. In transmission circuits reactive power is dispatched by altering the taps of transformers and this, in turn, controls the network voltages.

5.5.1 Use of Tap-Changing Transformers to Control Voltage in a Distribution System

Consider the 40 MVA 132/11 kV transformer with a reactance of 13% on its rating shown in Figure 5.11. It is equipped with an on-load tap-changer that is used to maintain constant voltage at an 11 kV busbar by compensating for variations in the voltage of the 132 kV network and for the voltage drop across the transformer. The variation of network voltage at the 132 kV transformer busbar for heavy and light loading conditions, and the loads of the transformer are given in Table 5.1. Active power losses in the transformer are ignored and it is assumed that the value of the reactance of the transformer is not influenced by the change in the turns ratio.

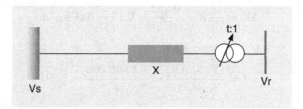

Figure 5.11 Tap changing transformer in a distribution circuit

t is the fraction of the nominal transformation ratios, that is the tap ratio/nominal ratio. For example, a transformer of nominal ratio 132 to 11 kV when tapped to give 144 to 11 kV has a t of $144/132 = 1.09$.

Choosing S_{BASE} of 40 MVA and V_{BASES} of 132 kV and 11 kV.

Under heavy loading conditions,

$V_r = 11$ kV, 1 p.u.

$V_s = 120$ kV, 0.909 p.u.

$Q = 13.94$ MVAr, 0.3485 p.u.

$$\Delta V = \frac{Q}{V_S}X = \frac{0.3485}{0.909} \times 0.13 = 0.05 \text{ p.u.}$$

$$V_r t = (V_s - \Delta V)$$

$$t = \frac{(V_s - \Delta V)}{V_r} = \frac{0.909 - 0.05}{1} = 0.86$$

Under light loading conditions,

$V_r = 11$ kV, 1 p.u.

$V_s = 145$ kV, 1.1 p.u.

$Q = 2.11$ MVAr, 0.053 p.u.

Table 5.1 Loading of transformer

	Load	Power Factor of Load	Voltage at 132 kV Busbar V_s	Desired Voltage at 11 kV Busbar V_r
Heavy loading conditions	32 MVA	0.90	120 kV	11 kV
Light loading conditions	4 MVA	0.85	145 kV	11 kV

$$\Delta V = \frac{Q}{V_S} X = \frac{0.053}{1.1} \times 0.13 = 0.006 \text{ p.u.}$$

$$V_r t = (V_s - \Delta V)$$

$$t = \frac{(V_s - \Delta V)}{V_r} = \frac{1.1 - 0.006}{1} = 1.094$$

A radial distribution circuit with two tap-changing transformers, is shown in the equivalent single-phase circuit of Figure 5.12. V_1 and V_2 are the nominal voltages; at

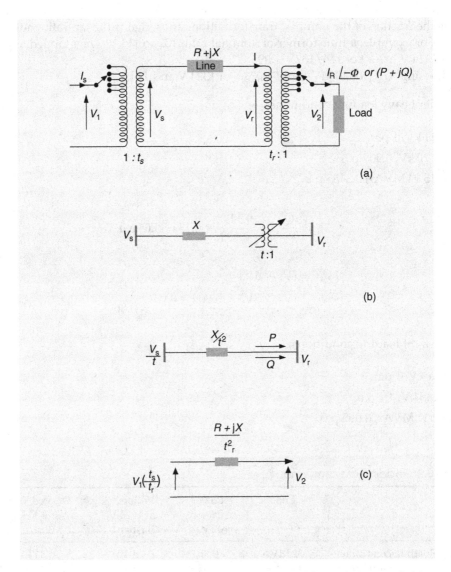

Figure 5.12 (a) Coordination of two tap-changing transformers in a radial transmission link (b) and (c) Equivalent circuits for dealing with off-nominal tap ratio, (b) Single transformer, (c) Two transformers

the ends of the circuit the actual voltages are t_sV_1 and t_rV_2. It is required to determine the tap-changing ratios needed to compensate completely for the voltage drop in the line. The product t_st_r will be made unity; this ensures that the overall voltage level remains in the same order and that the minimum range of taps on both transformers is used.

(Note that all values are in per unit; t is the off-nominal tap ratio.)

Transfer all quantities to the load circuit.

The line impedance becomes $(R + jX)/t_r^2$; $V_s = V_1t_s$ and, as the impedance has been transferred $V_r = V_1t_s$. The input voltage to the load circuit becomes V_1t_s/t_r and the equivalent circuit is as shown in Figure 5.10(c). The arithmetic voltage drop

$$= V_1\frac{t_s}{t_r} - V_2 \approx \frac{RP + XQ}{t_r^2V_2}$$

When $t_r = 1/t_s$

$$t_s^2V_1V_2 - V_2^2 = (RP + XQ)t_s^2$$

And

$$V_2 = 1/2\left[t_s^2V_1 \pm t_s\left(t_s^2V_1^2 - 4(RP + XQ)\right)^{\frac{1}{2}}\right] \tag{5.7}$$

If t_s is specified then t_r is defined. There are then two values of V_2 for a given V_1, *one low current, high voltage* and *one high current and low voltage*. Only the high voltage, low current solution is useful in a power system.

Example 5.3

A 132 kV line is fed through an 11/132 kV transformer from a constant 11 kV supply. At the load end of the line the voltage is reduced by another transformer of nominal ratio 132/11 kV. The total impedance of the line and transformers at 132 kV is $(25 + j66)\,\Omega$. Both transformers are equipped with tap-changing facilities which are arranged so that the product of the two off-nominal settings is unity. If the load on the system is 100 MW at 0.9 p.f. lagging, calculate the settings of the tap-changers required to maintain the voltage of the load busbar at 11 kV. Use a base of 100 MVA.

Solution

The line diagram is shown in Figure 5.13. As the line voltage drop is to be completely compensated, $V_1 = V_2 = 132\,\text{kV} = 1$ p.u. Also, $t_s \times t_r = 1$. The load is 100 MW, 48.3 MVAr., that is, $1 + j0.483$ p.u.

Using equation (5.7)

$$1 = 1/2\left[t_s^2 1 \pm t_s\left(t_s^2 1 - 4(0.14 \times 1 + 0.38 \times 0.48)\right)^{\frac{1}{2}}\right]$$

$$\therefore 2 = t_s^2 \pm t_s\left(t_s^2 - 1.28\right)^{\frac{1}{2}}$$

$$\therefore \left(2 - t_s^2\right)^2 = t_s^2\left(t_s^2 - 1.28\right)$$

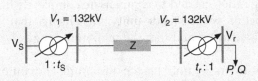

Figure 5.13 Schematic diagram of system for Example 5.3

Hence,

$$t_s = 1.21 \quad \text{and} \quad t_r = 1/1.21 = 0.83$$

These settings are large for the normal range of tap-changing transformers (usually not more than ±20% tap range). It would be necessary, in this system, to inject VArs at the load end of the line to maintain the voltage at the required value.

A transformer at the receiving end of a line does not improve the VAr flow in the circuit and the current in the supplying line is increased if the ratio is reduced. In countries with long and inadequate distribution circuits, it is often the practice to boost the received voltage by a variable ratio transformer so as to maintain rated voltage as the power required increases. Unfortunately, this has the effect of increasing the primary supply circuit current by the transformer ratio, thereby decreasing the primary voltage still further until voltage collapse occurs.

5.5.2 Use of Tap-Changing Transformers to Despatch VArs in a Transmission System

In transmission networks VArs may be dispatched by the adjustment of tap settings on transformers connecting busbars. Consider the situation in Figure 5.14(a), in which V_s and V_r are constant voltages representing the two connected systems. The circuit may be rearranged as shown in Figure 5.14(b), where t is the off-nominal (per unit) tap setting; resistance is zero. The voltage drop between busbars

$$\Delta V = \left(\frac{V_s}{t}\right) - V_r = \frac{X}{t^2} \cdot \frac{Q_T}{V_r}$$

Hence,

$$(V_s V_r t - V_r^2 t^2)1/X = Q_T$$

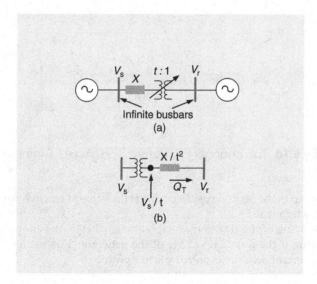

Figure 5.14 (a) Two power systems connected via a tap-change transformer, (b) Equivalent circuit with impedance transferred to receiver side

and

$$t(1-t)V^2/X = Q_T \quad \text{when} \quad V_s = V_r = V \tag{5.8}$$

When

$t < 1$, Q_T is positive, that is a flow of VArs into V_r
$t > 1$, Q_T is negative, a flow of VArs out of V_r

Also, $Q_T = t(1-t)S$, where S = short-circuit level, that is V^2/X.

Thus, by suitable adjustment of the tap setting, an appropriate injection of reactive power is obtained.

The idea can be extended to two transformers in parallel between networks. If one transformer is set to an off-nominal ratio of, say, 1.1 : 1 and the other to 0.8 : 1 (i.e. in opposite directions), then a circulation of reactive power occurs round the loop, resulting in a net absorption of VArs. This is known as 'tap stagger' and is a comparatively inexpensive method of VAr absorption.

Example 5.4

A synchronous generator (75 MVA, 0.8 p.f., 11.8 kV and $X_S = 1.1$ p.u.) is connected through an 11/275 kV tap changing transformer (75 MVA, $X_T = 0.15$ p.u., tap range = +/−20%) to a very large 275 kV power system, as shown in Figure 5.15.

a. What is the value of the internal emf and power angle of the generator when it exports 60 MW of active and zero MVAr of reactive power to the system? With the

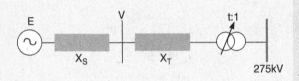

Figure 5.15 Tap changer in a transmission circuit – Example 5.4

transformer tap in the neutral position, what is the value of reactive power output at the generator terminals?

b. What is the value of the transformer tap at which 20 MVAr is imported from the 275 kV system, if the terminal voltage of the generator V is maintained at 1 p.u. when the generator does not export any active power.

Solution
Choosing a common base

$$S_{BASE} = 75\,\text{MVA}$$

$$V_{BASES} = 11\,\text{kV}, 275\,\text{kV}$$

Converting to the common base (Equation (2.7))

$$X_S = 1.1 \times \left(\frac{11.8}{11}\right)^2 = 1.26\,\text{p.u.}$$

$$X_T = 0.15\,p.u.$$

$$X = X_S + X_T = 1.41\,\text{p.u.}$$

At 60 MW and 0 MVAr exported into the 275 kV system,

$$I = 0.8 + j0\,\text{p.u.}$$

and the generator internal voltage and power angle are

$$E = 1 + IX = 1 + 0.8 \times j\,1.41 = 1 + j1.13$$

$$E = 1.5\angle 48°$$

With the transformer tap in the neutral position the reactive power output at the generator terminals is

$$V = 1 + IX = 1 + 0.8 \times j0.15 = 1 + j0.12$$

$$\mathbf{S} = \mathbf{VI}^* = (1 + j0.12)0.8 = 0.8 + j0.096\,\text{p.u.}$$

$$Q = 0.096 \times 75 = 7.2\,\text{MVAr}$$

If 20 MVAr is absorbed from the 275 kV system and V and the voltage of the 275 kV system are at 1 p.u. then using Equation (5.8)

$$t(1-t)V^2/X_T = Q_T$$

$$t(1-t)1/0.15 = \frac{-20}{75}$$

$$t(1-t) = -0.04$$

$$t^2 - t - 0.04 = 0$$

$$t = 1.04$$

The solution for t near 1 is chosen
Thus the transformer taps are set to +4%

5.6 Combined Use of Tap-Changing Transformers and Reactive-Power Injection

A common practical arrangement is shown in Figure 5.16, where the tertiary winding of a three-winding transformer is connected to a VAr compensator. For given load conditions it is proposed to determine the necessary transformation ratios with certain outputs of the compensator.

The transformer is represented by the equivalent star connection and any line impedance from V_1 or V_2 to the transformer can be lumped together with the transformer branch impedances. Here, V_N is the phase voltage at the star point of the equivalent circuit. The secondary impedance (X_S) is usually approaching zero and hence is neglected. Resistance and losses are ignored.

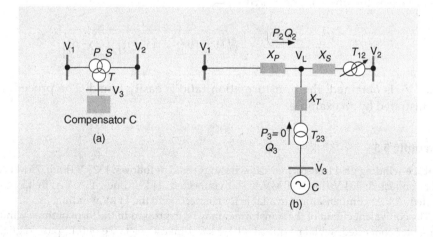

Figure 5.16 (a) Schematic diagram with combined tap-changing and synchronous compensation, (b) Equivalent network

The allowable ranges of voltage for V_1 and V_2 are specified and the values of the three-phase real and reactive power; P_2, Q_2, P_3, and Q_3 are given. P_3 is usually taken as zero.

The volt drop V_1 to V_L is given by

$$\Delta V_p \approx X_p \frac{Q_2/3}{V_N}$$

or

$$\Delta V_p \approx X_p \frac{Q_2}{V_L\sqrt{3}}$$

where V_L is the line voltage $= \sqrt{3}V_N$ and Q_2 is the total VArs.

Also,

$$\Delta V_q \approx X_p \frac{P_2}{V_L\sqrt{3}}$$

$$\therefore (V_N + \Delta V_p)^2 + (\Delta V_q)^2 = V_1^2$$

(see phasor diagram of Figure 2.24; phase values used)
and

$$\left(\frac{V_L}{\sqrt{3}} + X_p\frac{Q_2}{V_L\sqrt{3}}\right)^2 + X_p^2\left(\frac{P_2}{\sqrt{3}V_L}\right)^2 = V_1^2$$

$$\therefore (V_L^2 + X_p Q_2)^2 + X_p^2 P_2^2 = V_{1L}^2 V_L^2$$

where V_{1L} is the line voltage $= \sqrt{3}V_1$

$$\therefore V_L^2 = \frac{V_{1L}^2 - 2X_p Q_2}{2} \pm \frac{1}{2}\sqrt{\left[V_{1L}^2(V_{1L}^2 - 4X_p Q_2) - 4X_p^2 P_2^2\right]}$$

Once V_L is obtained, the transformation ratio is easily found. The procedure is best illustrated by an example.

Example 5.5

A three-winding grid transformer has windings rated as follows: 132 kV (line), 75 MVA, star connected; 33 kV (line), 60 MVA, star connected; 11 kV (line), 45 MVA, delta connected. A VAr compensator is available for connection to the 11 kV winding.

The equivalent circuit of the transformer may be expressed in the form of three windings, star connected, with an equivalent 132 kV primary reactance of 0.12 p.u., negligible secondary reactance, and an 11 kV tertiary reactance of 0.08 p.u. (both values expressed on a 75 MVA base).

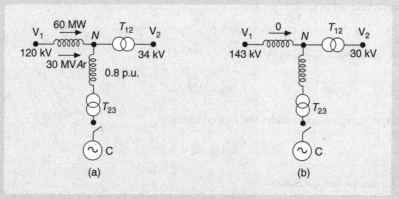

Figure 5.17 Systems for Example 5.4. (a) System with loading condition 1. (b) System with loading condition 2

In operation, the transformer must deal with the following extremes of loading:

1. Load of 60 MW, 30 MVAr with primary and secondary voltages governed by the limits 120 kV and 34 kV; compensator disconnected.
2. No load, primary and secondary voltage limits 143 kV and 30 kV; compensator in operation and absorbing 20 MVAr.

Calculate the range of tap-changing required. Ignore all losses.

Solution
The value of X_P, the primary reactance (in ohms)

$$= 0.12 \times \frac{132^2}{75} = 27.8\,\Omega$$

Similarly, the effective reactance of the tertiary winding is $18.5\,\Omega$. The equivalent star circuit is shown in Figure 5.17.
The first operating condition is as follows:

$$P_1 = 60\,\text{MW} \quad Q_1 = 30\,\text{MVAr} \quad V_{1L} = 120\,\text{kV}$$

Hence,

$$V_L^2 = \frac{1}{2}(120\,000^2 - 2 \times 27.8 \times 30 \times 10^6)$$

$$\pm \frac{1}{2}\sqrt{[120\,000^2\,(120\,000^2 - 4 \times 27.8 \times 30 \times 10^6) - 4 \times 27.8^2 \times 60^2 \times 10^{12}]}$$

$$= \left(63.6 \pm \frac{122}{2}\right)10^8 - 124.4 \times 10^8$$

$$\therefore V_L = 111\,\text{kV}$$

The second set of conditions are:

$$V_{1L} = 143\,\text{kV} \quad P_2 = 0 \quad Q_2 = 20\,\text{MVAr}$$

Again, using the formula for V_L,

$$V_L = 138.5\,\text{kV}$$

The transformation ratio under the first condition,

$$= 111/34 = 3.27$$

and, for the second condition

$$= 138.5/30 = 4.64$$

The actual ratio will be taken as the mean of these extremes, that is, 3.94, varying by ± 0.67 or $3.94 \pm 17\%$. Hence the range of tap-changing required is $\pm 17\%$.

Example 5.6

In the system shown by the line diagram in Figure 5.18, each of transformers T_A and T_B have tap ranges of $\pm 10\%$ in 10 steps of 1.0%. It is necessary to find the voltage boost needed on transformer T_A to share the power flow equally between the two lines.

The system data is as follows (on a common base):

All transformers: $X_T = 0.1$ p.u.
Transmission lines: $R = R' = 0$
$X = 0.20$ p.u.
$X' = 0.15$ p.u
$V_A = 1.1\angle 5°$
$V_B = 1.0\angle 0°$

Solution
We must first calculate the current sharing in the two parallel lines:

$$I_1 = \frac{1.1\angle 5 - 1.0\angle 0°}{j0.4} = 0.2397 - j0.2397$$

$$I_2 = \frac{1.1\angle 5 - 1.0\angle 0°}{j0.35} = 0.2740 - j0.2740$$

Any boost by transformer T_A will cause a current to circulate between the two busbars because the voltages V_A and V_B are assumed to be held constant by the voltage regulators on the generators.

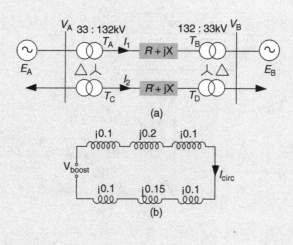

Figure 5.18 (a) Line diagram of system for Example 5.6. (b) Equivalent network with voltage boost V_{boost} acting

To equalize the currents, a circulating current is required, as in Figure 5.18(b), giving

$$I_{circ} = \frac{I_2 - I_1}{2}$$

$$I_{circ} = \frac{0.0343 - j0.0343}{2} = 0.0241\angle -45°$$

$$\therefore V_{Boost} = 0.0241\angle -45° \times j0.75 = 0.0180\angle 45°\,\text{V}$$

To achieve this boost, ideally T_A should be equipped with a phase changer of 45° and taps to give 1.8% boost. In practice, a tap of 2% would be used in either an in-phase boost (such as obtainable from a normal tapped transformer) or a quadrature boost (obtainable from a phase-shift transformer. In transmission networks it should be noted that because of the generally high X/R ratio, an in-phase boost gives rise to a quadrative current whereas a quadrature boost produces an in-phase circulating current, thereby adding to or subtracting from the real power flow.

Circulating reactive current by adjusting the taps in transformers in parallel circuits has been used to de-ice lines in winter by producing extra I^2R losses for heating. Two transformers in parallel can be tap-staggered to produce I^2X absorption under light-load, high-voltage conditions.

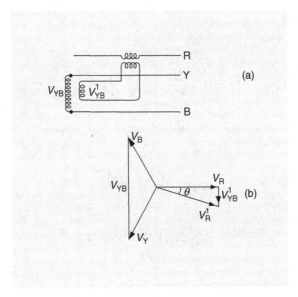

Figure 5.19 (a) Connections for one phase of a phase shift transformer. Similar connections to other two phases. (b) Corresponding phasor diagram

5.7 Phase-Shift Transformer

A quadrature phase shift can be achieved by the connections shown in Figure 5.19(a). The booster arrangement shows the injection of voltage into one phase only; it is repeated for the other two phases. In Figure 5.19(b), the corresponding phasor diagram is shown and the nature of the angular shift of the voltage boost V_{YB} indicated. By the use of tappings on the energizing transformer, several values of phase shift may be obtained.

Example 5.7

In the system shown in Figure 5.20, it is required to keep the 11 kV busbar at constant voltage. The range of taps is not sufficient and it is proposed to use shunt capacitors connected to the tertiary winding.

All impedances are referred to 33 kV. The impedance of the overhead line, Z_L referred to 33 kV $= (2.2 + j5.22)$ Ω. For the three-winding transformer the measured impedances between the windings and the resulting equivalent star impedances Z_1, Z_2 and Z_3 are given in Table 5.2.

Solution
The equivalent circuit referred to 33 kV is shown in Figure 5.20(b).
    The voltage at point C (referred to 33 kV) is

$$\frac{33\,000}{\sqrt{3}} - \Delta V_p$$

Table 5.2 Data for three-winding transformer

Winding MVA	MVA	Voltage kV	p.u. Z referred to nameplate VA	p.u. Z referred to 15 MVA	$Z(\Omega)$ referred to 33 kV side (Z_{BASE} = 72.6 Ω)	Equivalent $Z(\Omega)$ referred to 33 kV side. (Equation (3.14))
P-S	15	33/11	$0.008 + j0.1$	$0.008 + j0.1$	$0.58 + j7.26$	$Z_1 = 0.212 + j8.21$ $\frac{1}{2}(Z_{PS} + Z_{PT} - Z_{ST})$
P-T	5	33/1.5	$0.0035 + j0.0595$	$0.0105 + j.1785$	$0.76 + j12.96$	$Z_2 = 0.368 - j0.945$ $\frac{1}{2}(Z_{PS} + Z_{ST} - Z_{PT})$
S-T	5	11/1.5	$0.0042 + j0.0175$	$0.0126 + j0.0525$	$0.915 + j3.81$	$Z_3 = 0.547 + j4.76$ $\frac{1}{2}(Z_{PT} + Z_{ST} - Z_{PS})$

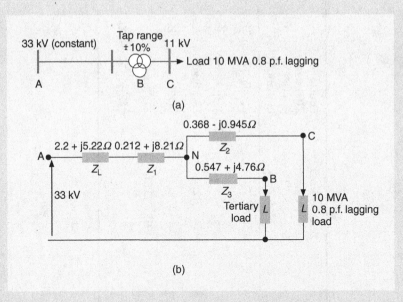

Figure 5.20 (a) Line diagram for Example 5.7. (b) Equivalent network – referred to 33 kV

where

$$\Delta V_p \approx \frac{RP + XQ}{V}$$

$$\therefore \Delta V_p \approx \frac{2.78 \times 8/3 \times 10^6 + 12.485 \times 6/3 \times 10^6}{33\,000/\sqrt{3}}$$

$$\Delta V_p = \frac{7.4 + 24.95}{19} = 1.703\,\text{kV} \quad \text{and} \quad V_C = 17.3\,\text{kV}$$

V_c referred to $11\,\text{kV} = 17.3/3 = 5.77\,\text{kV}$ (phase) or $10\,\text{kV}$ (line). In order to maintain $11\,\text{kV}$ at C, the voltage is raised by tapping down on the transformer. Using the full range of 10%, that is $t = 0.9$, the voltage at C is

$$\frac{17.3 \times \sqrt{3}}{((33 \times 0.9)/11)} = 11\,\text{kV}$$

The true voltage will be less than this as the primary current will have increased by (1/0.9) because of the change in transformer ratio. The tap-changing transformer is not able to maintain $11\,\text{kV}$ at C and the use of a static capacitor connected to the tertiary will be investigated.

Consider a shunt capacitor of capacity $5\,\text{MVAr}$ (the capacity of the tertiary).

Assume the transformer to be at its nominal ratio $33/11\,\text{kV}$. The voltage drop to point N

$$= \frac{2.412 \times 8/3 \times 10^6 + 13.42 \times 1/3 \times 10^6}{V_N \approx 19\,kV}$$

$$= 0.574\,kV$$

$$V_N = 19 - 0.574 = 18.43\,kV$$

Therefore the volt drop N to C

$$\Delta V_C = \frac{0.368 \times 8/3 \times 10^6 - 0.945 \times 6/3 \times 10^6}{18.43 \times 10^3} = -0.049\,kV$$

$$\therefore V_C = 18.43 - 0.049 = 18.381\,kV$$

Referred to 11 kV, $V_C = 10.65\,kV$ (line). Hence, to have 11 kV the transformer will tap such that $t = (1 - 0.35/11) = 0.97$, that is, a 3% tap change, which is well within the range and leaves room for load increases.

On no-load

$$\Delta V_p = \frac{2.959 \times 0 + 18.19 \times (-5/3)}{19}\,kV$$

$$= \frac{-30.3}{19} = -1.594\,kV$$

$$V_C = 19 + 1.6 = 20.6\,kV \text{ (phase)}$$

On the 11 kV side

$$V_C = 11.9\,kV \text{ (line)}$$

therefore the tap change will have to be at least 8.1%, which is well within the range.

5.8 Voltage Collapse

Voltage collapse is an important aspect of system stability.

Consider the circuit shown in Figure 5.21(a). If V_S is fixed (i.e. an infinite busbar), the graph of V_R against P for given power factors is as shown in Figure 5.21(b). In Figure 5.21(b), Z represents the series impedance of a 160 km long, double-circuit, 400 kV, 260 mm^2 conductor overhead line. The fact that two values of voltage exist for each value of power is easily understood by considering the analytical solution of this circuit. At the lower voltage a very high current is taken to produce the power.

The seasonal thermal ratings of the line are also shown, and it is apparent that for loads of power factor less than unity (lagging) the possibility exists that, before the thermal rating is reached, the operating power may be on that part of the character-istic where small changes in load cause large voltage changes and voltage instability will have occurred. In this condition the action of tap-changing transformers is

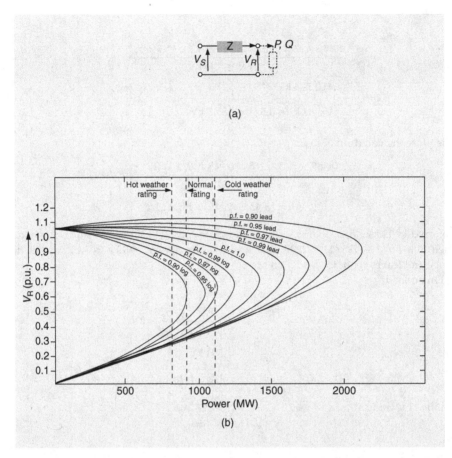

(a)

(b)

Figure 5.21 (a) Equivalent circuit of a line supplying a load $P + jQ$. (b) Relation between load voltages and received power at constant power factor for a 400 kV, $2 \times 260 \, mm^2$ conductor line, 160 km in length. Thermal ratings of the line are indicated

interesting. If the receiving-end transformers 'tap up' to maintain the load voltage, the line current increases, thereby causing further increase in the voltage drop. It would, in fact, be more profitable to 'tap down', thereby reducing the current and voltage drop. It is feasible therefore for a 'tapping-down' operation to result in increased secondary voltage, and vice versa.

The possibility of an actual voltage collapse depends upon the nature of the load. If this is stiff (constant power), for example induction motors, the collapse is aggravated. If the load is soft, for example heating, the power falls off rapidly with voltage and the situation is alleviated. Referring to Figure 5.21 it is evident that a critical quantity is the power factor; at full load a change in lagging power factor from 0.99 to 0.90 will precipitate voltage collapse. On long lines, therefore, for reasonable power transfers it is necessary to keep the power factor of transmission approaching unity, certainly above

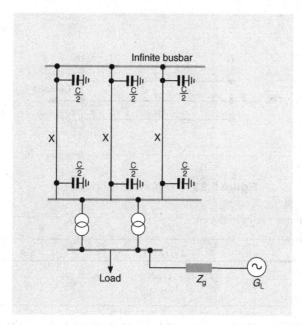

Figure 5.22 Line diagram of three long lines in parallel – effect of the loss of one line. G_L = local generators

0.97 lagging, and it is economically justifiable to employ VAr injection by static capacitors, synchronous compensators or Static VAr Compensators (SVCs) close by the load.

A problem arises with the operation of two or more lines in parallel, for example the system shown in Figure 5.22, in which the shunt capacitance has been represented as in a π section. If one of the three lines is removed from the circuit because of a fault, the system series reactance will increase from $X_L/3$ to $X_L/2$, and the capacitance, which normally improves the power factor, decreases to $2C$ from $3C$. Thus the overall voltage drop is greatly increased and, owing to the increased $I^2 X_L$ loss of the lines and the decreased generation of VArs by the shunt capacitances, the power factor decreases; hence the possibility of voltage instability. The same argument will, of course, apply to two lines in parallel.

Example 5.8

Figure 5.23 shows three parallel 400 kV transmission circuits each 250 km long. The parameters of the circuits are

$$
\begin{array}{ll}
\text{Resistance :} & 0.02\,\text{ohm/km} \\
\text{Inductance :} & 1.06\,\text{mH/km} \\
\text{Capacitance :} & 0.011\,\mu\text{F/km}
\end{array}
$$

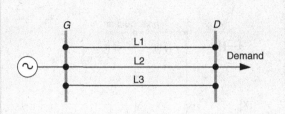

Figure 5.23 Diagram of Example 5.8

The load demand varies between

Demand	P_D (MW)	Q_D (MVAr)
Peak	1100	532.8
Off-peak	220	106.6

A power flow calculation for the two loading conditions shows the voltage at the demand busbar to be 0.894 during peak demand and 1.015 for off-peak demand conditions.

V_G (p.u.)	P_G (MW)	Q_G (MVAr)	V_D (p.u.)	P_D (MW)	Q_D (MVAr)
1	1117.5	451.7	0.894	1100	532.8
1	220.6	−304.4	1.015	220	106.6

To maintain the demand busbar voltage at 1 p.u. reactive compensation is needed

499.5 MVar (capacitive) during peak demand condition
83.3 MVAr (inductive) under off-peak condition

V_G (p.u.)	P_G (MW)	Q_G (MVAr)	V_D (p.u.)	P_D (MW)	Q_D (MVAr)	Compensation (MVAr)
1	1112.9	−166.2	1	1100	532.8	499.5
1	220.5	−216.3	1	220	106.6	−83.3

If one line is lost during peak demand (N-1 security requires that no load is shed if one circuit trips) the voltage at the receiving end would reduce to 0.745 p.u. and reactive compensation of 625.5 MVAr (capacitive) would be required to bring the voltage to 1 p.u.

V_G (p.u.)	P_G (MW)	Q_G (MVAr)	V_D (p.u.)	P_D (MW)	Q_D (MVAr)	Compensation (MVAr)
1	1139.9	983	0.745	1100	532.8	0
1	1119.7	40.6	1	1100	532.8	625.5

Usually, there will be local generation or compensation feeding the receiving-end busbars at the end of long lines. If this generation is electrically close to the load busbars, that is low connecting impedance Z_g, a fall in voltage will automatically increase the local VAr generation, and this may be sufficient to keep the reactive power transmitted low enough to avoid large voltage drops in the long lines. Often, however, the local generators supply lower voltage networks and are electrically remote from the high-voltage busbar of Figure 5.20, and Z_g is high. The fall in voltage now causes little change in the local generator VAr output and the use of static-controlled capacitors at the load may be required. As Z_g is inversely proportional to the three-phase short-circuit level at the load busbar because of the local generation, the reactive-power contribution of the local machines is proportional to this fault level. When a static or synchronous compensator reaches its rated limit, voltage can no longer be controlled and rapid collapse of voltage can follow because any VArs demanded by the load must now be supplied from sources further away electrically over the high-voltage system.

In the UK and some other countries, many generators are some distance from the load-centres. Consequently, the transmission system operator is required to install local flexible VAr controllers or compensators to maintain a satisfactory voltage at the delivery substations supplying the local distribution systems. Such flexible controllers, based on semiconductor devices which can vary the VAr absorption in a reactor or generation in a capacitor, are called FACTS (Flexible a.c. Transmission System).

Typical values of compensation required for a 400 kV or 500 kV network are:

$$\text{Peak load} = 0.3\,\text{kVAr/kW generating VArs}$$

$$\text{Light load} = 0.25\,\text{kVAr/kW absorbing VArs}$$

5.9 Voltage Control in Distribution Networks

Single-phase supplies to houses and other small consumers are tapped off from three-phase feeders connected between one phase and the neutral. Although efforts are made to allocate equal loads to each phase the loads are not applied at the same time and some unbalance occurs. In the distribution network (British practice) shown in Figure 5.24 an 11 kV distributor supplies a number of lateral feeders in which the voltage is approximately 400 V and then each phase, loaded separately.

The object of design is to keep the consumers' nominal 230 V supply within $-6/+10\%$ of the declared voltage. The main 33/11.5 kV transformer is controlled with an on-load tap changer to maintain the 11 kV busbar at a voltage approximately 5% above 11 kV. The distribution transformers have a secondary phase voltage of 433/250 V which is some 8.5% higher than the nominal value of 230 V. These transformers have taps of $+/-5\%$ that are only adjustable when the transformer is off-circuit (isolated).

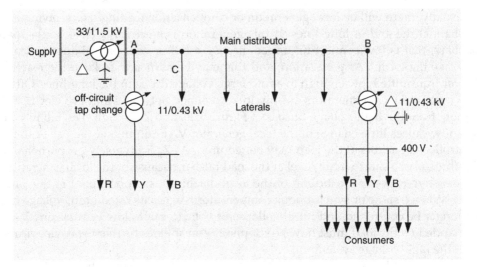

Figure 5.24 Line diagrams of typical radial distribution schemes

A typical distribution of voltage drops would be as follows: main 11 kV feeder distributor, 6%; 11/0.433 kV transformer, 3%; 400 V circuit, 7%; consumer circuit, 1.5%; giving a total of drop at full load of 17.5%. On very light load (10% of full load) the corresponding drop may be 1.5%.

To offset these drops, various voltage boosts are employed as follows: main transformer, +5%; distribution transformer, inherent boost of +4% (i.e. 433/250 V secondary) plus a 2.5% boost through off-circuit taps. These add to give a total boost of 11.5%. Hence the consumers' voltage varies between (+ 11.5 − 17.5), that is −6% and (+11.5 − 1.5) that is +10%, which is permissible. There will be a difference in consumer voltage depending upon the position of the lateral feeder on the main distributor; obviously, a consumer supplied from C will have a higher voltage than one supplied from B. In some circuits the voltage control of the 33/11.5 kV transformer is compounded with a measurement of current through the transformer. This is known as Line Drop Compensation and allows the voltage to be controlled at a remote point of the 11 kV feeder.

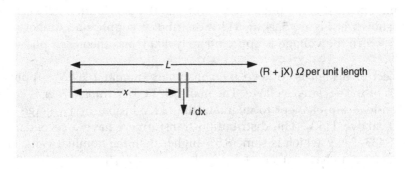

Figure 5.25 Uniformly loaded distribution line

5.9.1 Uniformly Loaded Feeder from One End

In areas with high load densities a large number of tappings are made from feeders and a uniform load along the length of a feeder may be considered to exist. Consider the voltage drop over a length dx of the feeder distant x metres from the supply end. Let i A be the current tapped per metre and R and X be the resistance and reactance per phase per metre, respectively. The length of the feeder is L (m) (see Figure 5.25).

The voltage across $dx = Rixdx \cos \phi + Xixdx \sin \phi$, where $\cos \phi$ is the power factor (assumed constant) of the uniformly distributed load.

The total voltage drop

$$= R \int_0^L ixdx \cos \phi + X \int_0^L ixdx \sin \phi$$

$$= Ri\frac{L^2}{2} \cos \phi + Xi\frac{L^2}{2} \sin \phi$$

$$= \frac{LR}{2}I \cos \phi + \frac{LX}{2}I \sin \phi$$

where $I = Li$, the total current load. Hence the uniformly distributed load may be represented by the total load tapped at the centre of the feeder (half its length).

5.10 Long Lines

On light loads the capacitive charging VArs of a line exceeds the inductive VArs consumed and the voltage rises, causing problems for generators. Shunt reactors are switched in circuit at times of light load to absorb the generated VArs.

A 500 km line can operate within ±10% voltage variation without shunt reactors. However, with, say, an 800 km line, shunt reactors are essential and the effects of these are shown in Figure 5.26. For long lines in general, it is usual to divide the system into sections with compensation at the ends of each section. This controls the voltage profile, helps switching, and reduces short circuit currents. Shunt compensation can be varied by switching discrete amounts of inductance. A typical 500 kV, 1000 km scheme uses compensation totalling 1200 MVAr.

Improvement in voltage profile may be obtained by compensation, using FACTS devices, at intermediate points, as well as at the ends of the line, as shown in Figure 5.26. If the natural load is transmitted there is, of course, constant voltage along the line with no compensation. If the various busbars of a sectioned line can be maintained at constant voltage regardless of load, each section has a theoretical maximum transmission angle of 90°. Thus, for a three section line a total angle of much greater than 90° would be possible. This is illustrated in Figure 5.27 for a three-section, 1500 km line with a unity power factor load.

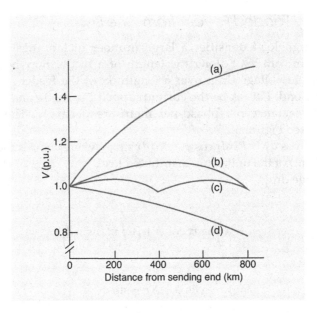

Figure 5.26 Voltage variation along a long line; (a) on no load with no compensation; (b) on no load with compensation at ends; (c) on no load with compensation at ends and at centre, (d) transmitting natural load, compensation at ends and centre

5.10.1 Sub-Synchronous Resonance

With very long lines the voltage drop from the series inductance can be very large. Series capacitors are then installed to improve the power transfer capacity and these effectively shorten the line electrically.

The combination of series capacitors and the natural inductance of the line (plus that of the connected systems) creates a resonant circuit of sub-synchronous resonant frequency. This resonance can interact with the generator-shaft critical torsional frequency, and a mechanical oscillation is superimposed on the rotating generator shaft that may have sufficient magnitude to cause mechanical failure.

Sub-synchronous resonance has been reported, caused by line-switching in a situation where trouble-free switching was normally carried out with all capacitors in service, but trouble occurred when one capacitor bank was out of service. Although this phenomena may be a rare occurrence, the damage resulting is such that, at the design stage, an analysis of possible resonance effects is required.

5.11 General System Considerations

Because of increasing voltages and line lengths, and also the wider use of underground circuits, the light-load reactive-power problem for an interconnected system becomes substantial, particularly with modern generators of limited VAr absorption

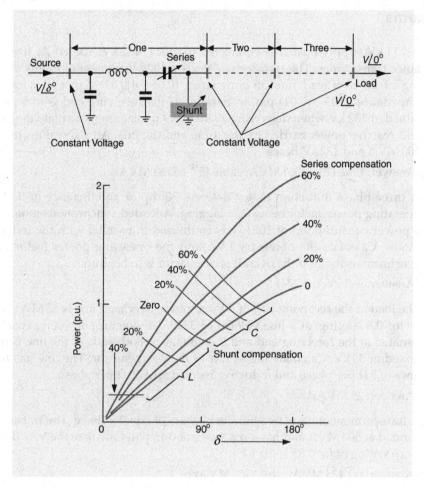

Figure 5.27 Power-angle curves for 1500 km line in three sections. Voltages at section-busbars maintained constant by variable compensation. Percentage of series and shunt compensation indicated (Figure adapted from IET)

capability. At peak load, transmission systems need to increase their VAr generation, and as the load reduces to a minimum (usually during the night) they need to reduce the generated VArs by the following methods, given in order of economic viability:

1. switch out shunt capacitors;
2. switch in shunt inductors;
3. run hydro plant on maximum VAr absorption;
4. switch out one cable in a double-circuit link;
5. tap-stagger transformers;
6. run base-load thermal generators at maximum VAr absorption.

Problems

5.1 An 11 kV supply busbar is connected to an 11/132 kV, 100 MVA, 10% reactance transformer. The transformer feeds a 132 kV transmission link consisting of an overhead line of impedance $(0.014 + j0.04)$ p.u. and a cable of impedance $(0.03 + j0.01)$ p.u. in parallel. If the receiving end is to be maintained at 132 kV when delivering 80 MW, 0.9 p.f. lagging, calculate the power and reactive power carried by the cable and the line. All p.u. values relate to 100 MVA and 132 kV bases.

(Answer: Line $(23 + j35)$ MVA; cable $(57 + j3.8)$ MVA)

5.2 A three-phase induction motor delivers 500 hp at an efficiency of 0.91, the operating power factor being 0.76 lagging. A loaded synchronous motor with a power consumption of 100 kW is connected in parallel with the induction motor. Calculate the necessary kVA and the operating power factor of the synchronous motor if the overall power factor is to be unity.

(Answer: 365 kVA, 0.274)

5.3 The load at the receiving end of a three-phase overhead line is 25 MW, power factor 0.8 lagging, at a line voltage of 33 kV. A synchronous compensator is situated at the receiving end and the voltage at both ends of the line is maintained at 33 kV. Calculate the MVAr of the compensator. The line has resistance of 5 Ω per phase and inductive reactance of 20 Ω per phase.

(Answer: 25 MVAr)

5.4 A transformer connects two infinite busbars of equal voltage. The transformer is rated at 500 MVA and has a reactance of 0.15 p.u. Calculate the VAr flow for a tap setting of (a) 0.85:1; (b) 1.1:1.

(Answer: (a) 425 MVAr; (b) −367 MVAr)

5.5 A three-phase transmission line has resistance and inductive reactance of 25 Ω and 90 Ω, respectively. With no load at the receiving end, a synchronous compensator there takes a current lagging by 90°; the voltage is 145 kV at the sending end and 132 kV at the receiving end. Calculate the value of the current taken by the compensator.
When the load at the receiving end is 50 MW, it is found that the line can operate with unchanged voltages at the sending and receiving ends, provided that the compensator takes the same current as before, but now leading by 90°. Calculate the reactive power of the load.

(Answer: 83.5 A; $Q_L = 24.2$ MVAr)

5.6 Repeat Problem 5.3 making use of $\partial Q / \partial V$ at the receiving end.

5.7 In Example 5.3, determine the tap ratios if the receiving-end voltage is to be maintained at 0.9 p.u. of the sending-end voltage.

(Answer: $t_s = 1.19$; $t_r = 0.84$)

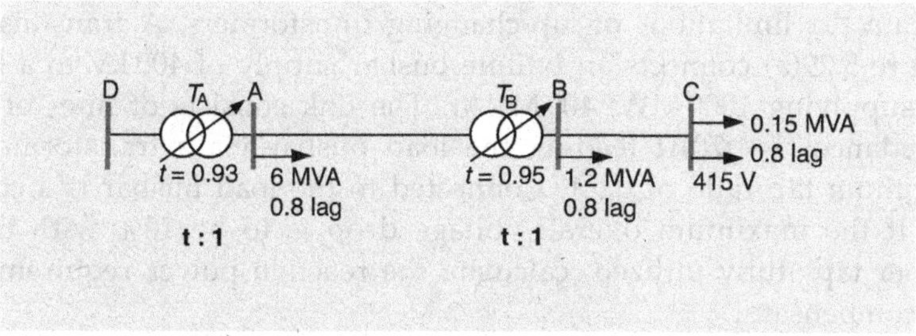

Figure 5.28 Line diagram for system in Problem 5.8

5.8 In the system shown in Figure 5.28, determine the supply voltage necessary at A to maintain a phase voltage of 240 V at the consumer's terminals at C. The data in Table 5.3 apply.

Table 5.3 Data for Problem 5.8

Line or Transformer	Rated Voltage kV	Rating MVA	Nominal Tap Ratio	Impedance (Ω)
BC	0.415			$0.0217 + j0.00909$
AB	11			$1.475 + j2.75$
DA	33			$1.475 + j2.75$
T_A	33/11	10	30.69/11	$1.09 + j9.8$ Referred to 33 kV
T_B	11/0.415	2.5	10.45/0.415	$0.24 + j1.95$ Referred to 11 kV

(Answer: 33 kV)

5.9 A load is supplied through a 275 kV link of total reactance 50 Ω from an infinite busbar at 275 kV. Plot the receiving-end voltage against power graph for a constant load power factor of 0.95 lagging. The system resistance may be neglected.

5.10 Describe two methods of controlling voltage in a power system.

a. Show how the scalar voltage difference between two nodes in a network is given approximately by:

$$\Delta V = \frac{RP + XQ}{V}$$

Each phase of a 50 km, 132 km overhead line has a series resistance of 0.156 Ω/km and an inductive reactance of 0.4125 Ω/km. At the receiving end the voltage is 132 kV with a load of 100 MVA at a power factor of 0.9 lagging. Calculate the magnitude of the sending-end voltage.

b. Calculate also the approximate angular difference between the sending-end and receiving-end voltages.

(Answer: (c) 144.1 kV; (d) 4.55°)

(*From Engineering Council Examination, 1997*)

5.11   Explain the limitations of tap-changing transformers. A transmission link
(Figure 5.29(a) connects an infinite busbar supply of 400 kV to a load busbar supplying 1000 MW, 400 MVAr. The link consists of lines of effective
impedance $(7+j70)\Omega$ feeding the load busbar via a transformer with a
maximum tap ratio of 0.9:1. Connected to the load busbar is a compensator. If the maximum overall voltage drop is to be 10% with the transformer taps fully utilized, calculate the reactive power requirement from
the compensator.

(Answer: 148 MVAr)

Note: Refer the voltage and line Z to the load side of transformer in Figure 5.29(b).

$$V_R = \frac{V_S}{t} - \left(\frac{\frac{RP}{t^2} + \frac{XQ}{t^2}}{V_R} \right)$$

5.12   A generating station consists of four 500 MW, 20 kV, 0.95 p.f. (generating
VArs) generators, each feeding through a 525 MVA, 0.1 p.u. reactance transformer onto a common busbar. It is necessary to transmit 2000 MW at 0.95
p.f. lagging to a substation maintained at 500 kV in a power system at a
distance of 500 km from the generating station. Design a suitable

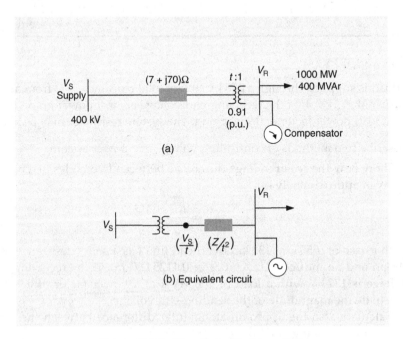

(a)

(b) Equivalent circuit

Figure 5.29   Circuits for Problem 5.11

transmission link of nominal voltage 500 kV to achieve this, allowing for a reasonable margin of stability and a maximum voltage drop of 10%. Each generator has synchronous and transient reactances of 2 p.u. and 0.3 p.u. respectively, and incorporates a fast-acting automatic voltage regulator. The 500 kV transmission lines have an inductive reactance per phase of 0.4 Ω/km and a shunt capacitive susceptance per phase of 3.3×10^{-6} S/km. Both series and shunt capacitors may be used if desired and the number of three-phase lines used should be not more than three – fewer if feasible. Use approximate methods of calculation, ignore resistance, and state clearly any assumption made. Assume shunt capacitance to be lumped at the receiving end only.

(Answer: Use two 500 kV lines with series capacitors compensating to 70% of series inductance)

5.13 It is necessary to transmit power from a hydroelectric station to a load centre 480 km away using two lines in parallel for security reasons.

Assume sufficient bundle conductors are used such that there are no thermal limitations, and the effective reactance per phase per km is 0.44 Ω and that the resistance is negligible. The shunt capacitive susceptance of each line is 2.27×10^{-6} S, per phase per km, and each line may be represented by the nominal π-circuit with half the capacitance at each end. The load is 2000 MW at 0.95 lagging and is independent of voltage over the permissible range.

Investigate, from the point of view of stability and voltage drop, the feasibility and performance of the link if the sending-end voltage is 345, 500, and 765 kV assuming the transmission angle is not to exceed 30°.

The lines may be compensated up to 70% by series capacitors and at the load-end compensators of 120 MVAr capacity are available. The maximum permissible voltage drop is 10%. As two lines are provided for security reasons, your studies should include the worst-operating case of only one line in use.

5.14 Explain the action of a variable-tap transformer, showing, with a phasor diagram, how reactive power may be despatched from a generator down a mainly reactive line by use of the taps. How is the level of real power despatch controlled?

Power flows down an H.V. line of impedance $0 + j0.15$ p.u. from a generator whose output passes through a variable-ratio transformer to a large power system. The voltage of the generator and the distant large system are both kept at 1.0 p.u. Determine the tap setting if 0.8 p.u. power and 0.3 p.u. VAr are delivered to a lagging load at the power system busbar. Assume the reactance of the transformer is negligible.

(Answer: $t = 1.052$)

(From Engineering Council Examination, 1995)

5.15 Two substations are connected by two lines in parallel, of negligible impedance, each containing a transformer of reactance 0.18 p.u. and rated at 120 MVA. Calculate the net absorption of reactive power when the transformer taps are set to 1:1.15 and 1:0.85, respectively (i.e. tap-stagger is used). The p.u. voltages are equal at the two ends and are constant in magnitude.

(Answer: 32 MVAr)

6

Load Flows

6.1 Introduction

A load flow (sometimes known as a power flow) is power system jargon for the steady-state solution of an electrical power network. It does not essentially differ from the solution of any other type of network except that certain constraints are peculiar to power systems and, in particular, the formulation is non-linear leading to the need for an iterative solution.

In previous chapters the manner in which the various components of a power system may be represented by equivalent circuits has been demonstrated. It should be stressed that the simplest representation of items of plant should always be used, consistent with the accuracy of the information available. There is no merit in using very complicated machine and line models when the load and other data are known only to a limited accuracy, for example, the long-line representation should only be used where absolutely necessary. Similarly, synchronous-machine models of more sophistication than those given in this text are needed only for very specialized purposes, for example in some stability studies. Usually, the size and complexity of the network itself provides more than sufficient intellectual stimulus without undue refinement of the components. Often, in high voltage networks, resistance may be neglected with little loss of accuracy and an immense saving in computation.

Load flow studies are performed to investigate the following features of a power system network:

1. Flow of MW and MVAr in the branches of the network.
2. Busbar (node) voltages.
3. Effect of rearranging circuits and incorporating new circuits on system loading.
4. Effect of temporary loss of generation and transmission circuits on system loading (mainly for security studies).
5. Effect of injecting in-phase and quadrature boost voltages on system loading.

Electric Power Systems, Fifth Edition. B.M. Weedy, B.J. Cory, N. Jenkins, J.B. Ekanayake and G. Strbac.
© 2012 John Wiley & Sons, Ltd. Published 2012 by John Wiley & Sons, Ltd.

6. Optimum system running conditions and load distribution.
7. Minimizing system losses.
8. Optimum rating and tap-range of transformers.
9. Improvements from change of conductor size and system voltage.

Planning studies will normally be performed for minimum-load conditions (examining the possibility of high voltages) and maximum-load conditions (investigating the possibility of low voltages and instability). Having ascertained that a network behaves reasonably under these conditions, further load flows will be performed to optimize voltages, reactive power flows and real power losses.

The design and operation of a power network to obtain optimum economy is of paramount importance and the furtherance of this ideal is achieved by the use of centralized automatic control of generating stations through system control centres. These control systems often undertake repeated load flow calculations in close to real time.

Although the same approach can be used to solve all load flow problems, for example the nodal voltage method, the object should be to use the quickest and most efficient method for the particular type of problem. Radial networks will require less sophisticated methods than closed loops. In very large networks the problem of organizing the data is almost as important as the method of solution, and the calculation must be carried out on a systematic basis and here the nodal-voltage method is often the most convenient. Methods such as network reduction combined with the Thevenin or superposition theorems are at their best with smaller networks. In the nodal method, great numerical accuracy is required in the computation as the currents in the branches are derived from the voltage differences between the ends. These differences are small in well designed networks so the method is ideally suited for computation using digital computers and the per unit system.

6.2 Circuit Analysis Versus Load Flow Analysis

The task of load flow analysis is conceptually similar to that of traditional circuit analysis, but there is a key difference, which is critical for understanding the specialized methods used for load flow calculations.

In circuit analysis, given all the values of impedances in the circuit and given the parameters of all voltage or current generators in the circuit, all nodal voltages and branch currents can be calculated directly. The key feature of this analysis is that the relationship between nodal voltages and branch current is linear (i.e. in the form $\mathbf{V} = \mathbf{Z} \times \mathbf{I}$).

In Load Flow analysis, loads and sources are defined in terms of powers not impedances or ideal voltage or current generators. All power network branches, transformers or overhead and underground circuits, are defined as impedances. The relationship between voltage, power and impedances is non-linear and appropriate methods for solving non-linear circuits need to be used.

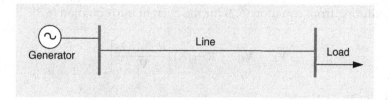

Figure 6.1　Simple two-busbar system

6.2.1 Power Flow in a Two-Busbar System

Consider a two-busbar (node) system of a generator that supplies a load over a transmission line (Figure 6.1).

Figure 6.2 shows the equivalent circuit of the system in Figure 6.1. The line resistance is R and reactance is X. The shunt susceptance of the line is neglected. In Figure 6.2 known variables are shown by solid lines (with an arrow showing the direction) and unknown variables are shown by dotted lines.

Initially it is assumed that the power the generator injects and the generator voltage are both specified. The equation that links the complex power of the generator with the current and the voltage at the generator busbar is:

$$\mathbf{S}_G = P_G + jQ_G = \mathbf{V}_G\mathbf{I}^* \tag{6.1}$$

The power and voltage are known at the same busbar, then from equation (6.1), the current through the line can be calculated directly:

$$\mathbf{I} = \frac{P_G - jQ_G}{\mathbf{V}_G^*} \tag{6.2}$$

Once the current is known, the voltage at the load busbar is given by:

$$\mathbf{V}_L = \mathbf{V}_G - (R + jX)\mathbf{I} \tag{6.3}$$

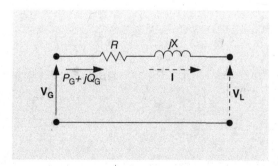

Figure 6.2　Power flow in a two-busbar system. The generator power and voltage are given

or by substituting from equation (6.2) for the current into equation (6.3):

$$\mathbf{V}_L = \mathbf{V}_G - (R + jX)\left[\frac{P_G - jQ_G}{\mathbf{V}_G^*}\right] \tag{6.4}$$

By defining $\mathbf{V}_G = V_G\angle\delta$:

$$\mathbf{V}_G^* = V_G\angle - \delta$$

Therefore:

$$\mathbf{V}_L = V_G\angle\delta - \left[\frac{RP_G + XQ_G}{V_G\angle - \delta}\right] - j\left[\frac{XP_G - RQ_G}{V_G\angle - \delta}\right] \tag{6.5}$$

As the current is known from the voltage and power at the generator busbar, the load voltage can be calculated directly from equation (6.5).

Now consider the situation shown in Figure 6.3. In this case, the voltage is specified at the generator busbar, while the power is known at the load busbar (and the voltage at the load is unknown).

The load power is

$$\mathbf{S}_L = P_L + jQ_L = \mathbf{V}_L\mathbf{I}^* \tag{6.6}$$

Although the complex power of the load is known, the load voltage is unknown and hence the current cannot be calculated directly.

The generator power is:

$$\mathbf{S}_G = P_G + jQ_G = \mathbf{V}_G\mathbf{I}^* \tag{6.7}$$

Again the current cannot be calculated, as the power at the generator is unknown. To calculate the power at the generator the losses in the line need to be known; for which the current is required.

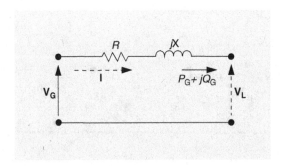

Figure 6.3 Power flow in a two-busbar system. The load power and generator voltage are given

In summary from equations (6.6) and (6.7), the current cannot be calculated directly, as power at the load, S_L, and the voltage at the generation, V_G (at opposite ends of the circuit) are given.

The expression that links the voltages at the generator and load is obtained by combining equations (6.3) and (6.6) as:

$$V_L = V_G - (R + jX)\left[\frac{P_L - jQ_L}{V_L^*}\right] \tag{6.8}$$

Equation (6.8) does not have a closed form solution. This equation, in relation to V_L is non-linear as it contains the product of the voltages at the load as shown below:

$$V_L V_L^* = V_G V_L^* - (R + jX)(P_L - jQ_L) \tag{6.9}$$

Solving Equation (6.8) requires an iterative method. The solution can start with an initial value of $V_L^{(0)}$, then find $V_L^{*(0)}$. This can then be substituted into equation (6.8) and a new value of $V_L^{(1)}$ calculated. The process is repeated for several iterations until the voltage of one iteration converges to the next iteration.

Once the value of voltage V_L at the load end is calculated, the current can be calculated from equation (6.6) and the losses can be calculated as:

Active losses $= I^2 R$,
The active power output of the generator is:

$$P_G = P_L + I^2 R$$

Similarly,
Reactive losses[1] $= I^2 X$,
The reactive power production of the generator is:

$$Q_G = Q_L + I^2 X$$

Example 6.1

The system shown in Figure 6.4 feeds a load of 20 MVA, 0.95 p.f. lagging (absorbing VArs). Draw a two busbar equivalent circuit for the network and carry out two iterations of the iterative solution.

---

[1] Active (real power) losses are dissipated as heat. Reactive power 'losses' are simply the difference in reactive power at the ends of the circuit. No energy is lost as heat.

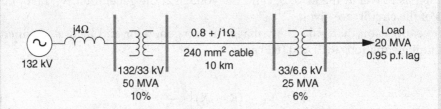

Figure 6.4 Line diagram of the system of Example 6.1

Solution

By choosing $S_{base} = 50\,\text{MVA}$, the different impedances in Figure 6.4 can be converted into p.u. values as:

	Base quantity	p.u. value
Source impedance	$132^2/50 = 348.5\,\Omega$	$j14/348.5 = j0.04$
132/33 kV transformer		$j0.1$
Cable	$33^2/50 = 21.78\,\Omega$	$0.037 + j0.046$
33/6.6 kV transformer		$0.06 \times 50/25 = j0.12$

For the load: $P = 20 \times 0.95 = 19\,\text{MW} = 0.38$ p.u. and $Q = 20 \times \sin(\cos^{-1}(0.95)) = 6.25\,\text{MVAr} = 0.125$ p.u.

The two busbar equivalent circuit is given in Figure 6.5. In the equivalent circuit the resistance is the resistance of the cable and equivalent reactance is the addition of source, transformer and cable reactances.

If the load busbar voltage is $\mathbf{V}_L$ then from equation (6.8):

$$\mathbf{V}_L = 1 - (0.037 + j0.306)\left[\frac{0.38 - j0.125}{\mathbf{V}_L^*}\right] \tag{6.10}$$

Initially assume that $\mathbf{V}_L^{(0)} = 1\angle 0°$ p.u. By substituting the first guess of $\mathbf{V}_L$ into the right hand side of equation (6.10):

$$\mathbf{V}_L^{(1)} = 1 - (0.037 + j0.306)(0.38 - j0.125)$$
$$= 0.95 - j0.11$$
$$= 0.956\angle - 6.6°$$

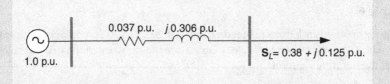

Figure 6.5 Two bus equivalent circuit of the system in Example 6.1

The complex conjugate of V_L is taken

$$V_L^{*(1)} = 0.95 + j0.11 = 0.956\angle 6.6°$$

The new value of V_L^* is substituted into the right hand side of equation (6.10) to find the second iteration of V_L:

$$V_L^{(2)} = 1 - (0.037 + j0.306)\left[\frac{0.38 - j0.125}{0.956\angle 6.6°}\right]$$

$$= 1 - (0.037 + j0.306)(0.418\angle - 24.8°)$$

$$= 1 - (0.037 + j0.306)(0.379 - j0.175)$$

$$= 0.933 - j0.109$$

$$= 0.94\angle - 6.7°$$

This iterative procedure is repeated until the difference between the nth iterative value of voltage and the $(n+1)^{th}$ iterative value of voltage becomes very small.

6.2.2 Relationship Between Power Flows and Busbar Voltages

In Chapter 2, the active and reactive power flow in a circuit was discussed. For a transmission circuit as $R \ll X$, $Z \approx X$ and $\theta \approx 90°$. Therefore from Equation (2.11):

$$P_G = \frac{V_G V_L}{X} \sin \delta$$

$$Q_G = \frac{V_G^2}{X} - \frac{V_G V_L}{X} \cos \delta = \frac{V_G(V_G - V_L)}{X} \cos \delta$$

The active power at the generator busbar is equal to the active power at the load as there are no active power losses in the circuit $(R = 0)$.

These expressions illustrate some key aspects of transporting active and reactive power across the network. Active power flow (P) requires a difference in phase angle between the busbar voltages while reactive power flow (Q) requires a difference in voltage magnitude between generator and load busbars.

Power systems are operated with relatively constant voltages and the differences in voltage magnitudes between various nodes are not allowed to be large. There are no such strict constraints on differences in phase angles across a line, but these are usually less than 30°. With $\delta \le 30°$, P is sensitive to changes in sin δ but cos δ remains close to 1.

It is possible to transmit a significant amount of active power over high voltage overhead AC transmission lines of several hundred kilometres. Reactive power cannot be transmitted over long distances, as this would require significant voltage drops that are unacceptable. We also observe that since $X \gg R$, the reactive losses are much larger than the active losses. Hence reactive power is supplied near the need usually by over-excited generators or shunt capacitive compensators.

6.3 Gauss-Seidel Method

The Gauss-Siedel method is a simple iterative technique for load flow calculations. The Gauss-Seidel method first approximates the load and generation by ideal current sources (converting powers into current injections using assumed values of voltages). The iteration process is then carried out using injected complex power until the voltages converge.

A three-busbar system is shown in Figure 6.6.

The admittances of circuits rather than their impedances are used. The complex admittance of the branch $i-j$ between nodes i and j is defined as:

$$\mathbf{y}_{ij} = \frac{1}{R_{ij} + jX_{ij}}$$

The current balance equations for each of the nodes are first derived. The injected current to Bus 1 is equal to the sum of the currents leaving this busbar through branches 1–2 and 1–3.

$$\mathbf{I}_1 = \mathbf{I}_{12} + \mathbf{I}_{13}$$

Then the branch currents are related to the busbar voltages using Ohm's law:

$$\mathbf{I}_1 = \mathbf{y}_{12} \times (\mathbf{V}_1 - \mathbf{V}_2) + \mathbf{y}_{13} \times (\mathbf{V}_1 - \mathbf{V}_3)$$

Rearranging this equation, the following is obtained:

$$\mathbf{I}_1 = (\mathbf{y}_{12} + \mathbf{y}_{13})\mathbf{V}_1 - \mathbf{y}_{12}\mathbf{V}_2 - \mathbf{y}_{13}\mathbf{V}_3 \tag{6.11}$$

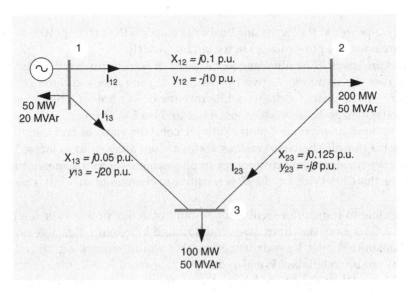

Figure 6.6 Three-busbar network for Gauss-Siedel load flow

The same procedure is repeated for Bus 2 and Bus 3

$$I_2 = -y_{12}V_1 + (y_{12} + y_{23})V_2 - y_{23}V_3 \tag{6.12}$$

$$I_3 = -y_{13}V_1 - y_{23}V_2 + (y_{13} + y_{23})V_3 \tag{6.13}$$

Equations (6.11)–(6.13) can be written in matrix form:

$$\begin{bmatrix} I_1 \\ I_2 \\ I_3 \end{bmatrix} = \begin{bmatrix} (y_{12} + y_{13}) & -y_{12} & -y_{13} \\ -y_{12} & (y_{12} + y_{23}) & -y_{23} \\ -y_{13} & -y_{23} & (y_{13} + y_{23}) \end{bmatrix} \begin{bmatrix} V_1 \\ V_2 \\ V_3 \end{bmatrix} \tag{6.14}$$

where:

$$I = \begin{bmatrix} I_1 \\ I_2 \\ I_3 \end{bmatrix} \text{ is the vector of nodal current injections}$$

$$V = \begin{bmatrix} V_1 \\ V_2 \\ V_3 \end{bmatrix} \text{ is the vector of nodal voltages}$$

$$Y_{BUS} = \begin{bmatrix} (y_{12} + y_{13}) & -y_{12} & -y_{13} \\ -y_{12} & (y_{12} + y_{23}) & -y_{23} \\ -y_{13} & -y_{23} & (y_{13} + y_{23}) \end{bmatrix}$$

is the admittance (or Y_{BUS}) matrix, with diagonal elements and off-diagonal elements defined as:

Diagonal elements

$$Y_{ii} = \sum_{\substack{j=1 \\ j \neq i}}^{N} y_{ij} \quad (N \text{ is the number of nodes})$$

Off diagonal elements

$$Y_{ij} = -y_{ji}$$

where Y_{ij} is the element at the ith row and j^{th} column of the Y_{BUS} matrix.

Substituting the values of admittances, the Y_{BUS} matrix of the example becomes:

$$Y_{BUS} = \begin{bmatrix} -j30 & j10 & j20 \\ j10 & -j18 & j8 \\ j20 & j8 & -j28 \end{bmatrix}$$

The three equations (6.11) to (6.13) are linearly dependent, that is by adding equations (6.12) and (6.13), equation (6.11) can be derived. This means that one of the equations can be eliminated and only two used to obtain the solution.

If the current injections were available, then the nodal voltages could be calculated simply from the matrix equation by finding the inverse of the reduced admittance matrix and then multiplying it by the current injection vector. However, in the formulation of the load flow the nodal currents are not available but real and reactive power injections are specified. Thus the Gauss-Siedel technique proceeds as follows.

A reference for the voltage magnitude and phase is set, and in this case it is assumed that the voltage at Bus 1 is known:

$$\mathbf{V}_1 = 1\angle 0°$$

The current injection at Bus 2 is:

$$\mathbf{I}_2 = \mathbf{Y}_{12}\mathbf{V}_1 + \mathbf{Y}_{22}\mathbf{V}_2 + \mathbf{Y}_{23}\mathbf{V}_3 \tag{6.15}$$

The injected current as a function of complex power and voltage at Bus 2 is:

$$\left(\frac{\mathbf{S}_2}{\mathbf{V}_2}\right)^* = \frac{P_2 - jQ_2}{\mathbf{V}_2^*} = \mathbf{Y}_{12}\mathbf{V}_1 + \mathbf{Y}_{22}\mathbf{V}_2 + \mathbf{Y}_{23}\mathbf{V}_3 \tag{6.16}$$

From equation (6.16) the voltage of Bus 2 is:

$$\mathbf{V}_2 = \frac{1}{\mathbf{Y}_{22}}\left[\frac{P_2 - jQ_2}{\mathbf{V}_2^*} - (\mathbf{Y}_{12}\mathbf{V}_1 + \mathbf{Y}_{23}\mathbf{V}_3)\right] \tag{6.17}$$

where P_2 and Q_2 are the active and reactive power flows into Bus 2.

The unknown complex voltage $\mathbf{V}_2$ appears on both sides of (6.17). The Gauss-Seidel method is to update the value of the voltage, in this case $\mathbf{V}_2$ on the left-hand side of (6.17), using the expression on the right-hand side, with values of the voltages already evaluated, in the present or previous iteration.

$$\mathbf{V}_2^{(p)} = \frac{1}{\mathbf{Y}_{22}}\left[\frac{P_2 - jQ_2}{\mathbf{V}_2^{(p-1)^*}} - (\mathbf{Y}_{12}\mathbf{V}_1 + \mathbf{Y}_{23}\mathbf{V}_3^{(p-1)})\right] \tag{6.18}$$

where p is the iteration number

Similarly, for Bus 3:

$$\mathbf{V}_3^{(p)} = \frac{1}{\mathbf{Y}_{33}}\left[\frac{P_3 - jQ_3}{\mathbf{V}_3^{(p-1)^*}} - (\mathbf{Y}_{13}\mathbf{V}_1 + \mathbf{Y}_{23}\mathbf{V}_2^{(p)})\right] \tag{6.19}$$

In general:

$$V_i^{(p)} = \frac{1}{Y_{ii}} \left[\frac{P_i - jQ_i}{V_i^{(p-1)^*}} - \sum_{j=1}^{i-1} Y_{ij} V_j^{(p)} - \sum_{j=i+1}^{N} Y_{ij} V_j^{(p-1)} \right] \tag{6.20}$$

The process can be started by initializing the value of all voltages to $1\angle 0°$. Hence for the iteration $p = 1$, the voltage at Bus 2 is given by:

$$V_2^{(1)} = \frac{1}{-j18} \left[\frac{-2 + j0.5}{1} - (j10 \times 1 + j8 \times 1) \right]$$

$$= 0.9722 - j0.1111 = 0.9785\angle - 6.5198°$$

Similarly, for Bus 3

$$V_3^{(1)} = \frac{1}{-j28} \left[\frac{-1 + j0.5}{1} - (j20 \times 1 + j8 \times (0.9722 - j0.1111)) \right]$$

$$= 0.9742 - j0.0675 = 0.9765\angle - 3.9612°$$

Table 6.1 shows how values of voltages change in this process.

The convergence of the iteration is shown in Figure 6.7.

Table 6.1 and Figure 6.7 show that, in this example, the iteration process converges after five or six iterations.

The changes in voltage magnitudes can be used as a criterion for convergence. In this case normally a maximum change in voltage from the previous iteration, for example 0.001 p.u is set. Alternatively, the power mismatches at each node, at each iteration can be calculated. The power mismatch is the difference between the complex power injected into a node and the power leaving a node through the network

Table 6.1 Changes of voltage magnitudes and phase angles in the Gauss-Siedel example

Iteration	Voltage Magnitude (pu)			Voltage Angle (Degrees)		
	Bus 1	Bus 2	Bus 3	Bus 1	Bus 2	Bus 3
1	1	0.978 551	0.976 539 249	0	−6.519 80	−3.96 121
2	1	0.957 666	0.967 134 203	0	−8.38 021	−4.45 358
3	1	0.949 463	0.96 419 005	0	−8.71 939	−4.55 110
4	1	0.947 275	0.963 412 677	0	−8.81 964	−4.58 133
5	1	0.946 675	0.963 197 267	0	−8.84 810	−4.58 974
6	1	0.946 508	0.963 137 161	0	−8.85 594	−4.59 206

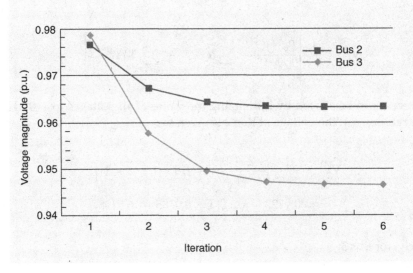

Figure 6.7 Convergence of Gauss-Siedel iteration

branches. When the power mismatch is less then a threshold, the iteration process is stopped.

6.4 Load Flows in Radial and Simple Loop Networks

Distribution systems are normally operated as radial networks and a simple iterative procedure can be used to solve load flows in them. One of the common methods used is the forward and backward method. The procedure used in this method is:

Step 1	Assume an initial nodal voltage magnitude and angle for each busbar ($1\angle 0°$ p.u. voltage is often used)
Step 2	Start from the source and move forward towards the feeder and lateral ends and calculate the current using equation (6.6)
Step 3	Start from the feeder and lateral ends and move towards the source while calculating the voltage at each busbar using equation (6.3)
Step 4	Repeat Steps 2 and 3 until the termination criterion ($\mathbf{V}_i^{(p)} - \mathbf{V}_i^{(p-1)} < \varepsilon$) is met at all busbars

Example 6.2

For the system shown in Figure 6.8 carry out two iterations of the forward and backward method.

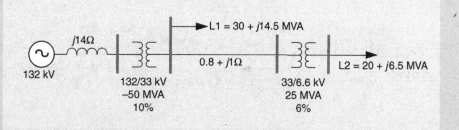

Figure 6.8 Line diagram of the system in Example 6.2

Solution

The equivalent circuit of the radial network shown in Figure 6.8 (in per unit on a 50 MVA base) is given in Figure 6.9. Neglecting losses in line 2–3, the current I_1 is given by:

$$I_1 = \frac{(0.6 - j0.29) + (0.38 - j0.125)}{V_2^*}$$
$$= \frac{0.98 - j0.415}{V_2^*}$$

(6.21)

Similarly, I_2 is given by:

$$I_2 = \frac{0.38 - j0.125}{V_3^*}$$

(6.22)

From equation (6.3), two voltage equations can be written as:

$$V_2 = 1 - j0.14 I_1$$

(6.23)

$$V_3 = V_2 - (0.037 + j0.166) I_2$$

(6.24)

Step 1:

Assume $V_2 = V_3 = 1\angle 0°$

Step 2:

From equations (6.21) and (6.22):

$$I_1^{(1)} = 0.98 - j0.415$$

$$I_2^{(1)} = 0.38 - j0.125$$

Step 3:

Substituting the values of I_1 and I_2 in equations (6.23) and (6.24), voltages can be found as:

$$V_2^{(1)} = 1 - j0.14(0.98 - j0.415)$$
$$= 0.942 - j0.137 = 0.952\angle - 8.29°$$

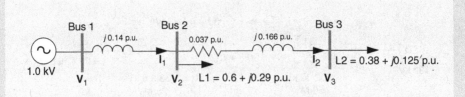

Figure 6.9 Equivalent circuit of the system in Example 6.2

$$\mathbf{V}_3^{(1)} = 0.942 - j0.137 - (0.037 + j0.166)(0.38 - j0.125)$$
$$= 0.907 - j0.196 = 0.928\angle - 12.15°$$

Repeating Step 2:

$$\mathbf{I}_1^{(2)} = \frac{0.98 - j0.415}{0.952\angle 8.29°} = 0.956 - j0.58$$

$$\mathbf{I}_2^{(2)} = \frac{0.38 - j0.125}{0.928\angle 12.15°} = 0.372 - j0.218$$

Repeating Step 3:

$$\mathbf{V}_2^{(2)} = 1 - j0.14(0.956 - j0.58)$$
$$= 0.92 - j0.133 = 0.93\angle - 8.23°$$

$$\mathbf{V}_3^{(2)} = 0.96 - j0.152 - (0.037 + j0.166)0.372 - j0.218$$
$$= 0.87 - j0.188 = 0.89\angle - 12.2°$$

Step 4:
Repeat Steps 2 and 3 until the termination criterion $(\mathbf{V}_i^{(n+1)} - \mathbf{V}_i^{(n)} < \varepsilon)$ is met

Example 6.3

Repeat Example 6.2 using the Gauss-Seidel method.

Solution:
The admittances of the branches are:

$$\mathbf{y}_{12} = \frac{1}{j0.14} = -j7.143$$

$$\mathbf{y}_{23} = \frac{1}{0.037 + j0.166} = 1.279 - j5.739$$

Therefore the $\mathbf{Y}_{BUS}$ matrix is:

$$\mathbf{Y}_{BUS} = \begin{bmatrix} -j7.143 & j7.143 & 0 \\ j7.143 & 1.279 - j12.881 & -1.279 + j5.739 \\ 0 & -1.279 + j5.739 & 1.279 - j5.739 \end{bmatrix}$$

From equation (6.18):

$$\mathbf{V}_2^{(p)} = \frac{1}{1.279 - j12.881}\left[\frac{-0.6 + j0.29}{\mathbf{V}_2^{(p-1)^*}} - (j7.143 \times 1 + (-1.279 + j5.739)\mathbf{V}_3^{(p-1)})\right]$$

From equation (6.19):

$$\mathbf{V}_3^{(p)} = \frac{1}{1.279 - j5.739}\left[\frac{-0.38 + j0.125}{\mathbf{V}_3^{(p-1)^*}} - (-1.279 + j5.739) \times \mathbf{V}_2^{(p)}\right]$$

These two equations are solved iteratively.

6.5 Load Flows in Large Systems

In large practical power systems, it is necessary to carry out many load flows for both planning and operation. These take into account the complex impedances of the circuits, the limits caused by circuit capacities, and the voltages that can be satisfactorily provided at all nodes in the system. Systems consisting of up to 3000 busbars, 6000 circuits, and 500 generators may have to be solved in reasonable time scales (e.g. 1–2 min) with accuracies requiring 32 or 64 bits for numerical stability. The many system states that are possible in a day's operation may have to be considered. Even though the Gauss-Seidel method could be employed for large networks, its convergence is slow (even with acceleration factors[2]) and sometimes it fails to converge. For large power systems the Newton-Raphson or fast decoupled methods are more commonly used. The speed of convergence of these methods is of extreme importance as the use of these methods in schemes for the automatic control of power systems requires very fast load-flow solutions.

For large meshed networks, the following types of busbar are specified for load-flow studies:

1. Slack, swing, or floating bus: A single busbar in the system is specified by a voltage, constant in magnitude and phase (the usual practice is to set the phase angle to 0). The generator at this node supplies the losses to the network; this is necessary because the magnitude of the losses will not be known until the calculation of currents is complete, and this cannot be achieved unless one busbar has no

---

[2] The correction in voltage from $\mathbf{V}^{(p)}$ to $\mathbf{V}^{(p+1)}$ is multiplied by a factor so that the number of iterations required to reach the specified convergence can be greatly reduced.

power constraint and can feed the required losses into the system. The location of the slack node can influence the complexity of the calculations; the node most closely approaching an infinite busbar should be used.

2. PQ bus: At a PQ bus the complex power $S = P \pm jQ$ is specified whereas the voltage magnitude and angle are unknown. All load buses are PQ buses.

3. PV bus: At a PV bus P and $|V|$ are specified; voltage angle and Q are unknown. Q is adjusted to keep $|V|$ constant. Most generator busbars are PV busbars.

In Section 6.3, the Gauss-Seidel method was illustrated on a network with slack and PQ busbars. To include PV Buses in the Gauss-Seidel method, Q needs to be guessed.

From equation (6.16):

$$S_i^* = V_i^* \sum_{j=1}^{N} Y_{ij} V_j = P_i - jQ_i$$

Hence, $Q_i^{(p)} = -Im\left[V_i^* \sum_{j=1}^{N} Y_{ij} V_j \right]$

During the iterative solution it is assumed $S_i^{(p)*} = P_i - jQ_i^{(p)}$
Tentatively solve for $V_i^{(p+1)}$

$$V_i^{(p+1)} = \frac{1}{Y_{ii}} \left[\frac{S_i^{(p)*}}{V_i^{(p)*}} - \sum_{\substack{j=1 \\ j \neq i}}^{N} Y_{ij} V_j^{(p)} \right]$$

But since $|V_i|$ is specified, replace $\left| V_i^{(p+1)} \right|$ by $|V_i|$.

6.5.1 Y_{BUS} *Matrix with Tap-Changing Transformers*

The formation of the **Y** matrix for a network having simple passive components was shown in Section 6.3. When the taps of a transformer are set to its nominal ratio the transformer is represented by a single series impedance. When the taps are off-nominal, adjustments have to be made as follows.

Consider a transformer of ratio $t{:}1$ connected between two nodes i and j; the series admittance of the transformer is Y_t. By defining an artificial node x between the voltage transforming element and the transformer admittance (Figure 6.10(a)), the following equations can be written:

$$S_i = V_i I_i^* \qquad\qquad (6.25)$$

$$S_x = \frac{V_i}{t} I_j^* \qquad\qquad (6.26)$$

From Figure 6.10(a),

$$I_j = \left[V_j - \frac{V_i}{t} \right] Y_t \tag{6.27}$$

As the $t:1$ transformer is ideal, $S_i = -S_x$. Therefore from (6.25)–(6.27):

$$I_i = -\frac{I_j}{t} = -\frac{1}{t}\left[V_j - \frac{V_i}{t} \right] Y_t$$

$$= \frac{Y_t}{t^2} V_i - \frac{Y_t}{t} V_j \tag{6.28}$$

From equations (6.27) and (6.28), the Y_{BUS} matrix is given by:

$$\begin{bmatrix} I_i \\ I_j \end{bmatrix} = \begin{bmatrix} \dfrac{Y_t}{t^2} & -\dfrac{Y_t}{t} \\ -\dfrac{Y_t}{t} & Y_t \end{bmatrix} \begin{bmatrix} V_i \\ V_j \end{bmatrix} \tag{6.29}$$

From equation (6.29) it is seen that for the node on the off-tap side of the transformer (that is, for node j), the following conditions apply:

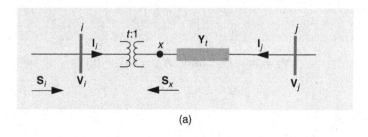

(a)

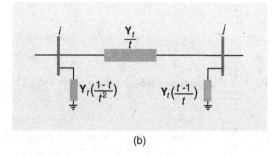

(b)

Figure 6.10 (a) Equivalent circuit of transformer with off-nominal tap ratio. Transformer series admittance on non-tap side. (b) The π section to represent transformer with off-nominal tap ratio

when forming Y_{jj} use Y_t, for the transformer, and when forming Y_{ij}, use $-Y_t/t$ for the transformer.

For the tap-side node (for node i) that the following conditions apply:

when forming Y_{ii} use Y_t/t^2, and when forming Y_{ij} use $-Y_t/t$.

These conditions may be represented by the π section shown in Figure 6.10(b), although it is probably easier to modify the mutual and self-admittances directly.

6.5.2 The Newton-Raphson Method

Although the Gauss-Seidel was the first popular method for load flow calculations, the Newton-Raphson method is now commonly used. The Newton-Raphson method has better convergence characteristics and for many systems is faster than the Gauss-Seidel method; the former has a much larger time per iteration but requires very few iterations (four is general), whereas the Gauss-Siedel requires up to 30 iterations, the number increasing with the size of system.

Consider an n-node power system. For the kth busbar, having links connecting node k to nodes j (depending on the number of links this could be from 1 to n $-$ 1) of admittance Y_{kj},

$$P_k + jQ_k = V_k I_k^* = V_k \sum_{j=1}^{n-1} \left(Y_{kj}V_j\right)^* \qquad (6.30)$$

Let

$$V_k = a_k + jb_k \quad \text{and} \quad Y_{kj} = G_{kj} + jB_{kj}$$

Then,

$$P_k + jQ_k = (a_k + jb_k)\sum_{j=1}^{n-1}\left[\left(G_{kj}+jB_{kj}\right)\left(a_j+jb_j\right)\right]^* \qquad (6.31)$$

from which,

$$P_k = \sum_{j=1}^{n-1}\left[a_k\left(a_jG_{kj}-b_jB_{kj}\right)+b_k\left(a_jB_{kj}+b_jG_{kj}\right)\right] \qquad (6.32)$$

$$Q_k = \sum_{j=1}^{n-1}\left[b_k\left(a_jG_{kj}-b_jB_{kj}\right)-a_k\left(a_jB_{kj}+b_jG_{kj}\right)\right] \qquad (6.33)$$

Hence, there are two non-linear simultaneous equations for each node. Note that $(n-1)$ nodes are considered because the slack node n is completely specified.

In order to explain the basic iterative procedure to solve these two non-linear simultaneous equations, equation (6.32) is defined as:

$$P_k = g_1(a_1, \ldots \ldots, a_{n-1}; b_1, \ldots \ldots, b_{n-1}) \tag{6.34}$$

Note that G_{kj} and B_{kj} are constants.

If the initial guess of variables are $a_1^*, \ldots \ldots, a_{n-1}^*; b_1^*, \ldots \ldots, b_{n-1}^*$ and errors are $\Delta a_1, \ldots \ldots, \Delta a_{n-1}; \Delta b_1, \ldots \ldots, \Delta b_{n-1}$, then equation (6.34) can be written as:

$$P_k^* + \Delta P_k = g_1\left(a_1^* + \Delta a_1, \ldots \ldots, a_{n-1}^* + \Delta a_{n-1}; b_1^* + \Delta b_1, \ldots \ldots, b_{n-1}^* + \Delta b_{n-1}\right) \tag{6.35}$$

By expanding equation (6.35) using Taylor's series and neglecting higher order terms:

$$\Delta P_k = \frac{\partial P_k}{\partial a_1} \Delta a_1 + \frac{\partial P_k}{\partial a_2} \Delta a_2 + \cdots \frac{\partial P_k}{\partial a_{n-1}} \Delta a_{n-1}$$
$$+ \frac{\partial P_k}{\partial b_1} \Delta b_1 + \frac{\partial P_k}{\partial b_2} \Delta b_2 + \cdots \frac{\partial P_k}{\partial b_{n-1}} \Delta b_{n-1}$$

Similar equations hold for ΔQ in terms of Δa and Δb.

These equations may be expressed generally using the Jacobian Matrix:

$$
\begin{bmatrix} \Delta P_1 \\ \vdots \\ \Delta P_{n-1} \\ \Delta Q_1 \\ \vdots \\ \Delta Q_{n-1} \end{bmatrix}
=
\begin{bmatrix}
\dfrac{\partial P_1}{\partial a_1} & \cdots & \dfrac{\partial P_1}{\partial a_{n-1}} & \dfrac{\partial P_1}{\partial b_1} & \cdots & \dfrac{\partial P_1}{\partial b_{n-1}} \\
& \cdots & & & \cdots & \\
& \cdots & & & \cdots & \\
\dfrac{\partial P_{n-1}}{\partial a_1} & \cdots & \dfrac{\partial P_{n-1}}{\partial a_{n-1}} & \dfrac{\partial P_{n-1}}{\partial b_1} & \cdots & \dfrac{\partial P_{n-1}}{\partial b_{n-1}} \\
\dfrac{\partial Q_1}{\partial a_1} & \cdots & \dfrac{\partial Q_1}{\partial a_{n-1}} & \dfrac{\partial Q_1}{\partial b_1} & \cdots & \dfrac{\partial Q_1}{\partial b_{n-1}} \\
& \cdots & & & \cdots & \\
& \cdots & & & \cdots & \\
\dfrac{\partial Q_{n-1}}{\partial a_1} & \cdots & \dfrac{\partial Q_{n-1}}{\partial a_{n-1}} & \dfrac{\partial Q_{n-1}}{\partial b_1} & \cdots & \dfrac{\partial Q_{n-1}}{\partial b_{n-1}}
\end{bmatrix}
\begin{bmatrix} \Delta a_1 \\ \vdots \\ \Delta a_{n-1} \\ \Delta b_1 \\ \vdots \\ \Delta b_{n-1} \end{bmatrix}
\tag{6.36}
$$

Jacobian matrix

For convenience, denote the Jacobian matrix by

$$
\begin{bmatrix} J_A & J_B \\ J_C & J_D \end{bmatrix}
$$

The elements of the matrix are evaluated for the values of P, Q and V at each iteration as follows.

For the submatrix J_A and from equation (6.32), diagonal elements are given by:

$$\frac{\partial P_k}{\partial a_k} = 2a_k G_{kk} - b_k B_{kk} + b_k B_{kk} + \sum_{\substack{j=1 \\ j \neq k}}^{n-1} \left(a_j G_{kj} - b_j B_{kj}\right) \tag{6.37}$$

This element may be more readily obtained by expressing some of the quantities in terms of the current at node k, $\mathbf{I}_k$, which can be determined separately at each iteration.

Let

$$\mathbf{I}_k = c_k + jd_k = (G_{kk} + jB_{kk})(a_k + jb_k) + \sum_{\substack{j=1 \\ j \neq k}}^{n-1} \left(G_{kj} + jB_{kj}\right)\left(a_j + jb_j\right)$$

from which,

$$c_k = a_k G_{kk} - b_k B_{kk} + \sum_{\substack{j=1 \\ j \neq k}}^{n-1} \left(a_j G_{kj} - b_j B_{kj}\right)$$

and

$$d_k = a_k B_{kk} + b_k G_{kk} + \sum_{\substack{j=1 \\ j \neq k}}^{n-1} \left(a_j B_{kj} + b_j G_{kj}\right)$$

So that,

$$\frac{\partial P_k}{\partial a_k} = a_k G_{kk} + b_k B_{kk} + c_k$$

Off-diagonal elements, where $k \neq j$, are given by:

$$\frac{\partial P_k}{\partial a_j} = a_k G_{kj} + b_k B_{kj} \tag{6.38}$$

For J_B,

$$\frac{\partial P_k}{\partial b_k} = -a_k B_{kk} + b_k G_{kk} + d_k$$

and

$$\frac{\partial P_k}{\partial b_j} = -a_k B_{kj} + b_k G_{kj} \quad (k \neq j)$$

For J_C,

$$\frac{\partial Q_k}{\partial a_k} = -a_k B_{kk} + b_k G_{kk} - d_k$$

and

$$\frac{\partial Q_k}{\partial a_j} = -a_k B_{kj} + b_k G_{kj} \quad (k \neq j)$$

For J_D,

$$\frac{\partial Q_k}{\partial b_k} = -a_k G_{kk} - b_k B_{kk} + c_k$$

and

$$\frac{\partial Q_k}{\partial b_j} = -a_k G_{kj} - b_k B_{kj} \quad (k \neq j)$$

The process commences with the iteration counter 'p' set to zero and all the nodes except the slack-bus being assigned voltages, usually 1 p.u.

From these voltages, P and Q are calculated from equations (6.32) and (6.33). The changes are then calculated:

$$\Delta P_k^p = P_k(\text{specified}) - P_k^p \quad \text{and} \quad \Delta Q_k^p = Q_k(\text{specified}) - Q_k^p$$

where p is the iteration number.

Next the node currents are computed as

$$I_k^p = \left(\frac{P_k^p + jQ_k^p}{V^p}\right)^* = c_k^p + jd_k^p$$

The elements of the Jacobian matrix are than formed, and from equation (6.36),

$$\begin{vmatrix} \Delta a \\ \Delta b \end{vmatrix} = \begin{vmatrix} J_A & J_B \\ J_C & J_D \end{vmatrix}^{-1} \begin{vmatrix} \Delta P \\ \Delta Q \end{vmatrix}$$

Hence, a and b are determined and the new values, $a_k^{p+1} = a_k^p + \Delta a_k^p$ and $b_k^{p+1} = b_k^p + \Delta b_k^p$, are obtained. The process is repeated ($p = p + 1$) until ΔP and ΔQ are less than a prescribed tolerance.

In the previous development of the Newton Raphson method the admittances and voltages were assumed to be in rectangular form. Sometimes the quantities are expressed in polar form. The polar form of the equations is:

$$P_k = P(V, \delta)$$
$$Q_k = Q(V, \delta)$$

(6.39)

For a link connecting nodes k and j of admittance, if $\mathbf{V}_k = V_k \angle \theta_k$, $\mathbf{V}_j = V_j \angle \theta_j$ and $\mathbf{Y}_{kj} = G_{kj} + jB_{kj}$, from equation (6.30) the power at a busbar is

$$\mathbf{S}_k = P_k + jQ_k = \mathbf{V}_k \mathbf{I}_k^*$$

$$= V_k \angle \delta_k \sum_{j=1}^{n-1} (G_{kj} - jB_{kj})(V_j \angle - \delta_j)$$

$$= V_k V_j \sum_{j=1}^{n-1} (G_{kj} - jB_{kj})(\cos \delta_{kj} + j\sin \delta_{kj})$$

where $\delta_{kj} = \delta_k - \delta_j$.

Therefore,

$$P_k = \sum_{j=1}^{n-1} V_k V_j (G_{kj} \cos \delta_{kj} + B_{kj} \sin \delta_{kj})$$

$$= V_k^2 G_{kk} + \sum_{\substack{j=1 \\ j \neq k}}^{n-1} V_k V_j (G_{kj} \cos \delta_{kj} + B_{kj} \sin \delta_{kj})$$

(6.40)

$$Q_k = \sum_{j=1}^{n-1} V_k V_j (G_{kj}\sin \delta_{kj} - B_{kj}\cos \delta_{kj})$$

$$= -V_k^2 B_{kk} + \sum_{\substack{j=1 \\ j \neq k}}^{n-1} V_k V_j (G_{kj}\sin \delta_{kj} - B_{kj}\cos \delta_{kj})$$

(6.41)

For a load busbar,

$$\Delta P_k = \sum_{j=1}^{n-1} \frac{\partial P_k}{\partial \delta_j} \Delta \delta_j + \sum_{j=1}^{n-1} \frac{\partial P_k}{\partial V_j} \Delta V_j$$

(6.42)

$$\Delta Q_k = \sum_{j=1}^{n-1} \frac{\partial Q_k}{\partial \delta_j} \Delta \delta_j + \sum_{j=1}^{n-1} \frac{\partial Q_k}{\partial V_j} \Delta V_j$$

(6.43)

For a generator PV busbar, only the ΔP_k equation is used as Q_k is not specified. The mismatch equation is:

$$
\begin{array}{|c|c|c|}
\hline
\Delta \mathbf{P}^{p-1} & J_A & J_B \\
\hline
\Delta \mathbf{Q}^{p-1} & J_C & J_D \\
\hline
\end{array}
\begin{array}{|c|}
\hline
\Delta \boldsymbol{\delta}^p \\
\hline
\left(\dfrac{\Delta \mathbf{V}^p}{\mathbf{V}^{p-1}} \right) \\
\hline
\end{array}
\tag{6.44}
$$

$\Delta \boldsymbol{\delta}^p$ is the correction to PQ and PV buses and $\Delta \mathbf{V}^p / \mathbf{V}^{p-1}$ is the correction to PQ buses.

For buses k and j, off-diagonal elements are given by:

$$
J_A : \frac{\partial P_k}{\partial \delta_j} = V_k V_j \left(G_{kj} \sin \delta_{kj} - B_{kj} \cos \delta_{kj} \right)
$$

$$
J_B : V_j \frac{\partial P_k}{\partial V_j} = V_k V_j \left(G_{kj} \cos \delta_{kj} + B_{kj} \sin \delta_{kj} \right)
$$

$$
J_C : \frac{\partial Q_k}{\partial \delta_j} = -V_k V_j \left(G_{kj} \cos \delta_{kj} + B_{kj} \sin \delta_{kj} \right)
$$

$$
J_D : V_j \frac{\partial Q_k}{\partial V_j} = V_k V_j \left(G_{kj} \sin \delta_{kj} - B_{kj} \cos \delta_{kj} \right)
$$

Also, diagonal elements are given by:

$$
J_A : \frac{\partial P_k}{\partial \delta_k} = \sum_{\substack{j=1 \\ j \neq k}}^{n-1} V_k V_j \left(-G_{kj} \sin \delta_{kj} + B_{kj} \cos \delta_{kj} \right)
$$

$$
= -Q_k - B_{kk} V_k^2
$$

Similarly,

$$
J_B : V_k \frac{\partial P_k}{\partial V_k} = P_k + G_{kk} V_k^2
$$

$$
J_C : \frac{\partial Q_k}{\partial \delta_k} = P_k - G_{kk} V_k^2
$$

$$
J_D : V_k \frac{\partial Q_k}{\partial V_k} = Q_k - B_{kk} V_k^2
$$

The computational process can be enhanced by pre-ordering and dynamic ordering, defined as

- Pre-ordering, in which nodes are numbered in sequence of increasing number of connections.

- Dynamic ordering, in which at each step in the elimination the next row to be operated on has the fewest non-zero terms.

6.5.3 Fast Decoupled Load Flow

In a power system, the changes in voltage angle mainly control flows of P and the voltage magnitudes of busbars mainly change Q. Therefore the coupling between the $P_k - V_j$ and $Q_k - \delta_j$ components is weak. Hence the submatrices J_B and J_C can be neglected. Equation (6.44) can then be reduced to two equations:

$$\left[\Delta \mathbf{P}^{p-1}\right] = [J_A][\Delta \boldsymbol{\delta}^p] \tag{6.45}$$

$$\left[\Delta \mathbf{Q}^{p-1}\right] = [J_D]\left[\frac{\Delta \mathbf{V}^p}{\mathbf{V}^{p-1}}\right] \tag{6.46}$$

The decoupled technique is further simplified by assuming that: under normal loading conditions, the angle difference across transmission lines is negligible, thus $\delta_{kj} \approx 0$ and for lines and cables it is reasonable to assume $B_{kj} \gg G_{kj}$, thus G_{kj} can be neglected. Therefore:

$$\text{Off-diagonal element of } J_A : \frac{\partial P_k}{\partial \delta_j} = -V_k V_j B_{kj} = -\frac{V_k V_j}{X_{kj}}$$

$$\text{Diagonal element of } J_A : \frac{\partial P_k}{\partial \delta_k} = \sum_{\substack{j=1 \\ j \neq k}}^{n-1} V_k V_j B_{kj} = \sum_{\substack{j=1 \\ j \neq k}}^{n-1} \frac{V_k V_j}{X_{kj}}$$

$$\text{Off-diagonal element of } J_D : V_j \frac{\partial Q_k}{\partial V_j} = -V_k V_j B_{kj} = -\frac{V_k V_j}{X_{kj}}$$

$$\text{Diagonal element of } J_D : V_k \frac{\partial Q_k}{\partial V_k} = \sum_{\substack{j=1 \\ j \neq k}}^{n-1} V_k V_j B_{kj} = \sum_{\substack{j=1 \\ j \neq k}}^{n-1} \frac{V_k V_j}{X_{kj}}$$

where X_{kj} is the branch reactance.
This method is called the fast decoupled load flow

Example 6.4

Using the fast decoupled method calculate the angles and voltages after the first iteration for the three-node network of Figure 6.11. Initially V_1 is 230 kV V_2 is 220 kV and V_3 is 228 kV all at $\theta = 0°$. Busbar 2 is a load consuming 200 MW, 120 MVAr; busbar 3 is a

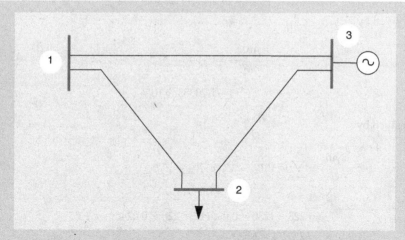

Figure 6.11 Three busbar network for fast decoupled load flow

generator node set at 70 MW and 228 kV. V_1 is an infinite busbar. The $\mathbf{Y}_{BUS}$ matrix is given by (all quantities are in Siemens):

$$\mathbf{Y}_{BUS} = \begin{bmatrix} 0.00819 - j0.049099 & -0.003196 + j0.019272 & -0.004994 + j0.030112 \\ -0.003196 + j0.019272 & 0.007191 - j0.043099 & -0.003995 + j0.02409 \\ -0.004994 + j0.030112 & -0.003995 + j0.02409 & 0.008989 - j0.053952 \end{bmatrix}$$

From equation (6.45)

$$\begin{bmatrix} \Delta P_2 \\ \Delta P_3 \end{bmatrix} = \begin{bmatrix} J_{A_{22}} & J_{A_{23}} \\ J_{A_{32}} & J_{A_{33}} \end{bmatrix} \begin{bmatrix} \Delta \delta_2 \\ \Delta \delta_3 \end{bmatrix} \tag{6.47}$$

$J_{A_{22}}$ is given by

$$\frac{\partial P_2}{\partial \delta_2} = \sum_{\substack{j=1 \\ j \neq 2}}^{3} V_k V_j B_{kj} = V_2(V_1 B_{12} + V_3 B_{23})$$

$$= 220 \times (230 \times 0.019272 + 228 \times 0.02409) \times 10^6$$
$$= 2.1835 \times 10^9$$

$J_{A_{23}}$ is given by

$$\frac{\partial P_2}{\partial \delta_3} = -V_2 V_3 B_{23}$$

$$= -220 \times 228 \times 0.02409 \times 10^6$$
$$= -1.20835 \times 10^9$$

$J_{A_{32}}$ is given by

$$\frac{\partial P_3}{\partial \delta_2} = -V_3 V_2 B_{32}$$

$$= -1.20835 \times 10^9$$

$J_{A_{33}}$ is given by

$$\frac{\partial P_3}{\partial \delta_3} = \sum_{\substack{j=1 \\ j \neq 3}}^{2} V_k V_j B_{kj} = V_3 (V_1 B_{13} + V_2 B_{32})$$

$$= 228 \times (230 \times 0.030112 + 220 \times 0.02409) \times 10^6$$
$$= 2.7874 \times 10^9$$

From (6.40):

$$P_2 = V_2 V_1 G_{21} + V_2^2 G_{22} + V_2 V_3 G_{23}$$
$$= 220 \times (-230 \times 0.003196 + 220 \times 0.007191 - 228 \times 0.003995) \times 10^6$$
$$= -14.0624 \, \text{MW}$$

$$P_3 = V_3 V_1 G_{31} + V_3 V_2 G_{32} + V_3^2 G_{33}$$
$$= 228 \times (-230 \times 0.004994 - 220 \times 0.003995 + 228 \times 0.008989) \times 10^6$$
$$= 5.0096 \, \text{MW}$$

Therefore,

$$\Delta P_2 = -200 + 14.0624 = -185.9376 \, \text{MW}$$
$$\Delta P_3 = 70 - 5.0096 = 64.99 \, \text{MW}$$

Now from equation (6.47):

$$\begin{bmatrix} -185.9376 \times 10^6 \\ 64.99 \times 10^6 \end{bmatrix} = \begin{bmatrix} 2.1835 \times 10^9 & -1.20835 \times 10^9 \\ -1.20835 \times 10^9 & 2.7874 \times 10^9 \end{bmatrix} \begin{bmatrix} \Delta \delta_2 \\ \Delta \delta_3 \end{bmatrix}$$

$$\begin{bmatrix} \Delta \delta_2 \\ \Delta \delta_3 \end{bmatrix} = \begin{bmatrix} -2.1835 \times 10^9 & -1.20835 \times 10^9 \\ -1.20835 \times 10^9 & -2.7874 \times 10^9 \end{bmatrix}^{-1} \begin{bmatrix} -185.9376 \times 10^6 \\ 64.99 \times 10^6 \end{bmatrix}$$

$$= \begin{bmatrix} -0.0951 \\ -0.0179 \end{bmatrix} \text{rad} = \begin{bmatrix} -5.45° \\ -1.026° \end{bmatrix}$$

$$\therefore \begin{bmatrix} \delta_2 \\ \delta_3 \end{bmatrix} == \begin{bmatrix} -5.45° \\ -1.026° \end{bmatrix}$$

From equation (6.41)

$$Q_2 = -V_2V_1[G_{21}\sin(\delta_2-\delta_1)-B_{21}\cos(\delta_2-\delta_1)]-V_2^2B_{22}-V_2V_3[G_{23}\sin(\delta_2-\delta_3)-B_{23}\cos(\delta_2-\delta_3)]$$

$$= -220\times230[-0.003196\times\sin(-5.45°)-0.019272\times\cos(-5.45°)]-220^2\times0.043099$$

$$-220\times228[-0.003995\times\sin(-5.45°+1.026°)-0.02409\times\cos(-5.45°+1.026°)]$$

$$Q_2 = 58.68\,\text{MVAr}$$

$$[\Delta Q_2]=-120+58.68=-61.32\,\text{MVAr}$$

From equation (6.46):

$$[\Delta Q_2] = [J_{D_2}]\left[\frac{\Delta V_2}{V_2}\right]$$

J_{D_2} is given by

$$V_k\frac{\partial Q_k}{\partial V_k} = \sum_{\substack{j=1\\j\neq2}}^{3}V_kV_jB_{kj} = (V_2V_1B_{21} + V_2V_3B_{23})$$

$$= 220 \times (230 \times 0.019272 + 228 \times 0.02409) \times 10^6$$

$$= 2.1835 \times 10^9$$

$$\Delta V_2 = \frac{\Delta Q_2 \times V_2}{J_{D_2}}$$

$$= \frac{-61.32 \times 10^6 \times 220 \times 10^3}{2.1835 \times 10^9} = -6.178\,\text{kV}$$

Hence

$$V_1 = 230\,\text{kV}$$
$$V_2 = 220 - 6.178 = 213.82\,\text{kV}$$
$$V_3 = 228\,\text{kV}$$

The process is repeated for the next iteration, and so on, until convergence is reached.

6.6 Computer Simulations

The line diagram of a part of a transmission system is shown in Figure 6.12, with details of the overhead lines given in Table 6.2.

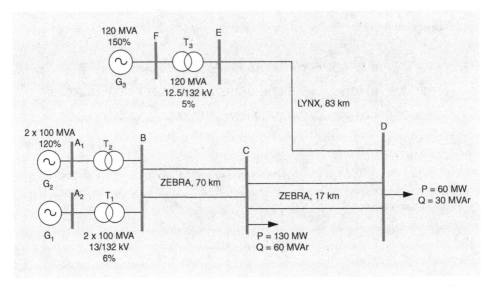

Figure 6.12 System used for simulation studies

Table 6.2 Overhead line data

Overhead line	Cross-section mm^2	Resistance mΩ/km	Reactance mΩ/km	Capacitance nF/km
ZEBRA (on a double circuit 132 kV tower)	400	76.0	387.0	9.50
LYNX (on a single circuit 132 kV tower)	175	178.0	401.0	8.90

The problem has been solved using a load flow program, IPSA[3]; a brief description of the arrangement of essential data is given. Most commercially available load flow programs require data to be entered in per unit. Therefore a base value of S was chosen as 100 MVA and voltage bases selected as the rated primary and secondary voltages of the transformers. Previously the capacitance of the lines has been neglected. However for HV cable circuits and EHV overhead lines, when the capacitance is known, it can be converted into susceptance for load flow programs. Tables 6.3-6.5 show the input data for the lines, busbars and transformers. For double circuit lines the per unit data is given for a single circuit. Transformer taps were changed to maintain the 132 kV busbar voltage at a reference.

---

[3] The IPSA load flow program uses a variant of the fast-decoupled load flow technique.

Table 6.3 Input line data (On 100 MVA, 132 kV base, the impedance base is: $Z_{base} = 132^2/100 = 174.24\,\Omega$)

Line	Resistance (p.u.) = R/Z_{base}	Reactance (p.u.) = X/Z_{base}	Susceptance (p.u.) = $2\pi f C \times Z_{base}$
B - C	$76 \times 10^{-3} \times 70/174.24$ $= 0.0305$	$387 \times 10^{-3} \times 70/174.24$ $= 0.1555$	$100\pi \times 9.5 \times 10^{-9} \times 70 \times 174.24$ $= 0.0364$
C - D	$76 \times 10^{-3} \times 17/174.24$ $= 0.0074$	$387 \times 10^{-3} \times 17/174.24$ $= 0.0378$	$100\pi \times 9.5 \times 10^{-9} \times 17 \times 174.24$ $= 0.0088$
E - D	$178 \times 10^{-3} \times 83/174.24$ $= 0.0848$	$401 \times 10^{-3} \times 83/174.24$ $= 0.1910$	$100\pi \times 8.9 \times 10^{-9} \times 83 \times 174.24$ $= 0.0404$

Table 6.4 Busbar data

Busbar	Type	V	δ	P_G (MW)	Q_G (MVAr)	P_L (MW)	Q_L (MVAr)	X_s(p.u.)
A1 and A2 (combined)	Slack	1.0	0	—	—	—	—	0.6
B	PQ (load)	—	—	—	—	130	60	—
C	PQ (load)	—	—	—	—	60	30	—
F	PV (generator)	1.05	—	120	+100 to −60	—	—	$1.5 \times 100/$ $120 = 1.25$

Table 6.5 Transformer data

Transformer	Resistance (p.u.) with X/R = 10	Reactance (p.u.)
T_1	0.006	0.06
T_2	0.006	0.06
T_3	0.0042	$0.05 \times 100/120 = 0.042$

The system shown in Figure 6.12 was implemented in the IPSA load flow software package. The two generators at busbars A_1 and A_2 were combined to create the slack busbar.

Figure 6.13 shows the load flow solution. P (upper value) and Q (lower value) flows are shown for each line and transformer. Table 6.6 shows the line flows and losses.

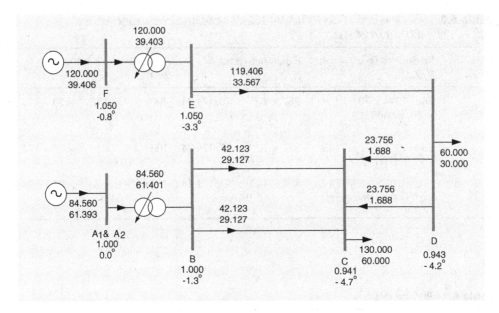

Figure 6.13 Load flow results

Table 6.6 Line and transformer flows

From bus	To bus	Flow		Losses	
		MW	MVAr	MW	MVAr
A_1 and A_2	B	84.560	61.401	0.315	3.147
B	A_1 and A_2	84.247	58.259	0.315	3.147
	C	42.123 (in each line)	29.127 (in each line)	1.666 (total)	1.632 (total)
C	B	41.290 (in each line)	28.314 (in each line)	1.666 (total)	1.632 (total)
	D	−23.756 (in each line)	−1.688 (in each line)	0.094 (total)	1.08 (total)
D	C	−23.757 (in each line)	−1.146 (in each line)	0.094 (total)	1.08 (total)
	E	107.514	32.292	11.901	1.271
F	E	120	39.403	0.584	5.837
E	D	119.406	33.567	11.901	1.271

Problems

6.1 Consider a simple power system composed of a generator, a 400 kV overhead transmission line and a load, as shown in Figure 6.14.

   If the desired voltage at the consumer busbar is 400 kV, calculate the:

   i. Active and reactive losses in the line.
   ii. Voltage magnitude and phase angle at busbar 1.
   iii. Active and reactive output of the generator at busbar 1.

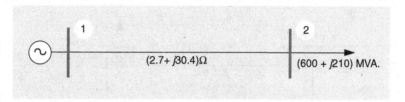

Figure 6.14 400 kV system for Problem 6.1

(Answer (i) 6.8 MW, 76.7 MVAr; (ii) 422.35 kV (L-L), 6.0°; (iii) 606.8 MW, 286.7 MVAr)

6.2 A load of $1 + j0.5$ p.u. is supplied through a transmission line by a generator G1 that maintains its terminal voltage at 1 p.u. (Figure 6.15).

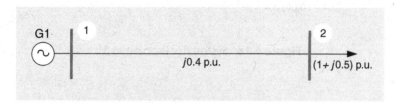

Figure 6.15 System for Problem 6.2

  i. Form the Y_{BUS} matrix for this system.
 ii. Perform two iterations of the Gauss-Seidel load flow.

(Answer: (i) $Y_{bus} = \begin{bmatrix} -2.5j & 2.5j \\ 2.5j & -2.5j \end{bmatrix}$; (ii) $V_1 = 1.0$ p.u., $V_2 = 0.6 - j0.3$ p.u.)

6.3 List the information which may be obtained from a load-flow study. Part of a power system is shown in Figure 6.16. The circuit reactances and values of real and reactive power (in the form $P \pm jQ$) at the various busbars are expressed as per unit values on a common MVA base. Resistance may be neglected. By the use of an iterative method, calculate the voltages at the stations after the first iteration.

(Answer: $V_2 = 1.0333 - j0.0333$ p.u., $V_3 = 1.1167 + j0.2333$ p.u., $V_4 = 1.0556 - j0.0556$ p.u.)

6.4 A 400 kV interconnected system is supplied from busbar A, which may be considered to be an infinite busbar. The loads and line reactances are as indicated in Figure 6.17.

Determine the flow of power in line AC using two iterations of voltages at each bus.

(Answer: $P_{AC} = 1.16$ GW)

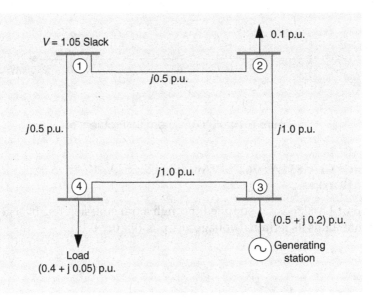

Figure 6.16 System for Problem 6.3

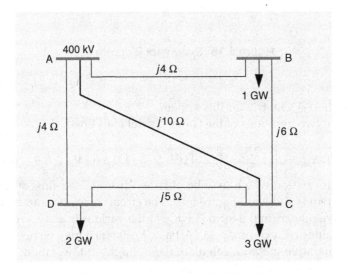

Figure 6.17 System for Problem 6.4

6.5 Determine the voltage at busbar (2), voltage at busbar (3) and the reactive power at bus (3) as shown in Figure 6.18, after the first iteration of a Gauss-Seidel load flow method. Assume the initial voltage to be $1\angle 0°$ p.u. All the quantities are in per unit on a common base.

(Answer: $0.99 - j0.0133$, $1.0 + j0.0015$ and $Q3 = 0.4$ p.u.)

6.6 Active power demand of the 132 kV system shown in Figure 6.19 is supplied by two generators G1 and G2. System voltage is supported by generator G2 and a

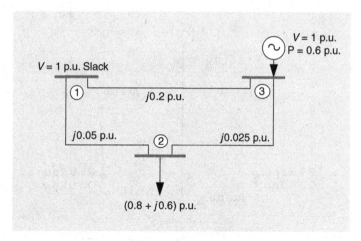

Figure 6.18 System for Problem 6.5

large synchronous compensator SC (see Figure 6.19), which both maintain the voltage at 1 p.u. at their respective nodes. Generator G1, connected at node 1, has no reactive power capacity available for voltage control.

  i. Form the Y_{bus} matrix for this system.
 ii. Perform two iterations of the Gauss-Seidel load flow.

(Answer: (i) $Y_{bus} = \begin{bmatrix} -7.5j & 2.5j & 5j \\ 2.5j & -6.5j & 4j \\ 5j & 4j & -9j \end{bmatrix}$; (ii) $V_1 = 0.9825 + j0.0520$ p.u., $V_2 = 1.0$ p.u., $V_3 = 0.9903 - j0.1386$)

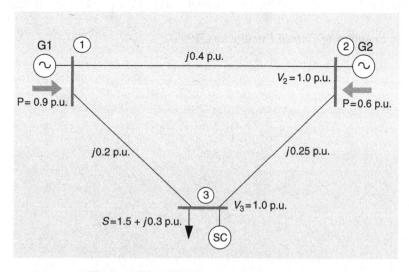

Figure 6.19 132 kV system for Problem 6.6

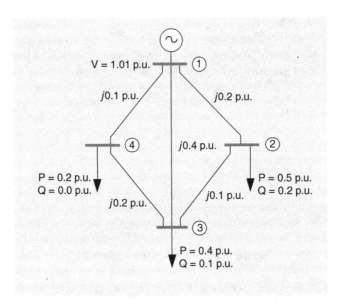

Figure 6.20 System for Problem 6.8

6.7   In Figure 6.20 the branch reactances and busbar loads are given in per unit on a common base. Branch resistance is neglected.

Explain briefly why an iterative method is required to determine the busbar voltages of this network.

Form the Y_{BUS} admittance matrix for this network.

Using busbar 1 as the slack (reference) busbar, carry out the first iteration of a Gauss-Seidel load-flow algorithm to determine the voltage at all busbars. Assume the initial voltages of all busbars to be 1.01 p.u.

(Answer: $V_2 = 0.997 - j0.03$, $V_3 = 0.9968 - j0.0415$, $V_4 = 1.0056 - j0.026$)

(From Engineering Council Examination, 1995)

7

Fault Analysis

7.1 Introduction

Calculation of the currents which flow when faults of various types occur is an essential part of the design of a power supply network. Typically, fault currents are obtained using computer packages by applying faults at various points in the network. The magnitudes of the fault currents give the engineer the current settings for the protection to be used and the ratings of the circuit breakers.

The types of fault commonly occurring in practice are illustrated in Figure 7.1, and the most common of these is the short circuit of a single conductor to ground or earth. Often, the path to earth contains resistance R_f in the form of an arc, as shown in Figure 7.1(f). Although the single-line-to-earth fault is the most common, calculations are frequently performed with the three-line, balanced short circuit (Figure 7.1(d) and (e)). This is usually the most severe fault and also the most amenable to calculation. The same currents flow for the faults shown in 7.1(d) and 7.1(e).

The causes of faults are summarized in Table 7.1, which gives the distribution of faults, due to various causes, on the England and Wales Transmission System in a typical year. Table 7.2 shows the components affected. In tropical countries the incidence of lightning is much greater than in the UK, resulting in larger numbers of faults.

As well as fault current I_f (in kA), fault MVA is frequently used as a rating; this is obtained from the expression $\sqrt{3}V_L I_f$, where V_L is the nominal line voltage of the faulted part in kV. The fault MVA is often referred to as the *fault level*. The calculation of fault currents can be divided into the following two main types:

1. Faults short-circuiting all three phases when the network remains balanced electrically. For these calculations, normal single-phase equivalent circuits may be used as in ordinary load-flow calculations.

Electric Power Systems, Fifth Edition. B.M. Weedy, J.B. Cory, N. Jenkins, J.B. Ekanayake and G. Strbac.
© 2012 John Wiley & Sons, Ltd. Published 2012 by John Wiley & Sons, Ltd.

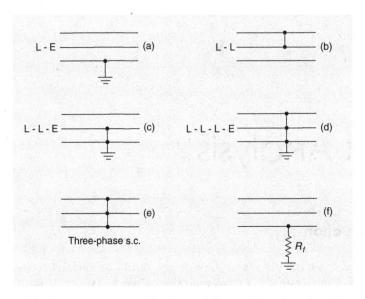

Figure 7.1 Common types of fault or short circuit (s.c.) L = line, E = earth

Table 7.1 Causes of overhead-line faults, England and Wales system 66 kV and above

	Faults/160 km of line/year
Lightning	1.59
Dew, fog, frost	0.15
Snow, ice	0.01
Gales	0.24
Salt spray	0.01
Total	2 faults per 160 km, giving a total of 232 faults on system/year

Table 7.2 Distribution of faults, England and Wales Transmission system

Type	Number of faults/year
Overhead lines	289
Cables	67
Switchgear	56
Transformers	59
Total	471

2. Faults other than three-phase short circuits when the network is electrically unbalanced. To facilitate these calculations a special method for dealing with unbalanced networks is used, known as the method of symmetrical components.

The main objects of fault analysis are:

1. to determine maximum and minimum three-phase short-circuit currents;
2. to determine the unsymmetrical fault current for single and double line-to earth faults, line-to-line faults, and sometimes for open-circuit faults;
3. investigation of the operation of protective relays;
4. determination of rated interrupting capacity of breakers;
5. to determine fault-current distribution and busbar-voltage levels during fault conditions.

7.2 Calculation of Three-Phase Balanced Fault Currents

The action of synchronous generators on three-phase short circuits has been described in Chapter 3. There it was seen that, depending on the time from the incidence of the fault, either the transient or the subtransient reactance should be used to represent the generator. For specifying switchgear, the value of the current flowing at the instant at which the circuit contacts open is required. It has been seen, however, that the initially high fault current, associated with the subtransient reactance, decays with the passage of time. Modern air-blast circuit breakers usually operate in 2 1/2 cycles and SF_6 breakers within 1 1/2 cycles of 60 or 50 Hz alternating current, and are operated through extremely fast protection. Older circuit breakers and those on lower voltage networks usually associated with relatively cruder protection can take in the order of eight cycles or more to operate. In calculations of fault currents on a transmission system it is usual to use the subtransient reactance of the generators and to ignore the effects of induction motors, which are connected to lower voltage networks.

The simple calculation of fault currents ignores the direct-current component, the magnitude of which depends on the instant in the cycle that the short circuit occurs. If the circuit breaker opens a reasonable time after the incidence of the fault, the direct-current component will have decayed considerably. With fast acting circuit breakers the actual current to be interrupted is increased by the direct-current component and it must be taken into account. In simple, manual calculations the symmetrical r.m.s. value may be modified by the use of multiplying factors to take account of the direct-current component of the fault current. Although these multipliers are determined by Standards and depend on the X/R ratio of the circuit to the fault, they are of the form:

- 8-cycle circuit breaker opening time, multiply by 1;
- 3-cycle circuit breaker opening time, multiply by 1.2;
- 2-cycle circuit breaker opening time, multiply by 1.4.

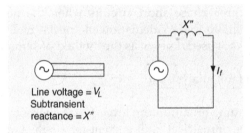

Figure 7.2 Voltage source with short circuit and single-phase equivalent circuit

Consider a generator with a short circuit across the three terminals, as shown in Figure 7.2. Defining the voltage base as V_L and the MVA base of S_b, the base impedance is,

$$Z_b = \frac{V_L^2}{S_b}$$

$$X''(\text{p.u}) = \frac{X''(\Omega)}{Z_b}$$

The short-circuit current

$$I_{s.c.} = \frac{1}{X''(p.u)} \tag{7.1}$$

The three-phase short-circuit level in p.u.

$$= V_{\text{nominal}} \times I_{s.c.} = 1 \times I_{s.c.} = \frac{1}{X''(p.u)}$$

The three-phase short-circuit level in volt-amperes

$$= \frac{S_b}{X''(p.u)} \tag{7.2}$$

Hence the short-circuit level contribution is immediately obtained if the subtransient reactance X''(p.u.) of the generator is known. Similarly if the impedance Z(p.u.) from an infinite busbar to the point of the fault on a network is known then the short-circuit level may be determined immediately.

Example 7.1

An 11.8 kV busbar is fed from three synchronous generators having the following ratings and subtransient reactances,

20 MVA, X''0.08 p.u.;   60 MVA, X''0.1 p.u.;   20 MVA, X''0.09 p.u.

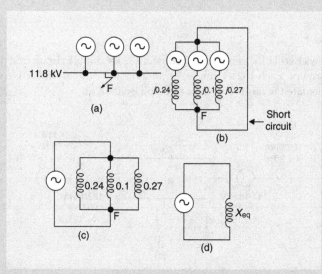

Figure 7.3 Line diagram and equivalent circuits for Example 7.1

Calculate the fault current and MVA if a three-phase symmetrical fault occurs on the busbars. Resistance may be neglected. The voltage base will be taken as 11.8 kV and the VA base as 60 MVA.

Solution
The subtransient reactance of the 20 MVA machine on the above base is $(60/20) \times 0.08$, that is, 0.24 p.u., and of the 20 MVA machine $(60/20) \times 0.09$, that is, 0.27 p.u. These values are shown in the equivalent circuit in Figure 7.3. As the generator e.m.f.s are assumed to be equal, one source may be used (Figure 7.3(c)). The equivalent reactance is

$$X_{eq} = \frac{1}{1/0.24 + 1/0.27 + 1/0.1} = 0.056 \, \text{p.u.}$$

Therefore the fault MVA

$$= 60/0.056 = 1071 \, \text{MVA}$$

and the fault current

$$\frac{1071 \times 10^6}{\sqrt{3} \times 11800} = 52\,402 \, \text{A}$$

Example 7.2

Figure 7.4 shows two 11 kV generators feeding a 132 kV double circuit line through generator transformers. Each 132 kV line is 40 km long and with an inductive reactance of 0.4 Ω/km. Calculate the fault level for a three-phase fault at:

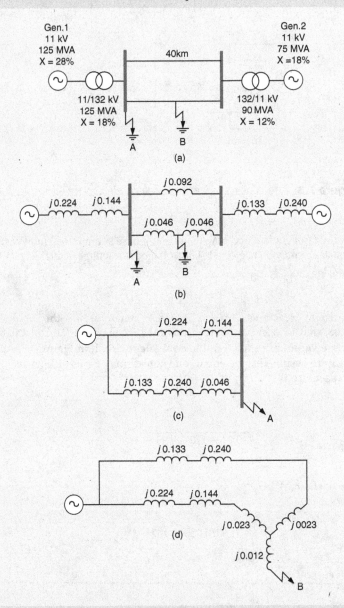

Figure 7.4 Line diagram and equivalent circuits for Example 7.2

a. Point A on the busbar close to generator 1.
b. Point B at the mid-point of one of the lines.

Use a 100 MVA base for the calculation, express your answer in Amperes and ignore the effect of any system loads.

Solution
On 100 MVA, 132 kV base

$$Z_b = \frac{132^2}{100} = 174.24\,\Omega$$

On 100 MVA base, the reactances of the generators, transformers and the line are:

		Reactance in p.u. on 100 MVA
Generator X	$j0.28 \times 100/125$	$j0.224$
Generator Y	$j0.18 \times 100/75$	$j0.240$
Generator transformer of X	$j0.18 \times 100/125$	$j0.144$
Generator transformer of Y	$j0.12 \times 100/90$	$j0.133$
Line	$j0.4 \times 40/174.24$	$j0.092$

The equivalent single-phase network of generator and line reactances is shown in Figure 7.4(b). For a fault at A, the equivalent circuit is reduced to the network shown in Figure 7.4(c) (note that the double circuit line is replaced by a single equivalent reactance).

The network in Figure 7.4(c) was reduced to give the final single equivalent reactance,

$$X_{eq} = 0.196 \text{ p.u.}$$

The fault level at point A

$$= \frac{100}{0.196} = 510.4 \text{ MVA}$$

Therefore fault current $= 510.4 \times 10^6 / \sqrt{3} \times 11 \times 10^3 = 26\,789$ A

In order to obtain the fault level at B, the equivalent circuit shown in Figure 7.4(b) was replaced by the network shown in Figure 7.4(d) by the use of the delta-star transformation. A further transformation is carried out on the network in Figure 7.4(d) to give the final single equivalent reactance,

$$X_{eq} = 0.209 \text{ p.u.}$$

The fault level at point B

$$= \frac{100}{0.209} = 479.1 \text{ MVA}$$

Therefore the fault current $= \dfrac{479.1 \times 10^6}{\sqrt{3} \times 11 \times 10^3} = 25\,146$ A

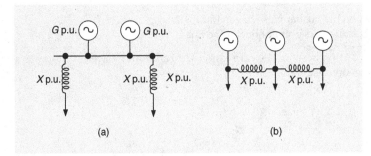

Figure 7.5 Connection of artificial reactors: (a) feeder connection; (b) ring system. Transient reactance of machines, G p.u.; reactance of artificial reactors X p.u

7.2.1 Current Limiting Reactors

The impedances presented to fault currents by transformers and machines when faults occur on substation or generating station busbars are low. To reduce the high fault current which would do considerable damage mechanically and thermally, artificial reactances are sometimes connected between bus sections. These current-limiting reactors usually consist of insulated copper strip embedded in concrete formers; this is necessary to withstand the high mechanical forces produced by the current in neighbouring conductors. The position in the circuit occupied by the reactor is a matter peculiar to individual designs and installations (see Figure 7.5(a) and (b)).

Example 7.3

A 400 kV power system contains three substations A, B, and C having fault levels (GVA) of 20, 20, and 30, respectively.

The system is to be reinforced by three lines each of reactance $j5\,\Omega$ connecting together the three substations as shown in Figure 7.6(a). Calculate the new fault level at C (three-phase symmetrical fault). Neglect resistance.

Solution
The equivalent circuit for the network is shown in Figure 7.6(b). The substations are represented by a voltage source (400 kV) in series with the value of reactance to give the specified initial fault level.

Select $S_b = 60$ GVA. Hence the effective reactance 'behind' A when subject to a three-phase short circuit $60/20 = 3$ p.u. Similarly, the effective reactance at $B = 3$ p.u. and at $C = 2$ p.u.

With $V_b = 400$ kV and $S_b = 60$ GVA,

$$Z_b = \frac{\left(400 \times 10^3\right)^2}{60 \times 10^9} = 2.67\,\Omega$$

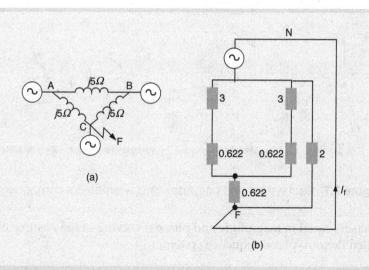

Figure 7.6 (a) Networks for Example 7.3 (b) Reduced network

The 5 Ω mesh is transformed into a star with arms of value 1.66 Ω and then converted into p.u. value by dividing by Z_b. The equivalent circuit with the fault on C after reinforcement is shown in Figure 7.6(b). Fault current at C with equivalent reactance 1.1 p.u. gives, fault level at C, 60/1.1 GVA = 54.66 GVA.

Note that this value is very high and that the result of reinforcement is always to increase the fault levels. The maximum rating of a 400 kV circuit breaker is about 50 GVA, hence either a busbar must be run split between two incoming feeders or a current limiting reactor must be installed.

7.3 Method of Symmetrical Components

This method formulates a system of three separate phasor systems which, when superposed, give the unbalanced conditions in the circuit. It should be stressed that the systems to be discussed are essentially artificial and used merely as an aid to calculation. The various sequence-component voltages and currents do not exist as physical entities in the network, although they could be monitored by special filters.

The method postulates that any three-phase unbalanced system of voltages and currents may be presented by the following three separate systems of phasors:

1. a balanced three-phase system in the normal a-b-c (red-yellow-blue) sequence, called the positive phase-sequence system;
2. a balanced three-phase system of reversed sequence, that is a-c-b (red-blue-yellow), called the negative phase-sequence system;

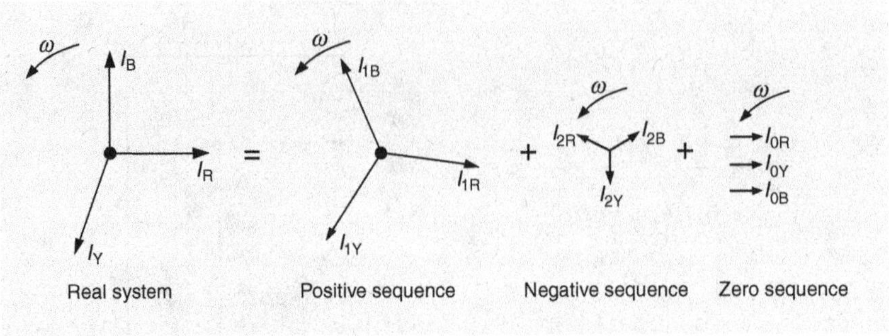

Figure 7.7 Real system and corresponding symmetrical components

3. three phasors equal in magnitude and phase revolving in the positive phase rotation, called the zero phase-sequence system.

In Figure 7.7 an unbalanced system of currents is shown with the corresponding system of symmetrical components. If each of the red-phase phasors are added, that is $I_{1R} + I_{2R} + I_{0R}$, the resultant phasor will be I_R in magnitude and direction; similar reasoning holds for the other two phases.

To express the phasors algebraically, use is made of the complex operator 'a' (sometimes denoted by λ or h) which denotes a phase-shift operation of $+120°$ and a multiplication of unit magnitude; that is.

$$V\angle\phi \times \mathbf{a} = V\angle\phi \times 1\angle 120°$$

$$= V\angle(\phi + 120°)$$

$$\mathbf{a} = e^{j2\pi/3} \quad \text{and} \quad \mathbf{a}^3 = e^{j3\times 2\pi/3} = 1$$

Also,

$$\mathbf{a}^2 + \mathbf{a} = (-0.5 - j0.866) + (-0.5 + j0.866) = -1$$

$$\therefore \mathbf{a}^3 + \mathbf{a}^2 + \mathbf{a} = 0$$

and

$$\therefore 1 + \mathbf{a} + \mathbf{a}^2 = 0$$

For positive-sequence phasors, taking the red phasor as reference,

$$I_{1R} = I_{1R}e^{j0} = \text{reference phase (Figure 7.8)}$$
$$I_{1Y} = I_{1R}(-0.5 - j0.866) = \mathbf{a}^2 I_{1R}$$

and

$$I_{1B} = I_{1R}(-0.5 + j0.866) = \mathbf{a}I_{1R}$$

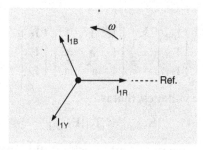

Figure 7.8 Positive-sequence phasors

For negative-sequence quantities,

$$\mathbf{I}_{2R} = \mathbf{I}_{2R}(1 + j0) = \text{reference phase (Figure 7.9)}$$
$$\mathbf{I}_{2Y} = \mathbf{I}_{2R}(-0.5 + j0.866) = \mathbf{aI}_{2R}$$
$$\mathbf{I}_{2B} = \mathbf{I}_{2R}(-0.5 - j0.866) = \mathbf{a}^2\mathbf{I}_{2R}$$

Returning to the original unbalanced system of currents, $\mathbf{I}_R$, $\mathbf{I}_Y$, and $\mathbf{I}_B$,

$$\mathbf{I}_R = \mathbf{I}_{1R} + \mathbf{I}_{2R} + \mathbf{I}_{0R}$$
$$\mathbf{I}_Y = \mathbf{I}_{1Y} + \mathbf{I}_{2Y} + \mathbf{I}_{0Y} = \mathbf{a}^2\mathbf{I}_{1R} + \mathbf{aI}_{2R} + \mathbf{I}_{0R}$$
$$\mathbf{I}_B = \mathbf{I}_{1B} + \mathbf{I}_{2B} + \mathbf{I}_{0B} = \mathbf{aI}_{1R} + \mathbf{a}^2\mathbf{I}_{2R} + \mathbf{I}_{0R}$$

Hence in matrix form,

$$\begin{bmatrix} \mathbf{I}_R \\ \mathbf{I}_Y \\ \mathbf{I}_B \end{bmatrix} = \begin{bmatrix} 1 & 1 & 1 \\ 1 & \mathbf{a}^2 & \mathbf{a} \\ 1 & \mathbf{a} & \mathbf{a}^2 \end{bmatrix} \begin{bmatrix} \mathbf{I}_{0R} \\ \mathbf{I}_{1R} \\ \mathbf{I}_{2R} \end{bmatrix} \tag{7.3}$$

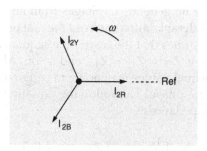

Figure 7.9 Negative-sequence phasors

Inverting the matrix,

$$
\begin{bmatrix} I_{0R} \\ I_{1R} \\ I_{2R} \end{bmatrix} = \frac{1}{3} \begin{bmatrix} 1 & 1 & 1 \\ 1 & a & a^2 \\ 1 & a^2 & a \end{bmatrix} \begin{bmatrix} I_R \\ I_Y \\ I_B \end{bmatrix}
\tag{7.4}
$$

The above also holds for voltages, that is.

$$
[E_{actual}] = [T_s][E_{1,2,0}]
$$

where $[T_s]$ is the symmetrical component transformation matrix,

$$
\begin{bmatrix} 1 & 1 & 1 \\ 1 & a^2 & a \\ 1 & a & a^2 \end{bmatrix}
$$

In a three-wire system the instantaneous voltages and currents add to zero. There is no neutral connection and hence no single phase currents. A fourth wire or connection to earth must be provided for single-phase currents to flow. In a three-wire system the zero phase-sequence components are replaced by zero in equations (7.3) and (7.4).

Where zero-phase currents flow,

$$
I_{0R} = \frac{I_R + I_Y + I_B}{3} = \frac{I_N}{3}
$$

where I_N is the neutral current.

$$
\therefore I_N = 3I_{0R} = 3I_{0Y} = 3I_{0B}
$$

In the application of this method it is necessary to calculate the symmetrical components of the current in each line of the network and then to combine them to obtain the actual values. The various phase-sequence values are obtained by considering a network derived from the actual network in which only a particular sequence current flows; for example, in a zero-sequence network, only zero-sequence currents and voltages exist. The positive-sequence network is identical to the real, balanced equivalent network, that is it is the same as used for three-phase symmetrical short-circuit studies. The negative sequence network is almost the same as the real one except that the values of impedance used for rotating machines are different and there are no generated voltages from an ideal machine. The zero-sequence network is considerably different from the real one.

The above treatment assumes that the respective sequence impedances in each of the phases are equal, that is $Z_{1R} = Z_{1Y} = Z_{1B}$, and so on. Although this covers most cases met in practice, unequal values may occur in certain circumstances, for example an open circuit on one phase. The following equation applies for the voltage drops across the phase impedances:

$$
\begin{bmatrix} V_R \\ V_Y \\ V_B \end{bmatrix} = [T_S] \begin{bmatrix} V_{0R} \\ V_{1R} \\ V_{2R} \end{bmatrix} = \begin{bmatrix} Z_R & 0 & 0 \\ 0 & Z_Y & 0 \\ 0 & 0 & Z_B \end{bmatrix} [T_S] \begin{bmatrix} I_{0R} \\ I_{1R} \\ I_{2R} \end{bmatrix}
$$

From which,

$$\begin{bmatrix} \mathbf{V}_R \\ \mathbf{V}_Y \\ \mathbf{V}_B \end{bmatrix} = \begin{bmatrix} \mathbf{Z}_R(\mathbf{I}_{0R} + \mathbf{I}_{1R} + \mathbf{I}_{2R}) \\ \mathbf{Z}_Y(\mathbf{I}_{0R} + a^2\mathbf{I}_{1R} + a\mathbf{I}_{2R}) \\ \mathbf{Z}_Y(\mathbf{I}_{0R} + a\mathbf{I}_{1R} + a^2\mathbf{I}_{2R}) \end{bmatrix}$$

Therefore

$$\mathbf{V}_{0R} = \frac{1}{3}(\mathbf{V}_R + \mathbf{V}_Y + \mathbf{V}_B)$$

$$= \frac{1}{3}\mathbf{I}_{1R}(\mathbf{Z}_R + a^2\mathbf{Z}_Y + a\mathbf{Z}_B) + \frac{1}{3}\mathbf{I}_{2R}(\mathbf{Z}_R + a\mathbf{Z}_Y + a^2\mathbf{Z}_B) + \frac{1}{3}\mathbf{I}_{0R}(\mathbf{Z}_R + \mathbf{Z}_Y + \mathbf{Z}_B)$$

and, similarly, expressions $\mathbf{V}_{1R}$ and $\mathbf{V}_{2R}$ may be obtained.

It is seen that the voltage drop in each sequence is influenced by the impedances in all three phases. If, as previously assumed, $\mathbf{Z}_R = \mathbf{Z}_Y = \mathbf{Z}_B = \mathbf{Z}$, then the voltage drops become $\mathbf{V}_{1R} = \mathbf{I}_{1R}\mathbf{Z}$, and so on, as before.

7.4 Representation of Plant in the Phase-Sequence Networks

7.4.1 The Synchronous Machine (see Table 3.1)

The positive-sequence impedance $\mathbf{Z}_1$ is the normal transient or subtransient value. Negative-sequence currents set up a magnetic field rotating in the opposite direction to that of the positive-sequence currents and which rotates round the rotor surface at twice the synchronous speed; hence the effective impedance ($\mathbf{Z}_2$) is different from $\mathbf{Z}_1$. The zero-sequence impedance $\mathbf{Z}_0$ depends upon the nature of the connection between the star point of the windings and the earth and the single-phase impedance of the stator windings in parallel. Resistors or reactors are frequently connected between the star point and earth for reasons usually connected with protective gear and the limitation of over-voltages. Normally, the only voltage sources appearing in the networks are in the positive-sequence one, as the generators only generate positive-sequence e.m.f.s.

7.4.2 Lines and Cables

The positive- and negative-sequence impedances are the normal balanced values. The zero-sequence impedance depends upon the nature of the return path through the earth if no fourth wire is provided. It is also modified by the presence of an earth wire on the towers that protect overhead lines against lightning strikes. In the absence of detailed information the following rough guide to the value of $\mathbf{Z}_0$ may be used. For a single-circuit line $\mathbf{Z}_0/\mathbf{Z}_1 = 3.5$ with no earth wire and $\mathbf{Z}_0/\mathbf{Z}_1 = 2$ with one earth wire. For a double-circuit line, $\mathbf{Z}_0/\mathbf{Z}_1 = 5.5$. For underground cables, $\mathbf{Z}_0/\mathbf{Z}_1 = 1 - 1.25$ for single core and 3–5 for three-core cables.

7.4.3 Transformers

The positive- and negative-sequence impedances are the normal balanced ones. The zero-sequence connection of transformers is, however, complicated, and depends on

Table 7.3　Zero sequence representation of transformers

Connections of Windings		Representation per Phase	Comments
Primary	Secondary		
(star, earthed)	(star, earthed)	Z_0	Zero-sequence currents free to flow in both primary and secondary circuits
(star)	(star, earthed)	Z_0	No path for zero-sequence currents in primary circuits
(delta)	(star, earthed)	Z_0	Single-phase currents can circulate in the delta but not outside it
(delta)	(star)	Z_0	No flow of zero-sequence currents possible
(delta)	(delta)	Z_0	No flow of zero-sequence currents possible
(star with tertiary T delta)	(star, earthed)	Z_0	Tertiary winding provides path for zero-sequence currents

the nature of the connection of the windings. Table 7.3 lists the zero-sequence representation of transformers for various winding arrangements. Zero-sequence currents in the windings on one side of a transformer must produce the corresponding ampere-turns in the other, but three in-phase currents cannot flow in a star connection without a connection to earth. They can circulate round a delta winding, but not in the lines outside it. Owing to the mutual impedance between the phases, $Z_0 \neq Z_1$. For a three-limb transformer, $Z_0 < Z_1$; for a five-limb transformer, $Z_0 > Z_1$. An example showing the nature of the three sequence networks for a small transmission link is shown in Figure 7.10.

7.5 Types of Fault

In the following, a single voltage source in series with an impedance is used to represent the power network as seen from the point of the fault. This is an extension of Thevenin's theorem to three-phase systems. It represents the general method used for manual calculation, that is the successive reduction of the network to a single

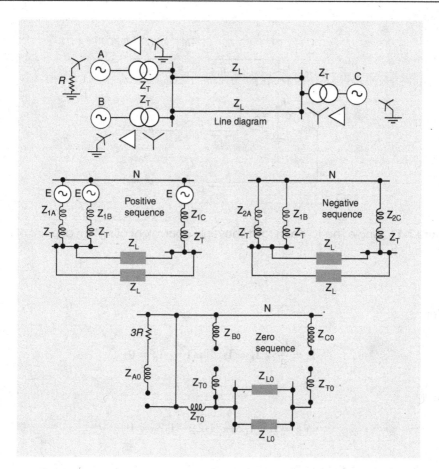

Figure 7.10 Typical transmission link and form of associated sequence networks

impedance and voltage or current source. The network is assumed to be initially on no-load before the occurrence of the fault, and linear, so that superposition applies.

7.5.1 Single-Line-To-Earth Fault

The three-phase circuit diagram is shown in Figure 7.11, where the three phases are on open-circuit at their ends.

Let I_1, I_2 and I_0 be the symmetrical components of I_R and let V_1, V_2 and V_0 be the components of V_R. For this condition, $V_R = 0$, $I_B = 0$, and $I_Y = 0$. Also, Z_R includes components Z_1, Z_2, and Z_0.

From equation (7.4)

$$I_0 = \frac{1}{3}(I_R + I_Y + I_B)$$

$$I_1 = \frac{1}{3}(I_R + aI_Y + a^2I_B)$$

$$I_2 = \frac{1}{3}(I_R + a^2I_Y + aI_B)$$

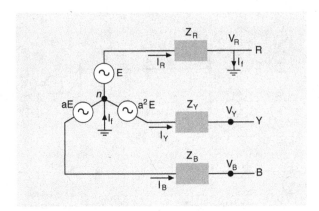

Figure 7.11 Single line-to-earth fault-Thevenin equivalent of system at point of fault

Hence,

$$\mathbf{I}_0 = \frac{\mathbf{I}_R}{3} = \mathbf{I}_1 = \mathbf{I}_2 \quad (as\ \mathbf{I}_Y + \mathbf{I}_B = 0)$$

Also,

$$\mathbf{V}_R = \mathbf{E} - \mathbf{I}_1\mathbf{Z}_1 - \mathbf{I}_2\mathbf{Z}_2 - \mathbf{I}_0\mathbf{Z}_0 = 0$$

Eliminating $\mathbf{I}_0$ and $\mathbf{I}_2$, we obtain

$$\mathbf{E} - \mathbf{I}_1(\mathbf{Z}_1 + \mathbf{Z}_2 + \mathbf{Z}_0) = 0$$

hence

$$\mathbf{I}_1 = \frac{\mathbf{E}}{\mathbf{Z}_1 + \mathbf{Z}_2 + \mathbf{Z}_0} \tag{7.5}$$

The fault current,

$$\mathbf{I}_f = \mathbf{I}_R = 3\mathbf{I}_1$$

So

$$\mathbf{I}_f = \frac{3\mathbf{E}}{\mathbf{Z}_1 + \mathbf{Z}_2 + \mathbf{Z}_0} \tag{7.6}$$

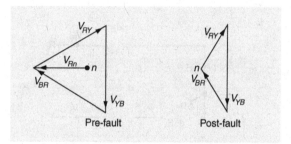

Figure 7.12 Pre- and post-fault phasor diagrams-single-line-to-earth fault

The e.m.f. of the Y phase $= a^2E$, and (from equation (7.4))

$$I_Y = I_0 + a^2I_1 + aI_2$$

$$V_Y = a^2E - I_0Z_0 - a^2I_1Z_1 - aI_2Z_2$$

The pre-fault and post-fault phasor diagrams are shown in Figure 7.12, where it should be noted that only V_{YB} remains at its pre-fault value.

It is usual to form an equivalent circuit to represent equation (7.5) and this can be obtained from an inspection of the equations. The circuit is shown in Figure 7.13 and it will be seen that $I_1 = I_2 = I_0$, and

$$I_1 = \frac{E}{Z_1 + Z_2 + Z_0}$$

7.5.2 Line-To-Line Fault

In Figure 7.14, $E =$ e.m.f. per phase and the R phase is again taken as the reference phasor. In this case, $I_R = 0$, $I_Y = -I_B$, and $V_Y = V_B$.

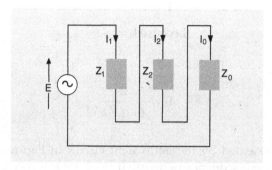

Figure 7.13 Interconnection of positive-, negative-, and zero-sequence networks for single-line-to-earth faults

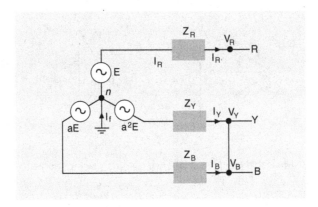

Figure 7.14 Line-to-line fault on phases Y and B

From equation (7.4),

$$\mathbf{I}_0 = 0$$

$$\mathbf{I}_1 = \frac{1}{3}\mathbf{I}_Y(\mathbf{a} - \mathbf{a}^2)$$

and

$$\mathbf{I}_2 = \frac{1}{3}\mathbf{I}_Y(\mathbf{a}^2 - \mathbf{a})$$

$$\therefore \mathbf{I}_1 = -\mathbf{I}_2$$

As $\mathbf{V}_Y = \mathbf{V}_B$ (but not equal to zero)

$$\mathbf{a}^2\mathbf{E} - \mathbf{a}^2\mathbf{I}_1\mathbf{Z}_1 - \mathbf{a}\mathbf{I}_2\mathbf{Z}_2 = \mathbf{a}\mathbf{E} - \mathbf{a}\mathbf{I}_1\mathbf{Z}_1 - \mathbf{a}^2\mathbf{I}_2\mathbf{Z}_2$$

$$\therefore \mathbf{E}(\mathbf{a}^2 - \mathbf{a}) = \mathbf{I}_1\left[\mathbf{Z}_1(\mathbf{a}^2 - \mathbf{a}) + \mathbf{Z}_2(\mathbf{a}^2 - \mathbf{a})\right]$$

$$\text{because } \mathbf{I}_1 = -\mathbf{I}_2$$

hence

$$\mathbf{I}_1 = \frac{\mathbf{E}}{\mathbf{Z}_1 + \mathbf{Z}_2} \tag{7.7}$$

This can be represented by the equivalent circuit in Figure 7.15, in which, of course, there is no zero-sequence network. If the connection between the two lines has an impedance $\mathbf{Z}_f$ (the fault impedance), this is connected in series in the equivalent circuit.

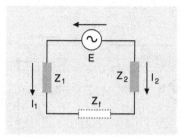

Figure 7.15 Interconnection of sequence networks for a line-to-line fault (including fault impedance Z_f, if present)

7.5.3 Line-To-Line-To-Earth Fault (Figure 7.16)

$$\mathbf{I}_R = 0 \quad \mathbf{V}_Y = \mathbf{V}_B = 0$$

and

$$\mathbf{I}_R = \mathbf{I}_1 + \mathbf{I}_2 + \mathbf{I}_0 = 0$$

$$a^2\mathbf{E} - a^2\mathbf{I}_1\mathbf{Z}_1 - a\mathbf{I}_2\mathbf{Z}_2 - \mathbf{I}_0\mathbf{Z}_0 = \mathbf{V}_Y = 0$$

and

$$a\mathbf{E} - a\mathbf{I}_1\mathbf{Z}_1 - a^2\mathbf{I}_2\mathbf{Z}_2 - \mathbf{I}_0\mathbf{Z}_0 = \mathbf{V}_B = 0$$

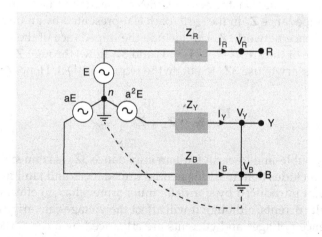

Figure 7.16 Line-to-line-to-earth fault

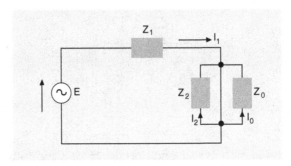

Figure 7.17 Interconnection of sequence networks–double line-to-earth fault

Hence,

$$I_1 = \frac{E}{Z_1 + [Z_2 Z_0 / (Z_2 + Z_0)]} \tag{7.8}$$

$$I_2 = -I_1 \frac{Z_0}{Z_2 + Z_0} \tag{7.9}$$

and

$$I_0 = -I_1 \frac{Z_2}{Z_2 + Z_0} \tag{7.10}$$

These can be represented by the equivalent circuit as shown in Figure 7.17.

The inclusion of impedances in the earth path, such as the star-point-to earth connection in a generator or transformer, modifies the sequence diagrams. For a line-to-earth fault an impedance Z_g in the earth path is represented by an impedance of $3Z_g$ in the zero-sequence network. Z_g can include the impedance of the fault itself, usually the resistance of the arc. As $I_1 = I_2 = I_0$ and $3I_1$ flows through Z_g in the physical system, it is necessary to use $3Z_g$ to obtain the required effect. Hence,

$$I_f = \frac{3E}{Z_1 + Z_2 + Z_0 + 3Z_g}$$

Again, for a double-line-to-earth fault an impedance $3Z_g$ is connected as shown in Figure 7.18. Z_g includes both machine neutral impedances and fault impedances.

The phase shift introduced by star-delta transformers has no effect on the magnitude of the fault currents, although it will affect the voltages at various points. If the positive-sequence voltages and currents are advanced by a certain angle then the negative-sequence quantities are retarded by the same angle for a given connection.

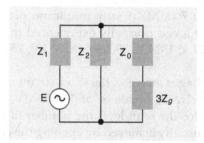

Figure 7.18 Modification of network in Figure 7.17 to account for neutral impedance **Z**g

7.6 Fault Levels in a Typical System

In Figure 7.19, a section of a typical system is shown. At each voltage level the fault level can be ascertained from the reactances given. It should be noted that the short-circuit level will change with network conditions, and there will normally be two extreme values: that with all plant connected and that with the minimum plant normally connected. The short-circuit MVA at 275 kV busbars in Britain is normally

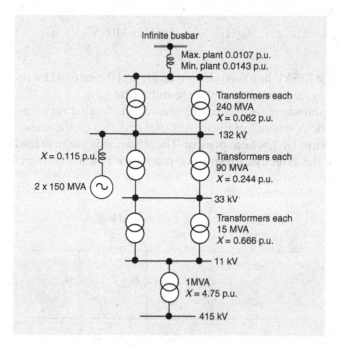

Figure 7.19 Typical transmission system. All reactances on a 100 MVA base

10 000 MVA, but drops to 7000 MVA with minimum plant connected. Maximum short-circuit (three-phase) levels normally experienced in the British system are as follows: 275 kV, 15 000 MVA; 132 kV, 3500 MVA; 33 kV, 750/1000 MVA; 11 kV, 150/250 MVA; 415 V, 30 MVA.

As the transmission voltages increase, the short-circuit currents also increase, and for the 400 kV system, circuit breakers of 35 000 MVA breaking capacity are required. In order to reduce the fault level the number of parallel paths is reduced by sectionalizing. This is usually achieved by opening the circuit breaker connecting two sections of a substation or generating station busbar. One great advantage of direct-current transmission links in parallel with the alternating-current system is that no increase in the short-circuit currents results.

7.6.1 Circuit Parameters with Faults

7.6.1.1 Fault Resistance

The resistance of the fault is normally that of the arc and may be approximated by

$$R_a(\Omega) = \frac{44\, V}{I_f} \text{ for } V < 110\,\text{kV and } V \text{ in kV and } I_f \text{ in A}$$

and

$$R_a(\Omega) = \frac{22\, V}{I_f} \text{ for } V > 110\,\text{kV}$$

For example, a 735 kV line would have an arc resistance of 4 Ω with a fault current of 4 kA, assuming no resistance in the ground return path.

The overall grounding resistance depends on the footing resistance (resistance of tower metalwork to ground) of the towers (R_T) and also on the resistance per section of the ground wires (R_S), where present. The situation is summarized in Figure 7.20, where usually the effective grounding resistance is smaller than the individual

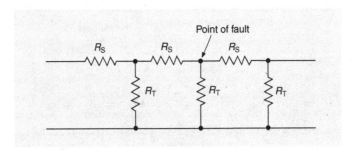

Figure 7.20 Equivalent network of towers with footing resistance (R_T) and ground wire sections (resistance R_S)

tower footing résistance. There is normally a spread in the values of R_T but normally R_T should not exceed $10\,\Omega$ (ground wires present).

Typical values of fault resistance with fault location are as follows: at source, $0\,\Omega$; on line with ground wires, $15\,\Omega$; on line without ground wires, $50\,\Omega$.

7.6.1.2 X/R Ratio

The range of X/R values for typical voltage class (Canadian) are as follows: 735 kV, 18.9–20.4; 500 kV, 13.6–16.5; 220 kV, 2–25; 110 kV, 3–26. The X/R value decreases with separation of the fault point from the source and can be substantially decreased by the fault resistance.

For distribution circuits, X/R is lower, and, although data is limited, typical values are 10 (at source point) and 2–4 on a medium voltage overhead line.

Example 7.4

A synchronous machine A generating 1 p.u. voltage is connected through a star-star transformer, reactance 0.12 p.u., to two lines in parallel. The other ends of the lines are connected through a star-star transformer of reactance 0.1 p.u. to a second machine B, also generating 1 p.u. voltage. For both transformers, $X_1 = X_2 = X_0$.

Calculate the current fed into a double-line-to-earth fault on the line-side terminals of the transformer fed from A.

The relevant per unit reactances of the plant, all referred to the same base, are as follows:

For each line: $X_1 = X_2 = 0.30$, $X_0 = 0.70$.
For generators:

	X_1	X_2	X_0
Machine A	0.30	0.20	0.05
Machine B	0.25	0.15	0.03

The star points of machine A and of the two transformers are solidly earthed.

Solution
The positive-, negative-, and zero-sequence networks are shown in Figure 7.21. All per unit reactances are on the same base. From these diagrams the following equivalent reactances up to the point of the fault are obtained: $Z_1 = j0.23$ p.u. $Z_2 = j0.18$ and $Z_0 = j0.17$ p.u.

The red phase is taken as reference phasor and the blue and yellow phases are assumed to be shorted at the fault point. From the equivalent circuit for a line-to-line fault,

$$I_1 = \frac{E}{Z_1 + [Z_2 Z_0 / (Z_2 + Z_0)]}$$

$$= \frac{1}{j[0.23 + [0.17 \times 0.18 / (0.17 + 0.18)]]} = -j3.15 \text{ p.u.}$$

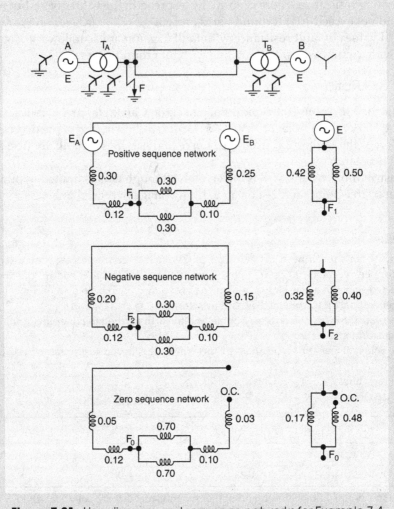

Figure 7.21 Line diagram and sequence networks for Example 7.4

$$\mathbf{I}_2 = -\mathbf{I}_1 \frac{\mathbf{Z}_0}{\mathbf{Z}_2 + \mathbf{Z}_0}$$

$$= -j3.15 \times \frac{0.17}{0.17 + 0.18} = -j1.53 \text{ p.u.}$$

$$\mathbf{I}_0 = -\mathbf{I}_1 \frac{\mathbf{Z}_2}{\mathbf{Z}_2 + \mathbf{Z}_0}$$

$$= -j3.15 \times \frac{0.18}{0.17 + 0.18} = -j1.62 \text{ p.u.}$$

$$\mathbf{I}_Y = \mathbf{I}_0 + \mathbf{a}^2\mathbf{I}_1 + \mathbf{a}\mathbf{I}_2$$

$$= j1.62 + (-0.5 - j0.866)(-j3.15) + (-0.5 + j0.866)(j1.53)$$

$$= -4.05 + j2.43 \text{ p.u.}$$

$$I_B = I_0 + aI_1 + a^2I_2$$
$$= j1.62 + (-0.5 + j0.866)(-j3.15) + (-0.5 - j0.866)(j1.53)$$
$$= 4.05 + j2.43 \text{ p.u.}$$

$$I_Y = I_B = 4.72 \text{ p.u.}$$

The correctness of the first part of the solution can be checked as

$$I_R = I_0 + I_1 + I_2$$
$$= j1.62 - j3.15 + j1.53 = 0$$

Example 7.5

An 11 kV synchronous generator is connected to a 11/66 kV transformer which feeds a 66/11/3.3 kV three-winding transformer through a short feeder of negligible imped-ance. Calculate the fault current when a single-phase-to-earth fault occurs on a termi-nal of the 11 kV winding of the three-winding transformer. The relevant data for the system are as follows:

Generator: $X_1 = j0.15$ p.u., $X_2 = j0.10$ p.u., $X_0 = j0.03$ p.u., all on a 10 MVA base; star point of winding earthed through a 3 Ω resistor.

11/66kV Transformer: $X_1 = X_2 = X_0 = j0.1$ p.u. on a 10 MVA base; 11 kV winding delta connected and the 66 kV winding star connected with the star point solidly earthed.

Three-winding transformer: A 66 kV winding, star connected, star point solidly earthed; 11 kV winding, star connected, star-point earthed through a 3 Ω resistor; 3.3 kV winding, delta connected; the three windings of an equivalent star connection to repre-sent the transformer have sequence impedances,

66 kV winding $X_1 = X_2 = X_0 = j0.04$ p.u.,
11 kV winding $X_1 = X_2 = X_0 = j0.03$ p.u.,
3.3 kV winding $X_1 = X_2 = X_0 = j0.05$ p.u.,

all on a 10 MVA base. Resistance may be neglected throughout.

Solution:
The line diagram and the corresponding positive-, negative-, and zero-sequence networks are shown in Figure 7.22. A 10 MVA base will be used. The 3 Ω earthing resistor has the following p.u. value:

$$\frac{3 \times 10 \times 10^6}{(11)^2 \times 10^6} \text{ or } 0.25 \text{ p.u.}$$

Much care is needed with the zero-sequence network owing to the transformer con-nections. For a line-to-earth fault, the equivalent circuit shown in Figure 7.13 is used, from which

$$I_1 = I_2 = I_0 \quad \text{and} \quad I_f = I_1 + I_2 + I_0$$

Hence

$$\mathbf{I}_f = \frac{3 \times 1}{\mathbf{Z}_1 + \mathbf{Z}_2 + \mathbf{Z}_0 + 3\mathbf{Z}_g}$$

$$= \frac{3}{j0.32 + j0.27 + j0.075 + 0.75}$$

$$= \frac{3}{0.75 + j0.66} = \frac{3}{1\angle 41°} \text{ p.u.}$$

$$\mathbf{I}_f = \frac{3 \times 10 \times 10^6}{\sqrt{3} \times 11000} = 1575 \text{ A}$$

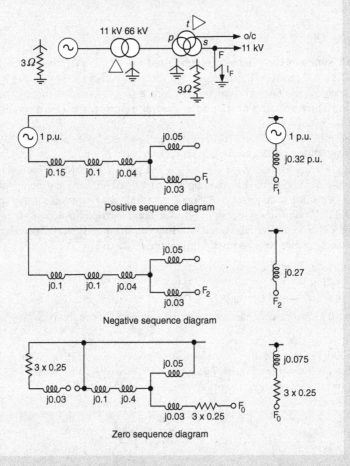

Figure 7.22 Line diagram and sequence networks for Example 7.5

7.7 Power in Symmetrical Components

The total power in a three-phase network

$$\mathbf{V}_a\mathbf{I}_a^* + \mathbf{V}_b\mathbf{I}_b^* + \mathbf{V}_c\mathbf{I}_c^*$$

where $\mathbf{V}_a$, $\mathbf{V}_b$, and $\mathbf{V}_c$ are phase voltages and $\mathbf{I}_a$, $\mathbf{I}_b$, and $\mathbf{I}_c$ are line currents. In phase (a),

$$P_a + jQ_a = (\mathbf{V}_{a0} + \mathbf{V}_{a1} + \mathbf{V}_{a2})(\mathbf{I}_{ao}^* + \mathbf{I}_{a1}^* + \mathbf{I}_{a2}^*)$$

$$= (\mathbf{V}_{a0}\mathbf{I}_{a0}^* + \mathbf{V}_{a1}\mathbf{I}_{a1}^* + \mathbf{V}_{a2}\mathbf{I}_{a2}^*)$$

$$+ (\mathbf{V}_{a0}\mathbf{I}_{a1}^* + \mathbf{V}_{a1}\mathbf{I}_{a2}^* + \mathbf{V}_{a2}\mathbf{I}_{a0}^*)$$

$$+ (\mathbf{V}_{a0}\mathbf{I}_{a2}^* + \mathbf{V}_{a1}\mathbf{I}_{a0}^* + \mathbf{V}_{a2}\mathbf{I}_{a0}^*)$$

with similar expressions for phases (b) and (c).

In extending this to cover the total three-phase power it should be noted that

$$\mathbf{I}_{b1}^* = (\mathbf{a}^2\mathbf{I}_{a1})^* = (\mathbf{a}^2)^*\mathbf{I}_{a1}^* = \mathbf{a}\mathbf{I}_{a1}^*$$

Similarly, $\mathbf{I}_{b2}^*$, $\mathbf{I}_{c1}^*$, and $\mathbf{I}_{c2}^*$ may be replaced.

The total power $= 3(\mathbf{V}_{a0}\mathbf{I}_{a0}^* + \mathbf{V}_{a1}\mathbf{I}_{a1}^* + \mathbf{V}_{a2}\mathbf{I}_{a2}^*)$ that is $3 \times$ (the sum of the individual sequence powers in any phase).

7.8 Systematic Methods for Fault Analysis in Large Networks

The methods described so far become unwieldy when applied to large networks and a systematic approach utilizing digital computers is used. The generators are represented by their voltages behind the transient reactance, and normally the system is assumed to be on no-load before the occurrence of the three-phase balanced fault. The voltage sources and transient reactances are converted into current sources and the admittance matrix is formed (including the transient reactance admittances). The basic equation $[\mathbf{Y}][\mathbf{V}] = [\mathbf{I}]$ is formed and solved with the constraint that the voltage at the fault node is zero. It is preferable, on grounds of storage and time, not to invert the matrix but to use Gaussian elimination methods. The computation efficiency may also be improved by utilizing the sparsity of the $\mathbf{Y}$ matrix. The mesh or loop (impedance matrix) method may be used, although the matrix is not so easily formed.

The following example illustrates a method suitable for determination of balanced three-phase fault currents in a large system by means of a digital computer.

Example 7.6

Determine the fault current in the system shown in Figure 7.23(a) for the balanced fault shown.

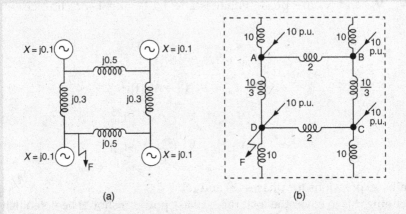

(a)   (b)

Figure 7.23 (a) Circuit for Example 7.6 and (b) equivalent circuit. All values are admittances (i.e. $-j$ Y). The generators, that is 1 p.u. voltage behind $-j10$ p.u. admittance, transform to $-j10$ p.u. current sources in parallel with $-j10$ p.u. admittance

Solution

The system in Figure 7.23(a) is replaced by the equivalent circuit shown in Figure 7.23(b) by converting a voltage source in series with transient reactance to a current source in parallel with the same reactance. The nodal admittance matrix is then formed. Finally, equation $[\mathbf{Y}][\mathbf{V}] = [\mathbf{I}]$ is formed with $\mathbf{V}_D = 0.0$ p.u.

$$
j\begin{bmatrix} 15.33 & -2.00 & 0 & -3.33 \\ -2.00 & 15.33 & -3.33 & 0 \\ 0 & -3.33 & 15.33 & -2.00 \\ -3.33 & 0 & -2.00 & 15.33 \end{bmatrix}\begin{bmatrix} V_A \\ V_B \\ V_C \\ 0 \end{bmatrix} = \begin{bmatrix} j10 \\ j10 \\ j10 \\ j10 - I_F \end{bmatrix}
$$

If the bottom row is eliminated and every element is multiplied by 1.5, we get:

$$
\begin{bmatrix} 23 & -3 & 0 \\ -3 & 23 & -5 \\ 0 & -5 & 23 \end{bmatrix}\begin{bmatrix} V_A \\ V_B \\ V_C \end{bmatrix} = \begin{bmatrix} 15 \\ 15 \\ 15 \end{bmatrix}
$$

from which by Gaussian elimination ($3 \times 1^{\text{st}}$ row $+ 23 \times 2^{\text{nd}}$ row):

$$\begin{bmatrix} 520 & -115 \\ -5 & 23 \end{bmatrix} \begin{bmatrix} V_B \\ V_C \end{bmatrix} = \begin{bmatrix} 390 \\ 15 \end{bmatrix}$$

Thus,

$$V_B = 0.9394\,\text{p.u.} \quad V_C = 0.8564\,\text{p.u.} \quad V_A = 0.7747\,\text{p.u.}$$

And

$$I_F = j(10 + 3.33 \times 0.7747 + 2 \times 0.8564) = j14.2926\,\text{p.u.}$$

Fault MVA $= 14.2926 \times 100 = 1429.26\,\text{MVA}$

Instead of the nodal admittance method, the bus impedance method may be used for computer fault analysis and has the following advantages:

1. Matrix inversion is avoided, resulting in savings in computer storage and time.
2. The matrices for the sequences quantities are determined only once and retained for later use; they are readily modified for system changes.
3. Mutual impedances between lines are readily handled.
4. Subdivisions of the main system may be incorporated.

The system is represented by the usual symmetrical component sequence networks and, frequently, the positive and negative impedances are assumed to be identical. Balanced phase impedances for all items of plant are assumed as are equal voltages for all generators.

In the bus impedance method the network loop matrix, that is $[\mathbf{E}] = [\mathbf{Z}][\mathbf{I}]$, is set up in terms of the various loop currents. First, the buses of interest are short-circuited to the neutral. Consider a fault on one of the buses (k) only, currents in all the other busses short circuited to the neutral will be zero and, from equation $[\mathbf{E}] = [\mathbf{Z}][\mathbf{I}]$, $1.0 = Z_{kk}I_k$, where $I_k =$ fault current with three-phase symmetrical fault on k. Similarly, the currents with balanced faults on each of the other buses may be easily determined.

7.8.1 Computer Simulations

Part of the transmission system shown in Figure 6.12 was used. For fault calculations, normally transient and subtransient reactances and time constants of the generators are required. Table 7.4 gives the data for each generator on the machine base.

Table 7.4 Machine data

Generator	X_s	R	X'	X''	T'	T''
G1 and G2	1.2	0.012	0.35	0.25	1.1	0.035
G3	1.5	0.015	0.27	0.20	1.8	0.035

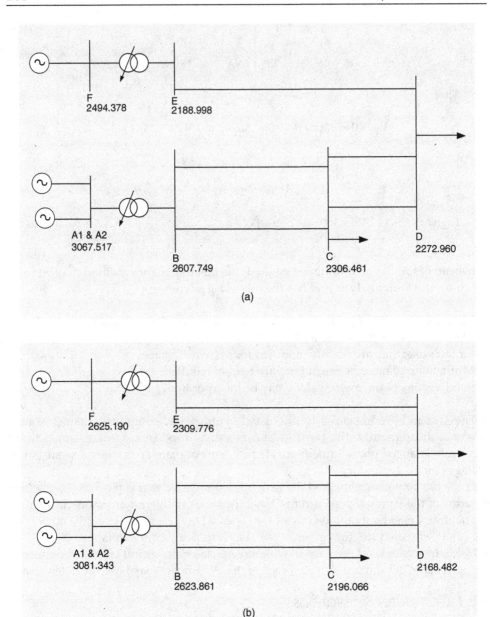

Figure 7.24 (a) Fault level in MVA at each busbar assuming all the busbar voltages are 1 p.u. (b) fault level calculated using actual pre-fault busbar voltages

In fault calculations so far it has been assumed that the busbar voltages are 1 p.u. Figure 7.24(a) shows the fault level of each busbar in MVA of the system shown in Figure 6.12 with 1 p.u. voltage on all the busbars. However pre-fault operating voltages of a network depend on the load flow and the resulting fault currents are

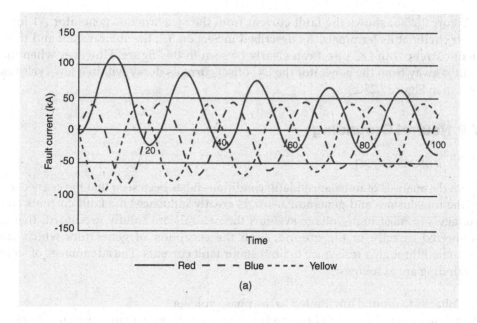

(a)

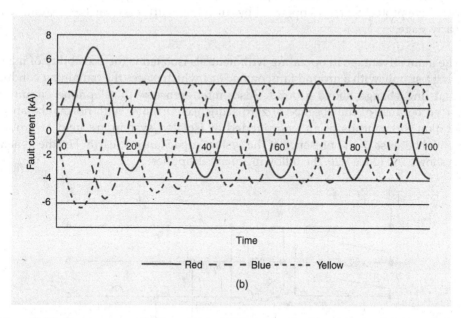

(b)

Figure 7.25 Three phase fault currents (a) for a fault at busbar A1 (b) for a fault at busbar D. Red phase has maximum DC offset

different when actual pre-fault operating voltages are used. Figure 7.24(b) shows the fault levels calculated using the actual busbar voltages.

Figure 7.25(a) shows the fault current from the synchronous generator A1 for a short circuit at its terminals. As described in Section 3.3, the subtransient and transient current with DC offset can clearly be seen in this figure. However when the fault is away from the generator the DC offset currents decay within a few cycles as shown in Figure 7.25(b).

7.9 Neutral Grounding

7.9.1 Introduction

From the analysis of unbalanced fault conditions it has been seen that the connection of the transformer and generator neutrals greatly influences the fault currents and voltages. In most high-voltage systems the neutrals are solidly grounded, that is connected directly to the ground, with the exception of generators which are grounded through a resistance to limit stator fault currents. The advantages of solid grounding are as follows:

1. Voltages to ground are limited to the phase voltage.
2. Intermittent ground faults and high voltages due to arcing faults are eliminated.
3. Sensitive protective relays operated by earth fault currents clear these faults at an early stage.

The main advantage in operating with neutrals isolated is the possibility of maintaining a supply with a ground fault on one line which places the remaining conductors at line voltage above ground. Also, interference with telephone circuits is reduced because of the absence of zero-sequence currents. With normal balanced operation the neutrals of an ungrounded or isolated system are held at ground potential because of the presence of the system capacitance to earth. For the general case shown in Figure 7.26, the following analysis applies:

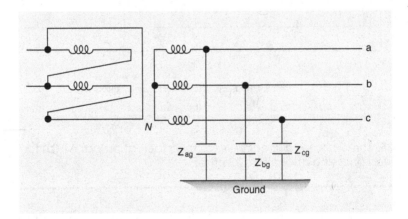

Figure 7.26 Line-to-ground capacitances in an ungrounded system

$$\frac{V_{ag}}{Z_{ag}} + \frac{V_{bg}}{Z_{bg}} + \frac{V_{cg}}{Z_{cg}} = 0 \tag{7.11}$$

Also,

$$V_{ag} = V_{an} + V_{ng}$$

where

V_{an} = voltage of line (a) to neutral
V_{ng} = voltage of neutral to ground

Similarly,

$$V_{bg} = V_{bn} + V_{ng}$$

and

$$V_{cg} = V_{cn} + V_{ng}$$

Substituting in equation (7.11) and separating terms,

$$\frac{V_{an}}{Z_{ag}} + \frac{V_{bn}}{Z_{bg}} + \frac{V_{cn}}{Z_{cg}} + V_{ng}\left(\frac{1}{Z_{ag}} + \frac{1}{Z_{bg}} + \frac{1}{Z_{cg}}\right) = 0 \tag{7.12}$$

The equation

$$\left(\frac{1}{Z_{ag}} + \frac{1}{Z_{bg}} + \frac{1}{Z_{cg}}\right) = Y_g$$

gives the ground capacitance admittance of the system.

7.9.2 Arcing Faults

Consider the single-phase system in Figure 7.27 at the instant when the instantaneous voltages are v on line (a) and $-v$ on line (b), where v is the maximum instantaneous voltage. The sudden occurrence of a fault to ground causes line (b) to assume a potential of $-2v$ and line (a) to become zero. Because of the presence of both L and C in the circuit, the sudden change in voltages by v produces a high-frequency oscillation of peak magnitude $2v$ superimposed on the power frequency voltages (see Chapter 10) and line (a) reaches $-v$ and line (b) $-3v$, as shown in Figure 7.28. These oscillatory voltages attenuate quickly due to the resistance present. The current in the arc to earth on line (a) is approximately 90° ahead of the fundamental voltage, and when it is zero the voltage will be at a maximum. Hence, if the arc extinguishes at the first current zero, the lines remain charged at $-v$ for (a) and $-3v$ for (b). The line potentials now change at power frequency until (a) reaches $-3v$, when the arc could restrike causing a voltage change of $-3v$ to 0, resulting in a transient

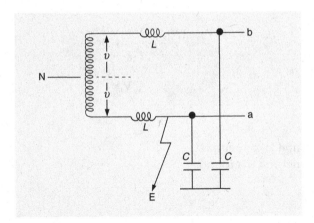

Figure 7.27 Single-phase system with arcing fault to ground

overvoltage of $+3v$ in line (a) and $+5v$ in line (b). This process could continue and the voltage build up further, but the resistance present usually limits the peak voltage to under $4v$. A similar analysis may be made for a three-phase circuit, again showing that serious overvoltages may occur with arcing faults because of the inductance and shunt capacitance of the system.

This condition may be overcome in an isolated neutral system by means of an arc suppression or Petersen coil. The reactance of this coil, which is connected between the neutral and ground, is made in the range 90–110% of the value required to neutralize the capacitance current. The phasor diagram for the network of Figure 7.29(a) is shown in Figure 7.29(b) if the voltage drop across the arc is neglected.

$$\mathbf{I}_a = \mathbf{I}_b = \sqrt{3}\mathbf{V}\omega C$$

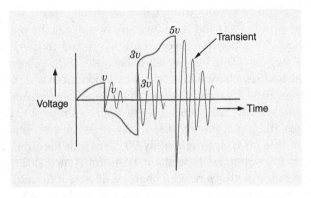

Figure 7.28 Voltage on line (a) of Figure 7.27

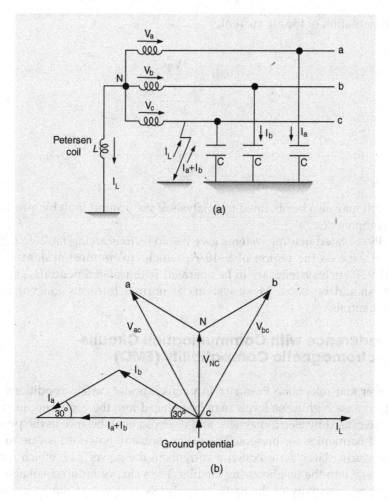

Figure 7.29 (a) System with arc suppression coil, (b) Phasor diagram of voltages and currents in part (a)

and

$$\mathbf{I}_a + \mathbf{I}_b = \sqrt{3} \times \sqrt{3}\mathbf{V}\omega C = 3\mathbf{V}\omega C$$

also,

$$\mathbf{I}_L = \frac{\mathbf{V}}{\omega L}$$

For compensation of the arc current,

$$\frac{\mathbf{V}}{\omega L} = 3\mathbf{V}\omega C$$

$$\therefore L = \frac{1}{3\omega^2 C}$$

And

$$X_L = \frac{1}{3\omega C}$$

This result may also be obtained by analysis of the ground fault by means of symmetrical components.

Generally, isolated neutral systems give rise to serious arcing-fault voltages if the arc current exceeds the region of 5–10 A, which covers most systems operating above 33 kV. If such systems are to be operated with isolated neutrals, arc-suppression coils should be used. Most systems at normal transmission voltages have grounded neutrals.

7.10 Interference with Communication Circuits– Electromagnetic Compatibility (EMC)

When power and telephone lines run in parallel, under certain conditions voltages sufficient to cause high noise levels may be induced into the communication circuits. This may be caused by electromagnetic and electrostatic unbalance in the power lines, especially if harmonics are present. The major problem, however, is due to faults to ground producing large zero-sequence currents in the power line, which inductively induce voltage into the neighbouring circuits. The value of induced voltage depends on the spacing, resistivity of the earth immediately below, and the frequency. If the telephone wires are a twisted pair or are situated close together and transposed, no voltage is induced between the communication conductors. However, a voltage can exist between the pair of wires and ground. These induced longitudinal voltages can be controlled by connecting the communication circuits to ground through an inductance which produces little attenuation at communication frequencies of 400–3500 Hz.

Capacitive coupling can occur if open communication circuits are run along the same route as power lines. Interference from underground cable circuits is much less (10%) than that from overhead lines.

Because of right-of-way constraints, telephone and power distribution lines run parallel along the same street in many urban areas. However, the interference in rural areas is often greater because communication lines and plant may be unshielded or have higher shield resistances, and unlike urban areas there is no extensive network of water and gas pipes to share the ground return currents.

Resistive coupling between power and communication circuits can exist:

- because of physical contact between them;
- via paths through the soil between telephone and power grounds.

Various formulae exist to calculate the value of mutual inductance-in H/m between circuits with earth return. These assume that ε_r (soil) is unity, displacement currents are much less than conduction currents, and the length of conductor is infinite.

During line-to-ground faults, induced voltages into communication circuits may be sufficient to be a shock hazard to personnel. Although in transmission circuits, equal current loading may be assumed in the phases, this is not the case in the lower voltage distribution circuits where significant residual currents may flow. Most telephone circuit standards now require up to 15 kV isolation if communication circuits are to be connected into substations for monitoring, control and communication purposes. Particular care must be taken with bonding the sheath of communication circuits brought into buildings where the power distribution system is also bonded to earth and to the building structure.

Problems

7.1 Four 11 kV generators designated A, B, C, and D each have a subtransient reactance of 0.1 p.u. and a rating of 50 MVA. They are connected in parallel by means of three 100 MVA reactors which join A to B, B to C, and C to D; these reactors have per unit reactances of 0.2, 0.4, and 0.2, respectively. Calculate the volt-amperes and the current flowing into a three-phase symmetrical fault on the terminals of machine B. Use a 50 MVA base.

  (Answer: 937.5 MVA; 49 206 A)

7.2 Two 100 MVA, 20 kV turbogenerators (each of transient reactance 0.2 p.u.) are connected, each through its own 100 MVA, 0.1 p.u. reactance transformer, to a common 132 kV busbar. From this busbar, a 132 kV feeder, 40 km in length, supplies an 11 kV load through a 132/11 kV transformer of 200 MVA rating and reactance 0.1 p.u. If a balanced three-phase short circuit occurs on the low-voltage terminals of the load transformer, determine, using a 100 MVA base, the fault current in the feeder and the rating of a suitable circuit breaker at the load end of the feeder. The feeder impedance per phase is $(0.035 + j0.14)$ Ω/km.

  (Answer: 431 MVA)

7.3 Two 60 MVA generators of transient reactance 0.15 p.u. are connected to a busbar designated A. Two identical machines are connected to another busbar B. Busbar A is connected to busbar B through a reactor, X. A feeder is supplied from A through a step-up transformer rated at 30 MVA with 10% reactance.

Calculate the reactance X, if the fault level due to a three-phase fault on the feeder side of the 30 MVA transformer is to be limited to 240 MVA. Calculate also the voltage on A under this condition if the generator voltage is 13 kV (line).

(Answer: $X = 0.075$ p.u.; $VA = 10.4$ kV)

7.4   A single line-to-earth fault occurs on the red phase at the load end of a 66 kV transmission line. The line is fed via a transformer by 11 kV generators connected to a common busbar. The line side of the transformer is connected in star with the star point earthed and the generator side is in delta. The positive-sequence reactances of the transformer and line are $j10.9 \, \Omega$ and $j44 \, \Omega$, respectively, and the equivalent positive and negative-sequence reactances of the generators, referred to the line voltage, are $j18 \, \Omega$ and $j14.5 \, \Omega$, respectively. Measured up to the fault the total effective zero sequence reactance is $j150 \, \Omega$. Calculate the fault current in the lines if resistance may be neglected. If a two-line-to-earth fault occurs between the blue and yellow lines, calculate the current in the yellow phase.

(Answer: 391 A; 1486 A)

7.5   A single-line-to-earth fault occurs in a radial transmission system. The following sequence impedances exist between the source of supply (an infinite busbar) of voltage 1 p.u. to the point of the fault: $Z_1 = 0.3 + j0.6$ p.u., $Z_2 = 0.3 + j0.55$ p.u., $Z_0 = 1 + j0.78$ p.u. The fault path to earth has a resistance of 0.66 p.u. Determine the fault current and the voltage at the point of the fault.

(Answer: $I_f = 0.74$ p.u.; $V_f = (0.43 - j0.23)$ p.u.)

7.6   Develop an expression, in terms of the generated e.m.f. and the sequence impedances, for the fault current when an earth fault occurs on phase (A) of a three-phase generator, with an earthed star point. Show also that the voltage to earth of the sound phase (B) at the point of fault is given by

$$V_B = \frac{-j\sqrt{3}E_A[Z_2 - aZ_0]}{Z_1 + Z_2 + Z_0}$$

Two 30 MVA, 6.6 kV synchronous generators are connected in parallel and supply a 6.6 kV feeder. One generator has its star point earthed through a resistor of 0.4 Ω and the other has its star point isolated. Determine: (a) the fault current and the power dissipated in the earthing resistor when an earth fault occurs at the far end of the feeder on phase (A); and (b) the voltage to earth of phase (B). The generator phase sequence is ABC and the impedances are as follows:

	Generator p.u.	Feeder Ω/p.h.
To positive-sequence currents	$j0.2$	$j0.6$

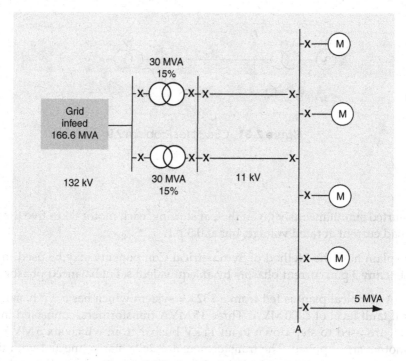

Figure 7.30 System for Problem 7.7

To negative-sequence currents	*j*0.16	*J*0.6
To zero-sequence currents	*j*0.06	*j*0.4

Use a base of 30 MVA.

(Answer: (a) 5000∠ − 58.45° A; 10 MW; (b) −2627 − *j*1351 V

7.7   An industrial distribution system is shown schematically in Figure 7.30. Each line has a reactance of *j*0.4 p.u. calculated on a 100 MVA base; other system parameters are given in the diagram. Choose suitable short-circuit ratings for the oil circuit breakers, situated at substation A, from those commercially available, which are given in the table below.

Short circuit (MVA)	75	150	250	350
Rated current (A)	500	800	1500	2000

The industrial load consists of a static component of 5 MVA and four large induction motors each rated at 6 MVA. Show that only three motors can be

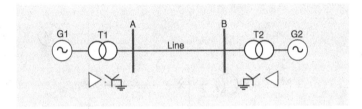

Figure 7.31 Circuit for Problem 7.9

started simultaneously given that, at starting, each motor takes five times full-load current at rated voltage, but at 0.3 p.f.

7.8 Explain how the Method of Symmetrical Components may be used to represent any 3 p.h. current phasors by an equivalent set of balanced phasors.

A chemical plant is fed from a 132 kV system which has a 3 p.h. symmetrical fault level of 4000 MVA. Three 15 MVA transformers, connected in parallel, are used to step down to an 11 kV busbar from which six 5 MVA, 11 kV motors are supplied. The transformers are delta-star connected with the star point of each 11 kV winding, solidly earthed. The transformers each have a reactance of 10% on rating and it may be assumed that $X_1 = X_2 = X_0$. The initial fault contribution of the motors is equal to five times rated current with 1.0 p.u. terminal voltage.
Using a base of 100 MVA,

a. calculate the fault current (in A) for a line-to-earth short circuit on the 11 kV busbar with no motors connected;
b. calculate the 3 p.h. symmetrical fault level (in MVA) at the 11 kV busbar if all the motors are operating and the 11 kV busbar voltage is 1.0 p.u.

(Answer: (a) 22 kA; (b) 555 MVA)

(From Engineering Council Examination, 1996)

7.9 Describe the effect on the output current of a synchronous generator following a solid three-phase fault on its terminals.

For the system shown in Figure 7.31 calculate (using symmetrical components):

a. the current flowing in the fault for a three-phase fault at busbar A;
b. the current flowing in the fault for a one-phase-to-earth fault at busbar B;
c. the current flowing in the faulted phase of the overhead line for a one-phase-to earth fault at busbar B.

Generators G1 and G2: $X_1'' = X_2'' = j0.1$ p.u.; 11 kV
Transformers T1 and T2: $X_1 = X_2 = X_0 = j0.1$ p.u.; 11/275 kV

(Earthed star-delta)

Line: $Z_1 = Z_2 = j0.05$ p.u., $Z_0 = j0.1$ p.u.; 275 kV

(All p.u. values are quoted on a base of 100 MVA)

Assume the pre-fault voltage of each generator is 1 p.u. and calculate the symmetrical fault currents (in amps) immediately after each fault occurs.

(Answer: (a) 1.89 kA; (b) 2.18 kA; (c) 0.89 kA)

(From Engineering Council Examination, 1997)

7.10 Why is it necessary to calculate short-circuit currents in large electrical systems?

A generator rated at 400 MW, 0.8 power factor, 20 kV has a star-connected stator winding which is earthed at its star point through a resistor of 1 Ω. The generator reactances, in per unit on rating, are:

$$X_1 = 0.2 \quad X_2 = 0.16 \quad X_0 = 0.14$$

The generator feeds a delta-star-connected generator transformer rated at 550 MVA which steps the voltage up to a 275 kV busbar. The transformer star-point is solidly earthed and the transformer reactance is 0.15 p.u. on its rating. The 275 kV busbar is connected only to the transformer. Assume that for the transformer $X_1 = X_2 = X_0$.

Using a base of 500 MVA calculate the base current and impedance of each voltage level.

Calculate the fault current in amperes for:

a. a 275 kV busbar three-line fault;
b. a 275 kV single-line-to earth fault on the busbar;
c. a 20 kV three-line fault on the generator terminals;
d. a 20 kV single-line-to-earth fault on the generator terminals.

(Answer: (a) 3.125 kA; (b) 4.1 kA; (c) 72.15 kA; (d) 11.44 kA)

(From Engineering Council Examination, 1995)

8

System Stability

8.1 Introduction

The stability of a power system is its ability to remain in an operating equilibrium when subjected to the disturbances that are inevitable in any network made up of generating plant supplying loads. The disturbance may be small (e.g. caused by changes in load) or large (e.g. due to fault). After the disturbance a stable system returns to a condition of acceptable voltages and power flows throughout the system.

Figure 8.1 shows how power system stability may be divided into several aspects in order to make the problem easier to address. These include the loss of synchronism between synchronous generators (angle stability) either due to faults and large disturbances (transient stability) or oscillations caused by changes in load and lack of damping (dynamic or small-signal stability). Voltage instability can be caused by large induction motor loads drawing excessive amounts of reactive power during network faults when the network voltage is depressed or by a lack of reactive power when excessive real power flows in a network.

When the rotor of a synchronous generator advances beyond a certain critical angle, the magnetic coupling between the rotor and the stator fails. The rotor, no longer held in synchronism with the rotating field of the stator currents, rotates relative to the field and pole slipping occurs. Each time the poles traverse the angular region within which it could be stable, synchronizing forces attempt to pull the rotor into synchronism. It is usual practice to disconnect the generator from the system if it commences to slip poles, as pole slipping causes large current to flow and high transient torques.

Synchronous, or angle, stability may be divided into: dynamic and transient stability. Dynamic stability is the ability of synchronous generators, when operating under given load conditions, to retain synchronism (without excessive angular oscillations) when subject to small disturbances, such as the continual changes in load or generation and the switching out of lines. It is most likely to result from the changes

Electric Power Systems, Fifth Edition. B.M. Weedy, B.J. Cory, N. Jenkins, J.B. Ekanayake and G. Strbac.
© 2012 John Wiley & Sons, Ltd. Published 2012 by John Wiley & Sons, Ltd.

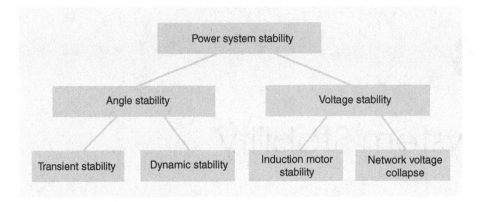

Figure 8.1 Main forms of power system stability

in source-to-load impedance resulting from changes in the network configuration or system state and is a consequence of lack of damping in the power system. Often, this is referred to as small-signal stability.

Transient stability is concerned with sudden and large changes in the network condition, such as those brought about by short-circuit faults. The maximum power that can be transmitted, the stability limit, taking into account fault conditions is usually less than the maximum stable steady-state load.

The stability of an asynchronous motor load is controlled by the voltage across it; if this becomes lower than a critical value, induction motors may become unstable and stall. This is, in effect, a voltage instability problem. When the voltage at the terminals of an asynchronous (induction) motor drops, perhaps due to a fault on the power system, its ability to develop torque is reduced and the motor slows down. If the fault on the network persists the motor stalls and draws very large amounts of reactive power. These reactive power flows depress the voltage at the motor terminals further and the section of network has to be isolated. This form of voltage instability is a well known hazard in oil refineries and chemical plants that have large induction motor loads. Voltage instability can also occur in large national power systems when the loading of transmission lines exceeds the stable (approximately horizontal) section of the P-V curve, see Figure 5.21. Once the load of the transmission line approaches the 'nose' of the P-V curve instability can occur.

In a power system it is possible for either angle or voltage instability to occur, and in practice one form of instability can influence the other. Angle or synchronous stability has traditionally been considered more onerous and hence has been given more attention in the past. Recently, with the increasing use of static VAr compensators and the experience of large national black-outs caused by a deficit of reactive power, the study of voltage collapse has become important.

8.2 Equation of Motion of a Rotating Machine

The kinetic energy of a rotating mass, such as a large synchronous generator, is

$$KE = \frac{1}{2}I\omega^2 \quad [J]$$

and the angular momentum is

$$M = I\omega \quad [Js/rad]$$

where ω is the synchronous speed of the rotor (rad/s) and I is the moment of inertia (kgm^2).

The inertia constant (H) is defined as the stored energy at synchronous speed divided by the rating of the machine G expressed in volt-amperes (VA). Hence

$$H = \frac{1}{2}\frac{I\omega^2}{G} \quad [Ws/VA] \tag{8.1}$$

and the stored kinetic energy is

$$KE = GH = \frac{1}{2}I\omega^2 = \frac{1}{2}M\omega \quad [J]$$

The angular velocity can be expressed in electrical degrees per second, ω_e, as

$$\omega_e = 360f \quad [\text{electrical degrees per second}]$$

where f is the system frequency in Hz.
Then

$$GH = \frac{1}{2}M(360f) \quad [J]$$

with

$$M = \frac{GH}{180f} \quad [Js/\text{electrical degree}]$$

The inertia constants (H) of generating sets tend to similar values. The inertia constant for steam or gas turbo-generator units decreases from $10\,Ws/VA$ (generator and turbine together) for machines up to 30 MVA to values in the order of $4\,Ws/VA$ for large machines, the value decreasing as the rating increases. For salient-pole hydro-electric units, H depends on the number of poles; for machines operating in the speed range 200–400 r.p.m. the value increases from about $2\,Ws/VA$ at 10 MVA rating to 3.5 Ws/VA at 60 MVA. A mean value for synchronous motors is 2 Ws/VA. Large wind turbines have an inertia constant in the range 2–5 Ws/VA.

The net accelerating torque on the rotor of a machine is

$$\Delta T = \text{mechanical torque input - electrical torque output}$$

$$= I\frac{d^2\delta}{dt^2}$$

$$\therefore \frac{d^2\delta}{dt^2} = \frac{\Delta T}{I} = \frac{(\Delta T\omega)\omega}{2\dfrac{I\omega^2}{2}} = \frac{\Delta P\omega}{2KE}$$

where ΔP = net power corresponding to ΔT,

$$\Delta P = P_{mech} - P_{elec}$$

$$\frac{d^2\delta}{dt^2} = \frac{\Delta P}{M} \tag{8.2}$$

In equation (8.2) a reduction in electrical power output results in an increase in δ.

Sometimes, the mechanical power input is assumed to be constant and the equation of motion becomes

$$\frac{d^2\delta}{dt^2} = -\frac{\Delta P}{M}$$

or

$$M\frac{d^2\delta}{dt^2} + \Delta P = 0$$

8.3 Steady-State Stability

The steady-state stability limit is the maximum power that can be transmitted in a network between sources and loads when the system is subject to small disturbances. The power system is, of course, constantly subjected to small changes as load variations occur. To calculate the limiting value of power, small increments of load are added to a model of the system; after each increment the generator excitations are adjusted to maintain constant terminal voltages and a load flow is carried out. Eventually, a condition of instability is reached.

The steady-state stability limits of synchronous machines have been discussed in Chapter 3 (Section 3.5). It was seen that provided the generator is connected to a busbar with a high fault level and operates within the 'safe area' of the performance chart (Figure 3.12), stability is assured; usually, a 20% margin of safety on the leading power factor, absorbing VArs side is allowed and the limit is extended by the use of automatic voltage regulators. Often, the performance charts are not used directly and the generator-equivalent circuit employing the synchronous impedance is used. The normal maximum operating load angle for modern machines is in the order of 60 electrical degrees, and for the limiting value of 90° this leaves 30° to cover the transmission network. In a complex system a reference point must be taken from which the load angles are measured; this is usually a point where the direction of power flow reverses.

The simplest criterion for steady-state synchronous stability is $\frac{\partial P}{\partial \delta} > 0$, that is the synchronizing coefficient must be positive. The use of this criterion involves the following assumptions:

1. generators are represented by constant impedances in series with their internal voltages;
2. the input torques from the turbines are constant;
3. changes in speed (frequency) are ignored;
4. electromagnetic damping in the generators is ignored;
5. the changes in load angle δ are small.

The degree of complexity to which the analysis is taken has to be decided, for example the effects of machine inertia, governor action and automatic voltage regulators can be included; these items, however, greatly increase the complexity of the calculations. The use of the criterion $\frac{\partial P}{\partial \delta} = 0$ alone gives a pessimistic or low result and hence an inbuilt factor of safety.

In a system with several generators and loads, the question as to where the increment of load is to be applied is important. A conservative method is to assume that the increment applies to one machine only, determine the stability, and then repeat for each of the other machines in turn. Alternatively, the power outputs from all but the two generators having the largest load angles are kept constant.

For calculations made without the aid of computers it is usual to reduce the network to the simplest form that will keep intact the generator nodes. The values of load angle, power and voltage are then calculated for the given conditions, $\frac{\partial P}{\partial \delta}$ determined for each machine, and, if positive, the loading is increased and the process repeated.

In a system consisting of a generator supplying a load through a network of lines and transformers of effective reactance, X_T

$$\frac{\partial P}{\partial \delta} = \frac{EV}{X_T} \cos \delta$$

where E and V are the supply- and receiving-end voltages and δ the total angle between the generator rotor and the phasor of V.

The power transmitted is obviously increased with higher system voltages and lower reactances, and it may be readily shown that adding series capacitance in the line increases the steady-state stability limit, although with an increased risk of instability due to resonance (a phenomenon known as sub-synchronous resonance).

The determination of $\dfrac{\partial P}{\partial \delta}$ is not very difficult if the voltages at the loads can be assumed to be constant or if the loads can be represented by impedances. Use can be made of the P-V, Q-V characteristics of the load if the voltages change appreciably with the redistribution of the power in the network; this process, however, is tedious to undertake manually.

Example 8.1

For the system shown in Figure 8.2, investigate the steady-state stability.

Solution
Using Equation (2.18) (Section 2.5.1) and working from the infinite busbar voltage, the voltage at Point A is given by

$$V_A = \sqrt{\left[\left(V + \frac{QX}{V}\right)^2 + \left(\frac{PX}{V}\right)^2\right]}$$

$$= \sqrt{\left[\left(1 + \frac{0.2 \times 0.5}{1}\right)^2 + \left(\frac{0.5 \times 0.5}{1}\right)^2\right]}$$

$$= 1.105 \, \text{p.u.}$$

at an angle of 5.19° to the infinite busbar.

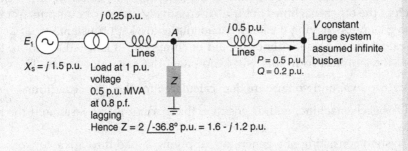

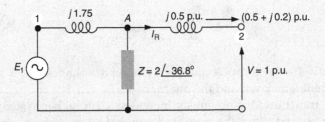

Figure 8.2 Circuit for Example 8.1

The reactive power absorbed by the line from point A to point 2 (the infinite busbar) is

$$I_R^2 X = \left(\frac{P^2 + Q^2}{V^2}\right) X = \left(\frac{0.5^2 + 0.2^2}{1^2}\right) 0.5$$

$$= 0.145 \, \text{p.u}$$

The actual load taken by A (if represented by an impedance) is given by

$$\frac{V_A^2}{Z} = \frac{1.105^2}{2\angle -36.8°} = 0.49 + j0.37 \, \text{p.u}$$

The total load supplied by link from generator to A

$$= (0.5 + 0.49) + j(0.2 + 0.145 + 0.37)$$

$$= 0.99 + j0.715 \, \text{p.u}$$

Internal voltage of generator E_1

$$= \sqrt{\left[\left(1.105 + \frac{0.715 \times 1.75}{1.105}\right)^2 + \left(\frac{0.99 \times 1.75}{1.105}\right)^2\right]}$$

$$= \sqrt{[5.006 + 2.458]} = 2.73 \angle 35.02°$$

Hence, the angle between E_1 and V is

$$35.02 + 5.19 = 40.21°$$

Since this angle is much less than 90°, the system is stable.

(Note that the formula $P = (EV/X) \sin \delta$ is not valid here as the load at A provides a resistance in the equivalent network. If the angle between E, and V by the approximate calculation shown above approaches 90°, a computer based calculation is called for.)

8.4 Transient Stability

Transient stability is concerned with the effect of large disturbances. These are usually due to faults, the most severe of which is the three-phase short circuit which governs the transient stability limits in the UK. Elsewhere, limits are based on other types of fault, notably the single-line-to-earth fault which is by far the most frequent in practice.

When a fault occurs at the terminals of a synchronous generator the real power output of the machine is greatly reduced as the voltage at the point of fault approaches zero and the only load on the machine is that of the inductive circuits of the generator. However, the input power to the generator from the turbine has not time to change during the short period of the fault and the rotor gains speed to store the excess energy. If the fault persists long enough the rotor angle will increase continuously and synchronism will be lost. Hence the time of operation of the protection and circuit breakers is all important.

An aspect of importance is the use of auto-reclosing circuit breakers. These open when the fault is detected and automatically reclose after a prescribed period (usually less than 1 s). If the fault persists the circuit breaker reopens and then recloses as before. This is repeated once more, when, if the fault still persists, the breaker remains open. Owing to the transitory nature of most faults, often the circuit breaker successfully recloses and the rather lengthy process of investigating the fault and restoring the line is avoided. The length of the auto-reclose operation must be considered when assessing transient stability limits; in particular, analysis must include the movement of the rotor over this period and not just the first swing.

If, in equation (8.2), both sides are multiplied by $2\dfrac{d\delta}{dt}$ then

$$\left(2\frac{d\delta}{dt}\right)\frac{d^2\delta}{dt^2} = \frac{d}{dt}\left(\frac{d\delta}{dt}\right)^2 = 2\frac{\Delta P}{M}\left(\frac{d\delta}{dt}\right)$$

$$\therefore \left(\frac{d\delta}{dt}\right)^2 = \frac{2}{M}\int_{\delta_0}^{\delta}\Delta P d\delta$$

(8.3)

If the machine remains stable during a system disturbance, the rotor swings until its angular velocity $\dfrac{d\delta}{dt}$ is zero; if $\dfrac{d\delta}{dt}$ does not become zero the rotor will continue to move and synchronism is lost. The integral of $\int\Delta P d\delta$ in Equation (8.3) represents an area on the P-δ diagram. Hence the criterion for stability is that the area between the P-δ curve and the line representing the power input P_0 must be zero. This is known as the *equal-area criterion*. It should be noted that this is based on the assumption that synchronism is retained or lost on the first swing or oscillation of the rotor, which may not always be the case. Physically, the criterion means that the rotor must be able to return to the system all the energy gained from the turbine during the acceleration period.

A simple example of the equal-area criterion may be seen by an examination of the switching out of one of two parallel lines which connect a generator to an infinite busbar (Figure 8.3). Initially the generator delivers P_0 at an angle δ_0 through both lines. When one line is switched out, the reactance of the circuit and hence the angle δ increases. If stability is retained, the two shaded areas (A_1 and A_2) are equal and the swinging rotor comes initially to rest at angle δ_2, after which the damped oscillation converges to δ_1. In this particular case the initial operating power and angle could be increased to such values that the shaded area between δ_0 and δ_1 (A_1) could be equal to the area between δ_1 and δ_3, where $\delta_3 = 180 - \delta_1$; this would be the condition for maximum input power. If it swings beyond δ_3, the rotor continues to accelerate and instability results.

The power-angle curves for the condition of a fault on one of two parallel lines are shown in Figure 8.4. The fault is cleared when the rotor has swung to δ_1 and the shaded area δ_0 to δ_1 (A_1) indicates the energy stored. The rotor continues to swing until it reaches δ_2 when the two areas A_1 and A_2 are equal. In this particular case P_0 is the maximum operating power for a fault clearance time corresponding to δ_1 and, conversely, δ_1 *is the critical clearing angle* for P_0. If the angle δ_1 is decreased (for

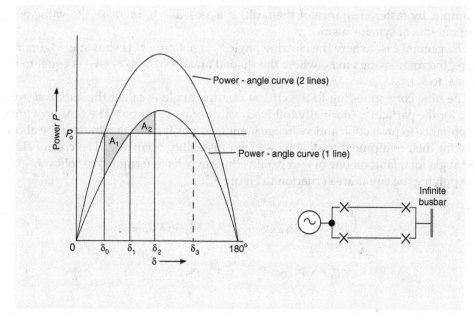

Figure 8.3 Power-angle curves for one line and two lines in parallel. Equal-area criterion. Resistance neglected

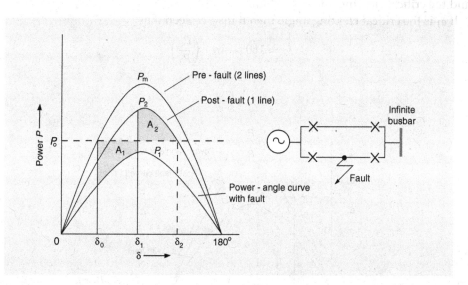

Figure 8.4 Fault on one line of two lines in parallel. Equal-area criterion. Resistance neglected. δ_1 is critical clearing angle for input power P_0

example, by faster clearance of the fault) it is possible to increase the value of P_0 without loss of synchronism.

The general case where the clearing angle δ_1 is not critical is shown in Figure 8.5. Here, the rotor swings to δ_2, where the shaded area from δ_0 to δ_1 (A_1) is equal to the area δ_1 to δ_2 (A_2).

The time corresponding to the critical clearing angle is called the *critical clearing time* for the particular (normally full-load) value of power input. The time is of great importance to protection and switchgear engineers as it is the maximum time allowable for their equipment to operate without instability occurring. The critical clearing angle for a fault on one of two parallel lines may be determined as follows:

Applying the equal-area criterion to Figure 8.5:

$$\int_{\delta_0}^{\delta_1} (P_0 - P_1\sin \delta)d\delta + \int_{\delta_1}^{\delta_2} (P_0 - P_2\sin \delta)d\delta = 0$$

$$\therefore [P_0\delta + P_1\cos \delta]_{\delta_0}^{\delta_1} + [P_0\delta + P_2\cos \delta]_{\delta_1}^{\delta_2} = 0$$

from which

$$\cos \delta_1 = \frac{P_0(\delta_0 - \delta_2) + P_1\cos \delta_0 - P_2\cos \delta_2}{P_1 - P_2} \tag{8.4}$$

and the critical clearing angle is obtained

If δ_1 is the critical clearing angle then it may be seen that

$$\delta_2 = 180 - \sin^{-1}\left(\frac{P_0}{P_2}\right)$$

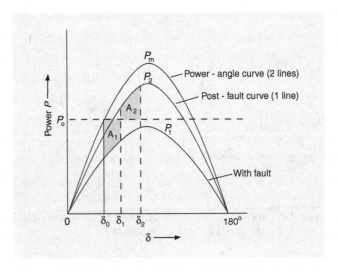

Figure 8.5 Situation as in Figure 8.4, but δ_1 not critical

It should be noted that a three-phase short circuit on the generator terminals or on a closely connected busbar absorbs zero power and prevents the generator outputting any real power to the system. Consequently, $P_1 = 0$ in Figures 8.4 and 8.5.

Example 8.2

A generator operating at 50 Hz delivers 1 p.u. power to an infinite busbar through a network in which resistance may be neglected. A fault occurs which reduces the maximum power transferable to 0.4 p.u., whereas before the fault this power was 1.8 p.u. and after the clearance of the fault it is 1.3 p.u. By the use of the equal-area criterion, determine the critical clearing angle.

Solution
The appropriate load-angle curves are shown in Figure 8.4. $P_0 = 1$ p.u., $P_1 = 0.4$ p.u., $P_2 = 1.3$ p.u., and $P_m = 1.8$ p.u.

$$\delta_0 = \sin^{-1}\left(\frac{1}{1.8}\right) = 33.8 \quad \text{electrical degrees}$$

$$\delta_2 = 180 - \sin^{-1}\left(\frac{1}{1.3}\right) = 180 - 50.28 = 129.72 \quad \text{electrical degrees}$$

Applying equation (8.4) (note that electrical degrees must be expressed in radians)

$$\cos\delta_1 = \frac{1(0.59 - 2.96) + 0.4 \times 0.831 - 1.3 \times (-0.64)}{0.4 - 1.3}$$

$$= 0.562$$

$$\therefore \delta_1 = 55.8°$$

In a large system it is usual to divide the generators and spinning loads into a single equivalent generator connected to an infinite busbar or equivalent motor. The main criterion is that the rotating machines should be electrically close when forming an equivalent generator or motor. If stability with faults in various places is investigated, the position of the fault will decide the division of machines between the equivalent generator and motor. A power system (including generation) at the receiving end of a long line would constitute an equivalent motor if not large enough to be considered an infinite busbar.

8.4.1 Reduction to Simple System

With a number of generators connected to the same busbar the inertia constant (H) of the equivalent machine is:

$$H_e = H_1 \frac{S_1}{S_b} + H_2 \frac{S_2}{S_b} \ldots + H_n \frac{S_n}{S_b}$$

where $S_1 \ldots S_n = $ MVA of the machines and $S_b = $ system base MVA.

For example, consider six identical machines connected to the same busbar, each having an H of 5 MWs/MVA and rated at 60 MVA. Making the system base MVA equal to the combined rating of the machines (360 MVA), the inertia constant of the equivalent coherent machine is:

$$H_e = 5 \times \frac{60}{360} \times 6 = 5 \quad \text{MWs/MVA}$$

It is important to remember that the inertia of the spinning loads must be included; normally, this will be the sum of the inertias of the induction motors and their mechanical loads.

Two synchronous machines connected by a reactance may be reduced to one equivalent machine feeding through the reactance to an infinite busbar system. The properties of the equivalent machine are found as follows.

The equation of motion for the two-machine system is:

$$\frac{d^2\delta}{dt^2} = \frac{\Delta P_1}{M_1} - \frac{\Delta P_2}{M_2} = \left(\frac{1}{M_1} + \frac{1}{M_2}\right)(P_0 - P_m \sin \delta)$$

where δ is the relative angle between the machines. Note that

$$\Delta P_1 = -\Delta P_2 = P_0 - P_m \sin \delta \tag{8.5}$$

where P_0 is the input power and P_m is the maximum transmittable power.

Consider a single generator of M_e with the same power transmitted to the infinite busbar system as that exchanged between the two synchronous machines. Then,

$$M_e \frac{d^2\delta}{dt^2} = P_0 - P_m \sin \delta$$

therefore,

$$M_e = \frac{M_1 M_2}{M_1 + M_2}$$

This equivalent generator has the same mechanical input as the first machine and the load angle δ it has with respect to the busbar is the angle between the rotors of the two machines.

Often, the maximum powers transferable before, during, and after a fault need to be calculated from the system configuration reduced to a network between the relevant generators. The use of network reduction by nodal elimination is most valuable in this context; it only remains then for the transfer reactances to be calculated, as any shunt impedance at the reduced nodes does not influence the power transferred.

With unbalanced faults more power is transmitted during the fault period than with three-phase short circuits and the stability limits are higher.

8.4.2 Effect of Automatic Voltage Regulators and Governors

These may be represented in the equation of motion as follows,

$$M\frac{d^2\delta}{dt^2} + K_d\frac{d\delta}{dt} = (P_{mech} - \Delta P_{mech}) - P_{elec} \tag{8.6}$$

where:

K_d = damping coefficient;
P_{mech} = power input;
ΔP_{mech} = change in input power due to governor action;
P_{elec} = electrical power output modified by the voltage regulator.

Equation (8.6) is best solved by digital computer.

8.5 Transient Stability–Consideration of Time

8.5.1 The Swing Curve

In the previous section, attention has been mainly directed towards the determination of the angular position of the rotor; in practice, the corresponding times are more important. The protection engineer requires allowable times rather than angles when specifying relay settings. The solution of equation (8.1) with respect to time is performed by means of numerical methods and the resulting time-angle curve is known as the swing curve. A simple step-by-step method will be given in detail and references will be made to more sophisticated methods used for digital computation.

In this method the change in the angular position of the rotor over a short time interval is determined. In performing the calculations the following assumptions are made:

1. The accelerating power ΔP at the commencement of a time interval is considered to be constant from the middle of the previous interval to the middle of the interval considered.
2. The angular velocity is constant over a complete interval and is computed for the middle of this interval.

These assumptions are probably better understood by reference to Figure 8.6. From Figure 8.6,

$$\omega_{n-\frac{1}{2}} - \omega_{n-\frac{3}{2}} = \frac{d^2\delta}{dt^2}\Delta t = \frac{\Delta P_{n-1}}{M}\Delta t$$

The change in δ over the $(n-1)$th interval, that is from times $(n-2)$ to $(n-1)$

$$= \delta_{n-1} - \delta_{n-2} = \Delta\delta_{n-1} = \omega_{n-\frac{3}{2}}\Delta t$$

as $\omega_{n-\frac{3}{2}}$ is assumed to be constant

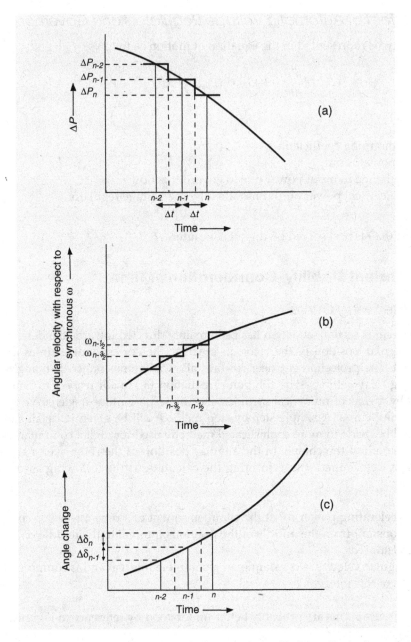

Figure 8.6 (a), (b), and (c) Variation of ΔP, ω and $\Delta\delta$ with time. Illustration of step-by-step method to obtain δ-time curve

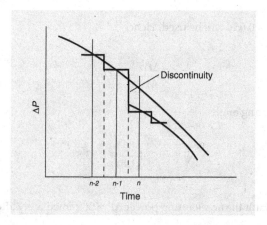

Figure 8.7 Discontinuity of ΔP in middle of a period of time

Over the nth interval,

$$\Delta\delta_n = \Delta\delta_n - \Delta\delta_{n-1} = \omega_{n-\frac{1}{2}}\Delta t$$

From the above,

$$\Delta\delta_n - \Delta\delta_{n-1} = \Delta t\left(\omega_{n-\frac{1}{2}} - \omega_{n-\frac{3}{2}}\right)$$

$$= \Delta t.\Delta t.\frac{\Delta P_{n-1}}{M}$$

$$\therefore \Delta\delta_n = \Delta\delta_{n-1} + \frac{\Delta P_{n-1}}{M}(\Delta t)^2 \tag{8.7}$$

It should be noted that $\Delta\delta_n$ and $\Delta\delta_{n-1}$ are the *changes* in angle.

Equation (8.7) is the basis of the numerical method. The time interval Δt used should be as small as possible (the smaller Δt, however, the larger the amount of labour involved), and a value of 0.05 s is frequently used. Any change in the operational condition causes an abrupt change in the value of ΔP. For example, at the commencement of a fault ($t = 0$), the value of ΔP is initially zero and then immediately after the occurrence it takes a definite value. When two values of ΔP apply, the mean is used. The procedure is best illustrated by an example.

Example 8.3

In the system described in Example 8.2 the inertia constant of the generator plus turbine is 2.7 p.u. Obtain the swing curve for a fault clearance time of 125 ms.

Solution

$$H = 2.7\,\text{p.u.}, f = 50\,\text{Hz}, G = 1\,\text{p.u.}$$

$$\therefore M = \frac{HG}{180f} = 3 \times 10^{-4}\,\text{p.u.}$$

A time interval $\Delta t = 0.05$ s will be used. Hence

$$\frac{(\Delta t)^2}{M} = 8.33$$

The initial operating angle

$$\delta_0 = \sin^{-1}\left(\frac{1}{1.8}\right) = 33.8°$$

Just before the fault the accelerating power $\Delta P = 0$. Immediately after the fault,

$$\begin{aligned} \Delta P &= 1 - 0.4 \sin \delta_0 \\ &= 0.78 \, \text{p.u.} \end{aligned}$$

The first value is that for the middle of the preceding period and the second is for the middle of the period under consideration. The value to be taken for ΔP at the commencement of this period is $(0.78/2)$, that is 0.39 p.u. At $t = 0$, $\delta = 33.8°$.

$$\Delta t_1 = 0.05 \, \text{s},$$

$$\Delta P = 0.39 \, \text{p.u.}$$

$$\Delta \delta_n = \Delta \delta_{n-1} + \frac{(\Delta t)^2}{M} \Delta P_{n-1}$$

$$\therefore \Delta \delta_n = 0 + 8.33 \times 0.39 = 3.25°$$

$$\therefore \delta_{0.05} = 33.8 + 3.25 = 37.05°$$

Δt_2

$$\Delta P = 1 - 0.4 \sin 37.05° = 0.76 \, \text{p.u.}$$

$$\Delta \delta_n = 3.25 + 8.33 \times 0.76 = 9.58°$$

$$\therefore \delta_{0.1} = 37.05 + 9.58 = 46.63°$$

Δt_3

$$\Delta P = 1 - 0.4 \sin 46.63° = 0.71 \, \text{p.u.}$$

$$\Delta \delta_n = 9.58 + 8.33 \times 0.71 = 15.49°$$

$$\therefore \delta_{0.15} = 46.63 + 15.49 = 62.12°$$

The fault is cleared after a period of 0.125 s. As this discontinuity occurs in the middle of a period (0.1–0.15 s), no special averaging is required (Figure 8.7).

If, on the other hand, the fault is cleared in 0.15 s, an averaging of two values would be required.

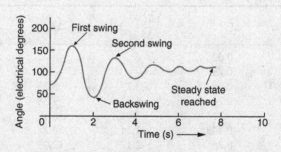

Figure 8.8 Typical swing curve for generator

From $t = 0.15$ s onwards,

$P = 1 - 1.3 \sin \delta$ (note change to P-δ curve of Figure 8.5)

Δt_4

$$\Delta P = 1 - 1.3 \sin 62.12° = -0.149 \, \text{p.u.}$$

$$\Delta \delta_n = 15.49 + 8.33 \times (-0.149) = 14.25°$$

$$\therefore \delta_{0.2} = 62.12 + 14.25 = 76.37°$$

Δt_5

$$\Delta P = 1 - 1.3 \sin 76.4° = -0.26 \, \text{p.u.}$$

$$\Delta \delta_n = 14.25 + 8.33 \times (-0.26) = 12.08°$$

$$\therefore \delta_{0.25} = 76.37 + 12.08 = 88.39°$$

Δt_6

$$\Delta P = 1 - 1.3 \sin 88.390 = -0.3 \, \text{p.u.}$$

$$\Delta \delta_n = 12.08 + 8.33 \times (-0.3) = 9.59°$$

$$\therefore \delta_{0.3} = 78.39 + 9.59 = 97.98°$$

If this process is continued, δ commences to decrease and the generator remains stable.

If computed by hand, a tabular calculation is recommended, as shown in Table 8.1

This calculation should be continued for at least the peak of the first swing, but if switching or auto-reclosing is likely to occur somewhere in the system, the calculation of δ needs to be continued until oscillations are seen to be dying away. A typical swing curve shown in Figure 8.8 illustrates this situation.

Different curves will be obtained for other values of clearing time. It is evident from the way the calculation proceeds that for a sustained fault, δ will continuously increase and stability will be lost. The critical clearing time should be calculated for conditions which allow the least transfer of power from the generator. Circuit breakers and the associated protection operate in times dependent upon their design; these times can be in the order of a few cycles of alternating voltage. For a given fault position a faster

Table 8.1 Tabular calculation of δn

t(s)	ΔP	$\dfrac{(\Delta t^2)}{M} \cdot \Delta P$	$\Delta \delta_n$	δ_n
0−	0.00	—	—	33.8
0+	0.78	—	—	33.8
0.05	0.39	3.25	3.25	37.05
0.1	0.76	6.33	9.58	46.63
0.15	0.71	5.91	15.49	62.12
0.2	−0.149	−1.24	14.25	76.37
0.25	−0.26	−2.17	12.08	88.39
. . .				

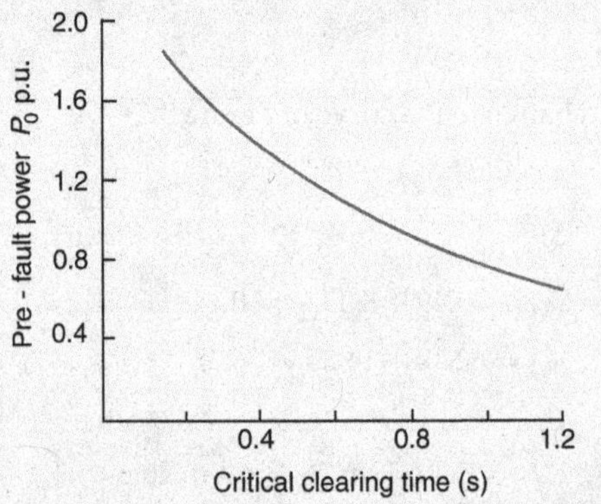

Figure 8.9 Typical stability boundary

clearing time implies a greater permissible value of input power P_0. A typical relationship between the critical clearing time and input power is shown in Figure 8.9 – this is often referred to as the stability boundary. The critical clearing time increases with increase in the inertia constant (H) of turbine generators. Often, the first swing of the machine is sufficient to indicate stability.

8.6 Transient Stability Calculations by Computer

It is obvious that a digital computer program can be readily written to carry out the simple studies of Section 8.5. If a load-flow program is readily available, then improved accuracy will be obtained if, for each value of δ_n, the actual power output of the generators is calculated. At the same time, the effect of the excitation system and the governor movement can be included. Such calculations make use of numerical integration packages based on mathematical concepts. Techniques such as trapezoidal integration provide fast and sufficiently accurate results for many stability studies; more accurate techniques, such as Runge-Kutta (fourth order), predictor-corrector routines, and so on, can be employed if the improved accuracy and longer run times can be economically justified. Most commercial stability programs offer various options for inclusion of generator controls, system switching and reclosing, compensator modelling, and transformer tap-change operation, according to some input criteria. Packages dealing with 1000 generators, 2000 lines, and 1500 nodes are available.

Example 8.4

Consider the network of Figure 8.10. Data for generators, transformers, lines and loads are given in Tables 8.2-8.5

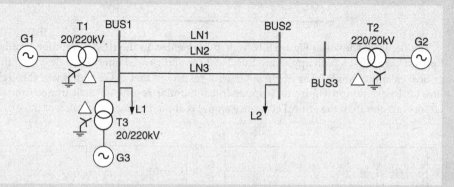

Figure 8.10 Network for Example 8.4

Table 8.2 Generator data

	G1	G2	G3
Rating (MVA)	750	750	250
X_s (p.u)	1.7	1.7	1.6
R_s (p.u)	0.05	0.05	0.06
X_d' (p.u)	0.35	0.35	0.3
X_d'' (p.u)	0.25	0.25	0.25
T_d' (s)	8.0	8.0	8.0
T_d'' (s)	0.03	0.03	0.03
H (s)	6.5	6.5	6.0

Table 8.3 Transformer data

	T1	T2	T3
Rating (MVA)	750	750	250
X_l (p.u)	0.15	0.15	0.12
R (p.u)	0	0	0

Table 8.4 Line data

	LN1	LN2	LN3	LN4
X (Ω)	115	115	115	9
R (Ω)	11	11	11	0.9
Y (µS)	1450	1450	1450	115

Table 8.5 Load data

	L1	L2
P (MW)	600	1000
Q (MVAr)	50	100

The network shown in Figure 8.10 was implemented in the IPSA computer simulation package. First a fault applied to Bus 3 at 2 sec and cleared after 80 ms (less than the critical clearance angle). The angle of G1 with respect to that of G2 and power through line LN1 is shown in Figure 8.11(upper trace). Similar results for a clearance time of 600 ms (greater than the critical clearance angle) is shown in Figure 8.11(lower trace).

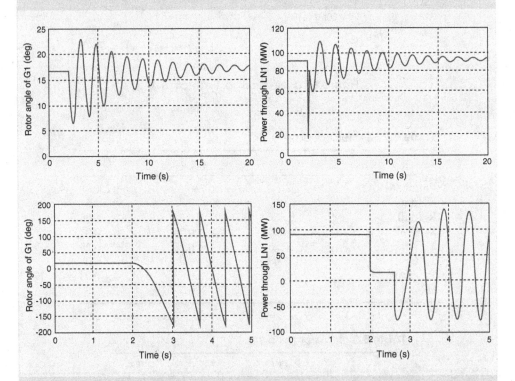

Figure 8.11 Angle of generator G1 and power through line LN1 before and after a fault at BUS 3. For 80 ms clearance time (upper trace). For 400 ms clearance time (lower trace)

As can be seen from Figure 8.11(upper trace) the angle between the two generators swings but comes back to a stable point after about 25 ms. When the fault clearance time is longer than the critical clearance time, the machines lose synchronism and pole slipping occurs, as shown in Figure 8.11(lower trace).

8.7 Dynamic or Small-Signal Stability

The power system forms a group of interconnected electromechanical elements, the motion of which may be represented by the appropriate differential equations. With large disturbances in the system the equations are non-linear, but with small changes the equations may be linearized with little loss of accuracy. The differential equations having been determined, the characteristic equation of the system is then formed, from which information regarding stability is obtained. The solution of the differential equation of the motion is of the form

$$\delta = k_1 e^{a_1 t} + k_2 e^{a_2 t} + \ldots \ldots k_n e^{a_n t}$$

where k_1, k_2,..k_n are constants of integration and a_1, a_2,..a_n are the roots of the characteristic equation as obtained through the well-known eigenvalue methods. If any of the roots have positive real terms then the quantity δ increases continuously with time and the original steady condition is not re-established. The criterion for stability is therefore that all the real parts of the roots of the characteristic equation, that is eigenvalues, be negative; imaginary parts indicate the presence of oscillation. Figure 8.12 shows the various types of motion. The determination

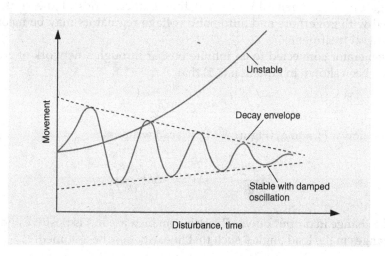

Figure 8.12 Types of response to a disturbance on a system

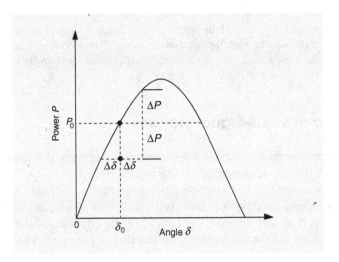

Figure 8.13 Small disturbance–initial operation on power-angle curve at $P_0 \delta_0$. Linear movement is assumed about $P_0 \delta_0$.

of the roots is readily obtained through an eigenvalue analysis package, but indirect methods for predicting stability have been established, for example the Routh-Hurwitz criterion in which stability is predicted without the actual solution of the characteristic equation. No information regarding the degree of stability or instability is obtained, only that the system is, or is not, stable. One advantage of using eigenvalues is that the characteristics of the control loops associated with governors and automatic voltage regulators may be incorporated in the general treatment.

For a generator connected to an infinite busbar through a network of zero resistance it has been shown in Equation (3.1) that

$$P = \frac{VE}{X} \sin \delta.$$

With operation at P_0 and δ_0 (Figure 8.13), we can write

$$M \frac{d^2 \Delta \delta}{dt^2} = -\Delta P = -\Delta \delta \left(\frac{\partial P}{\partial \delta} \right)_0$$

where the change in output power P causing an increase in δ is positive and refers to small changes in the load angle δ such that linearity may be assumed.

$$M s^2 \Delta \delta + \left(\frac{\partial P}{\partial \delta} \right)_0 \Delta \delta = 0 \qquad (8.8)$$

Where

$$s = \frac{d}{dt}$$

Here, $Ms^2 + (dP/d\delta)_0 = 0$ is the characteristic equation which has two roots

$$\pm\sqrt{\frac{-\left(\partial P / \partial \delta\right)_0}{M}}$$

When $(dP/d\delta)_0$ is positive, both roots are imaginary and the motion is oscillatory and undamped; when $(dP/d\delta)_0$ is negative, both roots are real, one positive and one negative and stability is lost. At $\delta = 90°$, $\left(\partial P / \partial \delta\right)_{90} = 0$, and the system is at the limit.

If damping is accounted for, the equation becomes

$$Ms^2\Delta\delta + Ks\Delta\delta + \left(\frac{\partial P}{\partial \delta}\right)_0 \Delta\delta = 0 \tag{8.9}$$

and the characteristic equation is

$$Ms^2 + Ks + \left(\frac{\partial P}{\partial \delta}\right)_0 = 0 \tag{8.10}$$

$$s_1, s_2 = \frac{-K \pm \sqrt{K^2 - 4M\left(\partial P / \partial \delta\right)}}{2M}$$

where K is the damping coefficient, assumed to be constant, independent of δ.

Again, if $\dfrac{\partial P}{\partial \delta}$ is negative, stability is lost. The frequency of the oscillation is given by the roots of the characteristic equation.

If the excitation of the generator is controlled by a fast-acting automatic voltage regulator without appreciable dead zone, the excitation voltage E is increased as increments of load are added. Hence the actual power-angle curve is no longer that for constant E (refer to Chapter 3) and the change of power may be obtained by linearizing the P-V characteristic at the new operating point (1), when

$$\Delta P = \left(\frac{\delta P}{\delta E}\right)_1 \Delta E$$

The complete equation of motion is now

$$Ms^2\Delta\delta + Ks\Delta\delta + \left(\frac{\partial P}{\partial \delta}\right)_1 \Delta\delta + \left(\frac{\partial P}{\partial E}\right)_1 \Delta E = 0 \tag{8.11}$$

Without automatic voltage control the stability limit is reached when $\delta = 90°$; with excitation control the criterion is obtained from the characteristic equation of (8.11).

Example 8.5

A synchronous generator of reactance 1.5 p.u. is connected to an infinite busbar system ($V = 1$ p.u.) through a line and transformers of total reactance 0.5 p.u. The no-load voltage of the generator is 1.1 p.u. and the inertia constant $H = 5$ MWs per MVA. All per unit values are expressed on the same base; resistance and damping may be neglected. Calculate the frequency of the oscillations set up when the generator operates at a load angle of 60° and is subjected to a small disturbance. The system frequency is 50 Hz.

Solution
The nature of the movement is governed by the sign of the quantity under the root sign in the equation for s_1 and s_2 (equation (8.10)). This changes when $K^2 = 4M\left(\frac{\partial P}{\partial \delta}\right)$; in this example $K = 0$ and the motion is undamped.

The roots of the characteristic equations give the frequency of oscillation; when $\delta_0 = 60°$,

$$\left(\frac{\partial P}{\partial \delta}\right)_{60} = \frac{1.1 \times 1}{2}\cos 60°$$
$$= 0.275 \text{ p.u.}$$

$$s_1 \text{ and } s_2 = \pm j \sqrt{\frac{\partial P}{\partial \delta} \cdot \frac{1}{M}}$$

$$= \pm j \sqrt{\frac{0.275}{5 \times \dfrac{1}{\pi \times 50}}}$$

$$= \pm j\sqrt{8.64} = 2.94 \text{ rad/s}$$

Therefore frequency of oscillation

$$= \frac{2.94}{2\pi} = 0.468 \text{ Hz}$$

and the periodic time

$$= \frac{1}{0.468} = 2.14 \text{ s}$$

8.7.1 Effects of Governor Action

In the above analysis the oscillations set up with small changes in load on a system have been considered and the effects of governor operation ignored. After a certain time has elapsed the governor control characteristics commence to influence the powers and oscillations, as explained in Section 4.3. It is now the practice to represent both the excitation system and the governor system with the dynamic equations of the generator in the state-space form, from which the eigenvalues of the complete system with feedback can be determined. Using well-established control design techniques, appropriate feedback paths and time constants can be established for a range of generating conditions and disturbances, thereby assuring adequate dynamic stability margins. Further

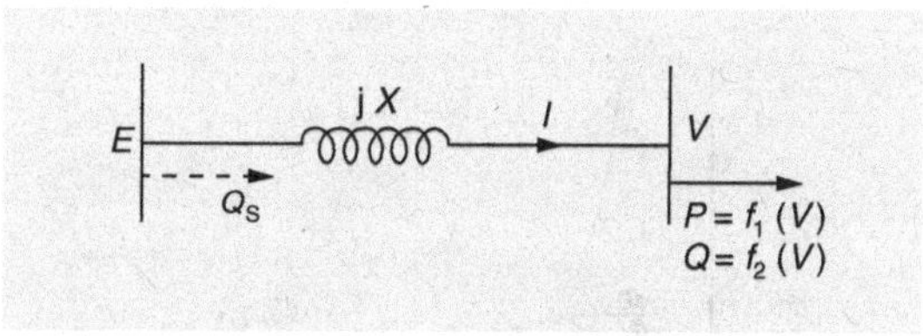

Figure 8.14 System with a load dependent on voltage as follows: $P = f_1(V)$ and $Q = f_2(V)$; Q_s = supply of VArs from $E = Q + I^2 X$; E = supply voltage

information on these design processes can be found in advanced control textbooks.

8.8 Stability of Loads Leading to Voltage Collapse

In Chapter 5 the power-voltage characteristics of a line supplying a load were considered. It was seen that for a given load power-factor, a value of transmitted power was reached, beyond which further decreases of the load impedance produce greatly reduced voltages, that is voltage instability (Figure 5.21). If the load is purely static, for example represented by an impedance, the system will operate stably at these lower voltages. Sometimes, in load-flow studies, this lower voltage condition is unknowingly obtained and unexpected load flows result. If the load contains non-static elements, such as induction motors, the nature of the load characteristics is such that beyond the critical point the motors will run down to a standstill or stall. It is therefore of importance to consider the stability of composite loads that will normally include a large proportion of induction motors.

The process of voltage collapse may be seen from a study of the V-Q characteristics of an induction motor under load, from which it is seen that below a certain voltage the reactive power consumed increases with decrease in voltage until $\frac{dQ}{dV} \to \infty$, when the voltage collapses. In the power system the problem arises owing to the impedance of the connection between the load and infinite busbar and is obviously aggravated when this impedance is high, that is connection is electrically weak. The usual cause of an abnormally high impedance is the loss of one line of two or more forming the connection. It is profitable, therefore, to study the process in its basic form – that of a load supplied through a reactance from a voltage source (Figure 8.14).

Already, two criteria for load instability have been given, that is $\frac{dP}{d\delta} = 0$ and $\frac{dQ}{dV} \to \infty$; from the system viewpoint, voltage collapse takes place when $\frac{dE}{dV} = 0$ or $\frac{dV}{dE} \to \infty$. Each value of E yields a corresponding value for V, and the plot of E-V is shown in Figure 8.15; also the plot of V-X for various values of E is shown in Figure 8.16. In these graphs the critical operating condition is clearly

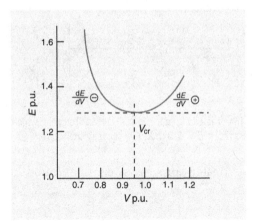

Figure 8.15 The E-V relationship per system in Figure 8.14. V_{cr}=critical voltage after which instability occurs

shown and the improvement produced by higher values of E is apparent, indicating the importance of the system-operating voltage from the load viewpoint.

In the circuit shown in Figure 8.14, and from Equation (2.17),

$$E = V + \frac{QX}{V} \quad if \quad \frac{PX}{V} \ll \frac{V^2 + QX}{V}$$

$$\therefore \frac{dE}{dV} = 1 + \left(\frac{dQ}{dV}.XV - QX\right)\frac{1}{V^2}$$

which is zero at the stability limit and negative in the unstable region.

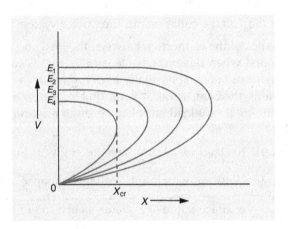

Figure 8.16 The V-X relationship. Effect of change in supply voltage E. X_{cr}=critical reactance of transmission link

At the limit,

$$\frac{dE}{dV} = 0 \quad and \quad \frac{dQ}{dV} = \left(\frac{QX}{V^2} - 1\right)\frac{V}{X} = \frac{Q}{V} - \frac{V}{X}$$

Also from Equation (2.17),

$$\frac{Q}{V} = \frac{E}{X} - \frac{V}{X}$$

$$\therefore \frac{dQ}{dV} = \frac{E}{X} - \frac{2V}{X} \tag{8.12}$$

Example 8.6

A load is supplied from a 275 kV busbar through a line of reactance 70 Ω phase-to-neutral. The load consists of a constant power demand of 200 MW and a reactive power demand Q which is related to the load voltage V by the equation:

$$(V - 0.8)^2 = 0.2(Q - 0.8)$$

$$Q = 5(V - 0.8)^2 + 0.8$$

This is shown in Figure 8.17, where the base quantities for V and P, Q are 275 kV and 200 MVA.

Examine the voltage stability of this system, indicating clearly any assumptions made in the analysis.

Solution
It has been shown that

$$E = \sqrt{\left[\left(V + \frac{QX}{V}\right)^2 + \left(\frac{PX}{V}\right)^2\right]}$$

If

$$\left(\frac{PX}{V}\right) \ll \left(V + \frac{QX}{V}\right)$$

then

$$E = \left(V + \frac{QX}{V}\right)$$

and

$$\frac{dE}{dV} = 1 + \left(\frac{dQ}{dV}\frac{X}{V} - \frac{QX}{V^2}\right) = 0$$

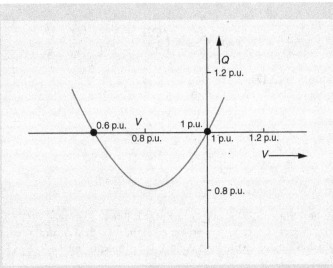

Figure 8.17 Example 8.6. Reactive power-voltage characteristic

In this problem

$P = 200\,\text{MW}, 1\,\text{p.u.}$
$Q = 200\,\text{MVAr}, 1\,\text{p.u.}$
$E = 275\,\text{kV}, 1\,\text{p.u.}$

And

$$X = \frac{70 \times 200}{275^2} = 0.185\,\text{p.u.}$$

Now

$$E = \left(V + \frac{QX}{V}\right)$$

$$\therefore 1 = V + \frac{0.185}{V}\left(\frac{(V-0.8)^2}{0.2} + 0.8\right)$$

$$V^2 = V - 0.925\,V^2 - 0.59 + 1.48\,V - 0.148$$

giving,

$$1.925\,V^2 - 2.48\,V + 0.74 = 0$$

$$\therefore V = \frac{+2.48 \pm \sqrt{2.48^2 - 4 \times 1.925 \times 0.74}}{2 \times 1.925}$$

$$= \frac{2.48 \pm 0.67}{3.85}$$

Taking the upper value, $V = 0.818$ p.u.,
Q at this voltage is

$$Q = \frac{(0.818 - 0.8)^2}{0.2} + 0.8 = 0.8016 \text{ p.u.}$$

$$\frac{dQ}{dV} = 10(0.818 - 0.8)$$

$$\frac{dE}{dV} = 1 + \frac{[10(0.818 - 0.8) \times 0.185 \times 0.818 - 0.8016 \times 0.185]}{0.818^2}$$

$$= 1 + \left(\frac{0.027 - 0.148}{0.818^2}\right)$$

which is positive, that is the system is stable.
(Note: $PX/V \approx 0.16$ and $(V^2 + QX)/V \approx 1$; therefore the approximation is reasonable.)

When the reactance between the source and load is very high, the use of tap-changing transformers is of no assistance. Large voltage drops exist in the supply lines and the 'tapping-up' of transformers increases these because of the increased supply currents. Hence the peculiar effects from tap-changing noticed when conditions close to a voltage collapse have occurred in practice, that is tapping-up reduces the secondary voltage, and vice versa. One symptom of the approach of critical conditions is sluggishness in the response to tap-changing transformers.

Studies into the voltage collapse phenomenon in interconnected power systems have shown how difficult it is to predict its occurrence or, more usefully, estimate the margin available to collapse at any given operating condition. One approach is to increase all loads at constant power factor until the load flow diverges. Failure of the load flow to converge is an indication that voltage collapse may be imminent. Unfortunately, this does not provide a realistic case since (1) any increase in real-power load will result in reserve generation being brought on line at various nodes of the system not previously used for generation input; and (2) as voltage falls, loads behave non-linearly (see Figure 8.15) but our knowledge of load behaviour below about 0.9 p.u. voltage is extremely sparse. Consequently, only worst-case situations can be assumed if voltage margins are to be estimated.

The voltage collapse phenomenon may be studied using the transient stability facility of many computer programs. In the network shown in Figure 8.10, the load on BUS 1 was increased in steps of $90 + j9$ MVA. Figure 8.18 shows the voltage collapse on that busbar as the load increases.

8.9 Further Aspects

8.9.1 Faults on the Feeders to Induction Motors

A common cause of the stalling of induction motors (or the low-voltage releases operating and removing them from the supply) occurs when the supply voltage is

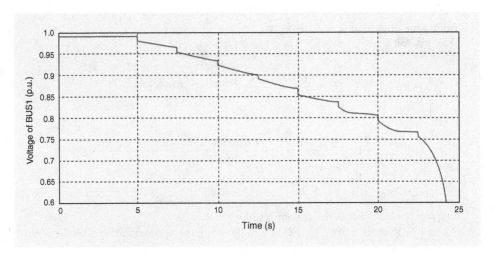

Figure 8.18 Voltage collapse on the network shown in Figure 8.10

either zero or very low for a brief period because of a fault on the supply system, commonly known as 'voltage dip' or 'voltage sag'. When the supply voltage is restored the induction motors accelerate and endeavour to attain their previous operating condition. In accelerating, however, a large current is taken, and this, plus the fact that the system impedance has increased due to the loss of a line, results in a depressed voltage at the motor terminals. If this voltage is too low the machines will stall or cut out of circuit.

8.9.2 Steady-State Instability Due to Voltage Regulators

Consider a generator supplying an infinite busbar through two lines, one of which is suddenly removed. The load angle of the generator is instantaneously unchanged and therefore the power output decreases due to the increased system reactance, thus causing the generator voltage to rise. The automatic voltage regulator of the generator then weakens its field to maintain constant voltage, that is decreases the internal e.m.f., and pole-slipping may result.

8.9.3 Dynamic Stability

The control circuits associated with generator AVRs, although improving steady-state stability, can introduce problems of poorly damped response and even instability. For this reason, dynamic stability studies are performed, that is steady-state stability analysis, including the automatic control features of the machines (see Section 8.7). The stability is assessed by determining the response to small step changes of rotor angle, and hence the machine and control-system equations are often linearized around the operating point, that is constant machine parameters and linear AVR characteristic. The study usually extends over several seconds of real-system time.

8.10 Multi-Machine Systems

In networks interconnecting many generators and dynamic loads, it is a complicated task, even with a fast digital computer, to solve a set of dynamic equations to establish if a given state of the system is stable or unstable following a credible disturbance. Obviously, many hundreds of studies may need to be run with different faults and the output data assessed in some way. It must be remembered that if each dynamic machine (generator or load) is to be represented by its swing trajectory, starting from an initial steady-state angle (as in Figure 8.8), then a multitude of trajectories up to 5 s will be presented by the computer output. Some machines will probably not be affected by the disturbance and their angles remain within a few degrees of their initial angle. Others will show oscillations but their mean angle could gradually diverge from the more or less unaffected machines – these diverging machines would indicate that the system is unstable and, if continued, would split up, by action of the interconnecting circuit-protection systems, into two or more 'islands'. A common feature of such studies often shows that machine angles oscillate at different frequencies but that they all gradually change their angles in unison from the initial angles, implying that the system is stable but that, subsequent to the initiating disturbance, the frequency of the whole system is either increasing or decreasing.

To enable a digital study to assess its own results, criteria need to be built into the software. One of these is whether or not the system angles are within a norm. Another is, in the case of instability, which parts of the system are splitting away from other parts. To achieve this assessment, a concept known as 'Centre of Inertia' (COI) is employed. This is similar to determining the centre of gravity of a mechanical system by writing

$$M_{tot}\delta_{COI} = \sum_{i=1}^{i=n} M_i\delta_i$$

where M_{tot} is the total area angular momentum and δ_{COI} is the average angle of the machines in the system.

Consequently, we now have, for an area of the system, that

$$M_{tot}\frac{d^2\delta_{COI}}{dt^2} + \Delta P_{area} = 0 \quad \text{(see equation (8.1))}$$

where ΔP_{area} is the combined power being input (negative if output) into that area. By use of conditional statements in the software used for post-analysis of a stability study, areas of the system which swing together, that is whose angles are within specified limits around the δ_{COI} (known as *coherency*), can be identified. Note also that the power transfer across the area boundary ΔP_{area} before the disturbance can also be established for each area, once the vulnerable areas are known. To ensure stability, further studies need to be undertaken with revised generator scheduling such that critical ΔP_{area} flows are reduced to a sustainable level. System operators should then observe these area flow limits to ensure stability under credible system contingencies. In most systems, by proper design of transient machine controllers or reinforcement of the transfer capability of the vulnerable circuits, only a few critical areas remain where power transfer is limited by stability considerations. For example, in the UK, the Scotland-England transfer over two double-circuit lines is often limited in this way; similarly under some conditions, circuits from North Wales to the rest of the National Grid system are flow-restricted.

8.11 Transient Energy Functions (TEF)

A fast stability study in large systems is essential to establish viable operating conditions. The use of Lyapunov-type functions for this purpose is left for advanced study. However, the equal-area criterion (EAC) is a form of energy function which can be used as a screening tool to enable a fast assessment of multi-machine stability to be made.

Figure 8.19 shows three curves similar to those of Figure 8.5.

From equation (8.5) we have that

$$M_e \frac{d^2\delta}{d\delta^2} = P_0 - P_2 \sin\delta$$

We have also that

$$\frac{d\delta}{dt} = \omega$$

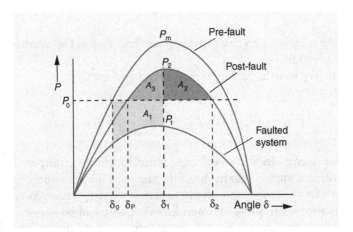

Figure 8.19 Equal-area criterion applied to an equivalent machine connected to infinite busbar through a system

And

$$\frac{d\omega}{dt} = \frac{P_0 - P_2 \sin \delta}{M_e}$$

noting that ω is the incremental speed from the steady-state speed ω_s. At any moment, the energy in this system consists of two components, namely:

$M_e\omega$ the incremental kinetic energy stored in the rotating masses of the machines at any instant;
$(P_0 - P_2 \sin \delta)$ the potential energy due to the excess electrical power at any instant.

   The energy balance in the system during the disturbance period can be obtained by integrating the two energy components over the period. However, we must remember that prior to the disturbance under steady-state conditions the energy put into the system equalled the energy being taken out. Our development of the EAC only took into account the energy increments or decrements from steady-state conditions; therefore we need only concern ourselves with the incremental kinetic and potential energy functions. We can compute a value for the total transient energy in the system, TEF, by using

$$\text{TEF} = \int_0^{\omega_1} M_e\omega \, d\omega - \int_{\delta_P}^{\delta_1} (P_0 - P_2 \sin \delta)d\delta \quad [J]$$

where ω_1 is the incremental angular speed at angle δ_1 and δ_P is the angle at which the system settles down following the disturbance; both angles are shown in Figure 8.19.
   In fact, for any intermediate value of ω or δ the value of TEF is an indication that there is a surplus or deficiency of incremental energy and that the system is still in a dynamic or oscillation state. Only when $\delta = \delta_P$ and $\omega = 0$, that is no increment on the steady-state synchronous speed, will the oscillations have died away and the system be stable.
   If the system is to regain a steady-state condition, the area A_2 must equal A_1, and the maximum angle attained at the limiting conditions is δ_2, where the incremental speed ω would again be zero. Under these conditions we see that

$$\text{TEF}_{\text{limit}} = -\int_{\delta_P}^{\delta_1} (P_0 - P_2 \sin \delta)d\delta \quad \text{since } \omega = 0$$

   $\text{TEF}_{\text{limit}}$ represents the maximum energy that the system can gain due to a disturbance and subsequently dissipate by transfer over the system, if it is to remain stable. Knowing the value of $\text{TEF}_{\text{limit}}$ from the above equation enables a quick determination of stability by calculating TEF as the disturbance proceeds. This can be done readily as the step-by-step integration proceeds because values of ω and δ will be available. Provided that the value of TEF remains less than $\text{TEF}_{\text{limit}}$ the system is stable and will settle down to a new angle δ_P. It is usual to compute TEF at fault clearance and assume afterwards that no more energy is added (rather, it is dissipated by system damping and losses), thereby checking that $\text{TEF} < \text{TEF}_{\text{limit}}$

It is worth noting that in Figure 8.19:

1. The kinetic energy-like term $\int_0^{\omega_1} M_e \omega \, d\omega$ is the area A_1.
2. The potential energy-like term $\int_{\delta_P}^{\delta_1} (P_0 - P_2 \sin\delta)d\delta$ is the area A_3.
3. TEF is equal to areas $A_1 + A_3$.
4. $\text{TEF}_{\text{limit}}$ is the area $A_2 + A_3$.
5. Equating TEF and $\text{TEF}_{\text{limit}}$ produces,

$$A_1 + A_3 = A_2 + A_3 \text{ i.e. } A_1 = A_2$$

which is the Equal Area Criterion for the system.

This shows that for stability the total energy gained during a disturbance must equal the capability of the system to transfer that energy away from the system afterwards. Considerable effort has been put into the determination of dynamic system boundaries through improved calculations of the TEF.

8.12 Improvement of System Stability

Apart from the use of fast-acting AVRs the following techniques are in use:

1. Reduction of fault clearance times, 80 ms is now the norm with SF_6 circuit breakers and high-speed protection (see Chapter 11).
2. Turbine fast-valving by bypass valving-this controls the accelerating power by closing steam valves. Valves which can close or open in 0.2 s are available. CCGTs can reduce power by fuel control within 0.2 to 0.5 s.
3. Dynamic braking by the use of shunt resistors across the generator terminals; this limits rotor swings. The switching of such resistors can be achieved by thyristors.
4. High-speed reclosure or independent pole tripping in long (point-to-point) lines. In highly interconnected systems the increase in overall clearance times on unsuccessful reclosure makes this technique of dubious value. Delayed auto-reclose (DAR) schemes with delays of 12–15 s are preferable if the voltage sag can be tolerated.
5. Increased use of II.V. direct-current links using thyristors and GTOs also alleviates stability problems.
6. Semiconductor-controlled static compensators enabling oscillations following a disturbance to be damped out.

7. Energy-storage devices, for example batteries, superconducting magnetic energy stores (SMES) with fast control, providing the equivalent of a UPS.

Problems

8.1 A round-rotor generator of synchronous reactance 1 p.u. is connected to a transformer of 0.1 p.u. reactance. The transformer feeds a line of reactance 0.2 p.u. which terminates in a transformer (0.1 p.u. reactance) to the LV side of which a synchronous motor is connected. The motor is of the round-rotor type and of 1 p.u. reactance. On the line side of the generator transformer a three-phase static reactor of 1 p.u. reactance per phase is connected via a switch. Calculate the steady-state power limit with and without the reactor connected. All per unit reactances are expressed on a 10 MVA base and resistance may be neglected. The internal voltage of the generator is 1.2 p.u. and of the motor 1 p.u.

(Answer: 5 MW and 3.13 MW for shunt reactor)

8.2 In the system shown in Figure 8.20, investigate the steady state stability. All per unit values are expressed on the same base and the resistance of the system (apart from the load) may be neglected. Assume that the infinite busbar voltage is 1 p.u.

8.3 A generator, which is connected to an infinite busbar through two 132 kV lines in parallel, each having a reactance of 70 Ω/phase, is delivering 1 p.u. to the infinite busbar. Determine the parameters of an equivalent circuit, consisting of a single machine connected to an infinite busbar through a reactance, which represents the above system

a. Pre-fault.
b. When a three-phase symmetrical fault occurs halfway along one line.
c. After the fault is cleared and one line isolated.

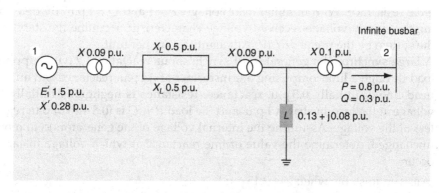

Figure 8.20 Line diagram of system in Problem 8.2

If the generator internal voltage is 1.05 p.u. and the infinite busbar voltage is 1.0 p.u, what is the maximum power transfer pre-fault, during the fault and post-fault?

Determine the swing curve for a fault clearance time of 125 ms.

The generator data are as follows:

Rating 60 MW at power factor 0.9 lagging.
Transient reactance 0.3 p.u.
Inertia constant 3 kWs/kVA.

(Answer: 1.62 p.u., 0.6 p.u., 1.24 p.u.)

8.4   An induction motor and a generator are connected to an infinite busbar. What is the equivalent inertia constant of the machines on 100 MVA base? Also calculate the equivalent angular momentum.

Data for the machines are:

Induction motor	Rating 40 MVA;
	Inertia constant 1 kWs/kVA.
Generator:	Rating 30 MVA;
	Inertia constant 10 kWs/kVA.

(Answer 3.4 Ws/VA, 0.00038 p.u.)

8.5   The P-V, Q-V characteristics of a substation load are as follows:

V	1.05	1.025	1	0.95	0.9	0.85	0.8	0.75
P	1.03	1.017	1	0.97	0.94	0.92	0.98	0.87
Q	1.09	1.045	1	0.93	0.885	0.86	0.84	0.85

The substation is supplied through a link of total reactance 0.8 p.u. and negligible resistance. With nominal load voltage, $P = 1$ and $Q = 1$ p.u. By determining the supply voltage-received voltage characteristic, examine the stability of the system by the use of dE/dV. All quantities are per unit.

8.6   A large synchronous generator, of synchronous reactance 1.2 p.u., supplies a load through a link comprising a transformer of 0.1 p.u. reactance and an overhead line of initially 0.5 p.u. reactance; resistance is negligible. Initially, the voltage at the load busbar is 1 p.u. and the load $P + jQ$ is $(0.8 + j0.6)$ p.u. regardless of the voltage. Assuming the internal voltage of the generators is to remain unchanged, determine the value of line reactance at which voltage instability occurs.

(Answer: unstable when $X = 2.15$ p.u.)

8.7   A load is supplied from an infinite busbar of voltage 1 p.u. through a link of series reactance 1 p.u. and of negligible resistance and shunt admittance. The load consists of a constant power component of 1 p.u. at 1 p.u. voltage and a per unit reactive power component (Q) which varies with the received voltage (V) according to

$$(V - 0.8)^2 = 0.2(Q - 0.8)$$

All per unit values are to common voltage and MVA bases.

Determine the value of X at which the received voltage has a unique value and the corresponding magnitude of the received voltage.

Explain the significance of this result in the system described. Use approximate voltage-drop equations.

(Answer: $X = 0.25$ p.u.; $V = 0.67$ p.u.)

8.8 Explain the criterion of stability based on the equal-area diagram.

A synchronous generator is connected to an infinite busbar via a generator transformer and a double-circuit overhead line. The transformer has a reactance of 0.15 p.u. and the line an impedance of $0 + j0.4$ p.u. per circuit. The generator is supplying 0.8 p.u. power at a terminal voltage of 1 p.u. The generator has a transient reactance of 0.2 p.u. All impedance values are based on the generator rating and the voltage of the infinite busbar is 1 p.u.

a. Calculate the internal transient voltage of the generator.
b. Determine the critical clearing angle if a three-phase solid fault occurs on the sending (generator) end of one of the transmission line circuits and is cleared by disconnecting the faulted line.

(Answer: (a) 1.035 p.u. (b) 64°

(From Engineering Council Examination, 1995)

8.9 A 500 MVA generator with 0.2 p.u. reactance is connected to a large power system via a transformer and overhead line which have a combined reactance of 0.3 p.u. All p.u. values are on a base of 500 MVA. The amplitude of the voltage at both the generator terminals and at the large power system is 1.0 p.u. The generator delivers 450 MW to the power system.

Calculate

a. the reactive power in MVAr supplied by the generator at the transformer input terminals;
b. the generator internal voltage;
c. the critical clearing angle for a 3 p.h. short circuit at the generator terminals.

(Answer: (a) 62 MVAr; (b) 1.04 p.u.; (c) 84°)

(From Engineering Council Examination, 1996)

9

Direct-Current Transmission

9.1 Introduction

The established method of transmitting large quantities of electrical energy is to use three-phase alternating-current. However, there is a limit to the distance that bulk a.c. can be transmitted unless some form of reactive compensation is employed. For long overhead lines either alternating current with reactive compensation (connected in shunt or series) or direct current may be used. If undersea crossings greater than around 50 km are required, then, because of the capacitive charging current of a.c. cables, d.c. is the only option.

Figure 9.1 shows the distances at which d.c. becomes cheaper than a.c for overhead line and submarine cable transmission. The terminal converter stations of a d.c. scheme are more expensive than a.c. substations but the overhead line is cheaper. The choice of whether to use a.c. or d.c. is usually made on cost. With the increasing use of high-voltage, high-current semiconductor devices, converter stations and their controls are becoming cheaper and more reliable, so making d.c. more attractive at shorter distances.

The main technical reasons for high-voltage direct-current (h.v.d.c.) transmission are for the:

1. interconnection of two large a.c. systems without having to ensure synchronism and be concerned over stability between them (for example, the UK-France cross-channel link of 2000 MW);
2. interconnection between systems of different frequency (for example, the connections between north and south islands in Japan, which use 50 and 60 Hz systems);
3. long overland transmission of high powers where a.c. transmission towers, insulators, and conductors are more expensive than using h.v.d.c. (for example, the Nelson River scheme in Manitoba – a total of 4000 MW over more than 600 km).

Electric Power Systems, Fifth Edition. B.M. Weedy, B.J. Cory, N. Jenkins, J.B. Ekanayake and G. Strbac.
© 2012 John Wiley & Sons, Ltd. Published 2012 by John Wiley & Sons, Ltd.

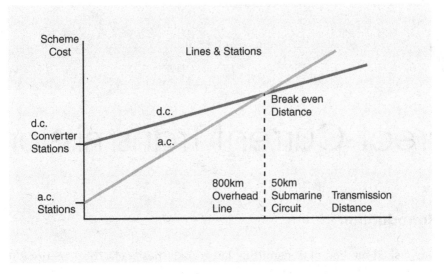

Figure 9.1 Costs of d.c. and a.c. transmission

The main advantages of h.v.d.c. compared with h.v.a.c. are:

1. two conductors, positive and negative to ground, are required instead of three, thereby reducing tower or cable costs;
2. the direct voltage can be designed equivalent to the peak of the alternating voltage for the same amount of insulation to ground (i.e. $V_{d.c.} = \sqrt{2}V_{a.c.}$);
3. the voltage stress at the conductor surface can be reduced with d.c., thereby reducing corona loss, audible emissions, and radio interference;
4. h.v.d.c. infeeds do not increase significantly the short-circuit capacity required of switchgear in the a.c. networks;
5. fast control of converters can be used to damp out oscillations in the a.c. system to which they are connected.

Disadvantages of h.v.d.c. are:

1. the higher cost of converter stations compared with an a.c. transformer substation;
2. the need to provide filters and associated equipment to ensure acceptable waveform and power factor on the a.c. networks;
3. limited ability to form multi-terminal d.c. networks because of the need for coordinated controls and the present lack of commercially available d.c. circuit breakers.

9.2 Current Source and Voltage Source Converters

Traditionally high power h.v.d.c. schemes have used thyristors with the d.c. current always flowing in the same direction (current source converters). These are used extensively for point-point transmission of bulk power and are available at d.c. voltage ratings of up to ±800 kV and are able to transmit 6500 MW over a single overhead

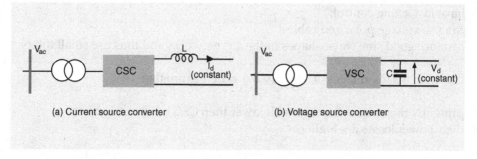

(a) Current source converter   (b) Voltage source converter

Figure 9.2 Commonly used converters for h.v.d.c. schemes

h.v.d.c. link. These schemes use converters employing thyristors and a large inductor is connected on the d.c. side. As the inductor maintains the d.c. current more or less constant (other than a small ripple) these converters are called current source converters (CSCs) (Figure 9.2(a)). In these converters while the thyristors are triggered on by a gate pulse, they turn off when the current through them falls to zero. Thus current source converters are also called line (or naturally) commutated converters.

Current source converter h.v.d.c. has a number of advantages. It can be made up to very high power and d.c. voltage ratings and the thyristors are comparatively robust with a significant transient overload capability. As the thyristors switch off only when the current through them has dropped to zero, switching losses are low.

However, it suffers from a number of disadvantages. Both the rectifier and inverter always draw reactive power from the a.c. networks and a voltage source, usually synchronous generation, is required at each end of the d.c. link to ensure commutation of the valves. When the a.c. voltage drops, due to a fault on the a.c. network, the valves may stop operating and experience commutation failure. The a.c. current contains significant harmonics and, although these can be reduced by 12 or even 24 pulse connection of the converters, large harmonic filters are required (which also provide some of the reactive power). Conventionally, the filters use open terminal switchgear and air-insulated busbars and so both the valve hall and the filters occupy a large area. Although there have been several three terminal CSC h.v.d.c. schemes constructed, this technology is mainly applied for bulk transfer of power between two stable a.c. power systems.

Recently large transistors called Insulated Gate Bipolar Transistors (IGBTs) have been used in h.v.d.c. systems where the d.c. voltage is always the same polarity and the current reverses to change the direction of power flow. These schemes are used in underground and submarine cable links of up to 1000 MW. The converters used for these schemes have a large capacitor on the d.c. side thus maintaining the d.c. voltage more or less constant. They are referred to as voltage source converters (VSCs) (Figure 9.2(b)). As IGBTs can be turned on and off by their gate voltage, these converters are also called forced commutated converters.

VSC h.v.d.c. schemes offer the following advantages. They:

1. can operate at any combination of active and reactive power;
2. have the ability to operate into a weak grid and even black-start an a.c. network;

3. have fast acting control;
4. can use voltage polarized cables;
5. produce good sine wave-shapes in the a.c. networks and thus use small filters;

The main disadvantages of VSC h.v.d.c. schemes are that:

1. presently their rating is very much lower than CSC h.v.d.c. schemes.
2. their power losses are higher.

9.3 Semiconductor Valves for High-Voltage Direct-Current Converters

The rapid growth in the use of h.v.d.c. since about 1980 has been due to the development of high-voltage, high-current semiconductor devices. These superseded the previously used complex and expensive mercury arc valves that employed a mercury pool as cathode and a high-voltage graded column of anodes, with the whole enclosed in steel and ceramic to provide a vacuum tight enclosure. Nowadays, the semiconductor devices are stacked to form a group which is able to withstand the design voltages and to pass the desired maximum currents – this group is termed a 'valve'.

9.3.1 Thyristors

Thyristors are manufactured from silicon wafers and are four-layer versions of the simple rectifier p-n junction, as shown in Figure 9.3(a). The p layer in the middle is connected to a gate terminal biased such that the whole unit can be prevented from passing current, even when a positive voltage exists on the anode. By applying a positive pulse to the gate, conduction can be started, after which the gate control has no effect until the main forward current falls below its latching value

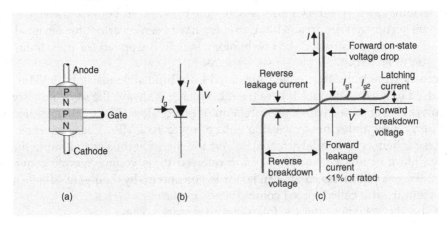

Figure 9.3 (a) Structure of a four-layer thyristor. (b) Symbol, (c) Thyristor characteristic: I_g gate current to switch thyristor on at forward voltage

(see Figure 9.3(c)). This current must be kept below the latching value for typically 100 µs before the thyristor is able to regain its voltage hold-off properties. (Note that in forward conduction there is still a small voltage across the p-n junctions, implying that power is being dissipated – hence the semiconductor devices must be cooled and their losses accounted for.)

In practice, many devices, each of rating 8.5 kV and up to 4000 A, are stacked in a valve to provide a rating of, say, 200 kV, 4000 A. Valves are connected in series to withstand direct voltages up to 800 kV to earth on each 'pole'. Each thyristor can be 15 cm in diameter and 2 cm depth between its anode and cathode terminals. A typical device is shown in Figure 9.4(a) and a valve in Figure 9.4(b).

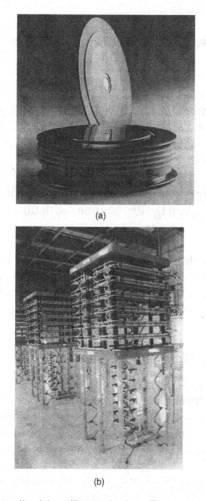

(a)

(b)

Figure 9.4 (a) High-power thyristor silicon device (Reproduced with permission from the Electric Power Research Institute, Inc.). (b) Thyristor valves in converter station (Reproduced with permission from IEEE.)

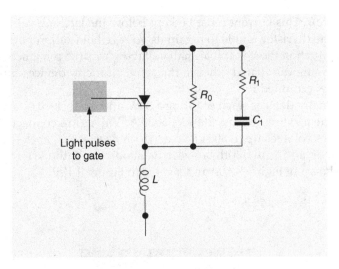

Figure 9.5 Circuitry associated with each thyristor

In order to turn on a thyristor a pulse of current from a gate circuit that is at the same potential as the cathode is required. Alternatively, a light trigged thyristor is triggered by a light pulse sent through a fibre-optic channel.

When many thyristors are connected in series to form a valve, it is necessary to:

1. obtain a uniform voltage distribution across each thyristor.
2. retain uniform transient voltage distributions with time.
3. control the rate of rise of current.

These are achieved by the auxiliary electrical circuitry shown in Figure 9.5. The inductor, L, limits the rate of rise of current during the early stage of conduction. The chain R_1, C_1 bypasses the thyristor, thus controlling the negative recovery voltage. The d.c. grading resistor, R_0, ensures uniform distribution of voltage across each thyristor in a valve.

9.3.2 Insulated Gate Bipolar Transistors

Recently the insulated gate bipolar thyristor (IGBT) has been used in h.v.d.c. schemes. The IGBT is a development of the MOSFET, in which removal of the voltage from the gate switches off the through current, thereby allowing power to be switched on or off at any point of an a.c. cycle. IGBTs have ratings up to 4 kV and 1000 A, although the on-state voltage drop and switching losses are larger than with a thyristor.

When a sufficient voltage is applied to the Gate with respect to the Emitter, it inverts the p region below the Gate (shaded area) thus forming a diode between the Emitter (n region) and Collector (p substrate). As shown in Figure 9.6(c), then when

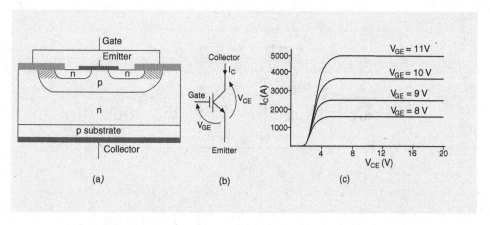

Figure 9.6 (a) structure of an IGBT (b) symbol, (c) characteristic

in addition the Emitter to Collector voltage is greater than 0.7 V, conduction between Emitter and Collector commences.

9.4 Current Source Converter h.v.d.c.

A converter is required at each end of a d.c. line and operates as a rectifier (a.c. to d.c.) or an inverter (power transfer from d.c. to a.c.). The valves at the sending end of the link rectify the alternating current, providing direct current which is transmitted to the inverter. Here, it is converted back into alternating current which is fed into the connected a.c. system (Figure 9.7(a)). If a reversal of power flow is required, the inverter and rectifier exchange roles and the direct voltage is reversed (Figure 9.7 (b)). This is necessary because the direct current can flow in one direction only (anode to cathode in the valves), so to reverse the direction (or sign) of power the voltage polarity must be reversed.

The alternating-current waveform injected by the inverter into the receiving end a.c. system, and taken by the rectifier, is roughly trapezoidal in shape, and thus produces not only a fundamental sinusoidal wave but also harmonics of an order dependent on the number of valves. For a six-valve bridge the harmonic order is $6n \pm 1$, that is 5, 7, 11, 13, and so on. Filters are installed at each converter station to tune out harmonics up to the 25th.

9.4.1 Rectification

For an initial analysis of the rectifier, a three-phase arrangement is shown in Figure 9.8(a). Figure 9.8(b) shows the currents and voltages in the three phases of the supply transformer. With no gate control, conduction will take place between the cathode and the anode of highest potential. Hence the output-voltage wave is the thick line and the current output is continuous. As the waveform segment from

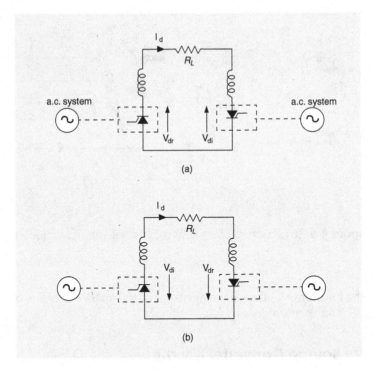

Figure 9.7 (a) Symbolic representation of two alternating current systems connected by a direct-current link; V_{dr} = direct voltage across rectifier, V_{di} = direct voltage across inverter, (b) System as in part (a) but power flow reversed

$\frac{\pi}{2} - \frac{\pi}{3}$ to $\frac{\pi}{2} + \frac{\pi}{3}$ repeats, the mean value of the direct-output voltage, V_0, can be obtained by integrating the sine wave from $\frac{\pi}{2} - \frac{\pi}{3}$ to $\frac{\pi}{2} + \frac{\pi}{3}$. Therefore:

$$V_0 = \frac{1}{2\pi/3} \int_{\frac{\pi}{2} - \frac{\pi}{3}}^{\frac{\pi}{2} + \frac{\pi}{3}} \hat{V} \sin(\omega t) d(\omega t)$$

$$= \frac{\hat{V} \sin\left(\frac{\pi}{3}\right)}{\frac{\pi}{3}} = \hat{V} \frac{3\sqrt{3}}{2\pi} = 0.83\hat{V} \tag{9.1}$$

where $\hat{V}$ is the peak a.c. voltage.

Defining the r.m.s. line-to-line voltage as V_L, from equation (9.1) V_0 can be obtained as:

$$V_0 = \hat{V} \frac{3\sqrt{3}}{2\pi} = V_L \frac{\sqrt{2}}{\sqrt{3}} \times \frac{3\sqrt{3}}{2\pi} = \frac{3}{\sqrt{2}} \frac{V_L}{\pi} = 0.675 V_L \tag{9.2}$$

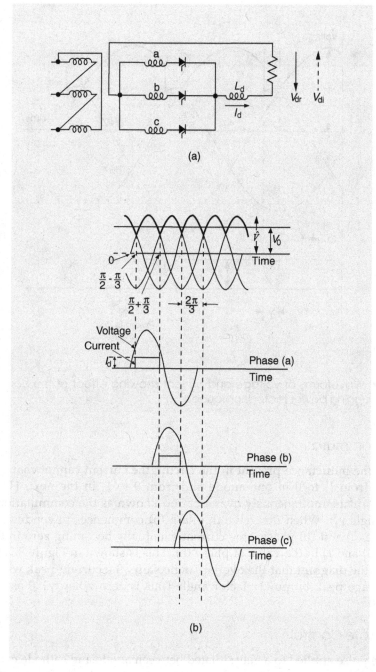

Figure 9.8 (a) Three-phase rectifier; V_{di} = voltage of inverter, (b) Waveforms of anode voltage and rectified current in each phase

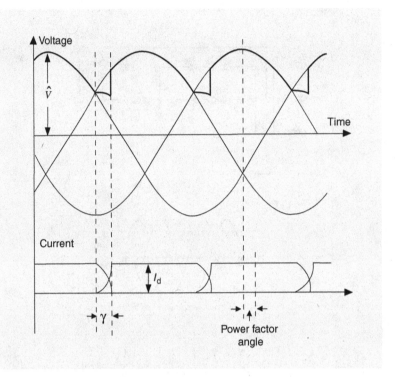

Figure 9.9 Waveforms of voltage and current showing effect of the commutation angle γ. A lagging power factor is produced

9.4.1.1 Commutation

Owing to the inductance present in the circuit, the current cannot change instantaneously from I_d to 0 in one anode and from 0 to I_d in the next. Hence, two anodes conduct simultaneously over a period known as the commutation time or overlap angle (γ). When the valve in phase (b) commences to conduct, it short-circuits the (a) and (b) phases, the current eventually becoming zero in the valve of phase (a) and I_d in the valve of phase (b). This is shown in Figure 9.9. It can be seen from the diagram that the overlap angle shifts the current peak with respect to the voltage peak by power factor angle (this is zero when γ=0 as shown in Figure 9.8).

9.4.1.2 Gate Control

A positive pulse applied to a gate situated between anode and cathode controls the instant at which conduction commences, and once conduction has occurred the gate exercises no further control. In the voltage waveforms shown in Figure 9.10 the conduction in the valves has been delayed by an angle α by suitably delaying the application of positive voltage to the gates. Ignoring the commutation angle γ, the new direct-output voltage with a delay angle of α is,

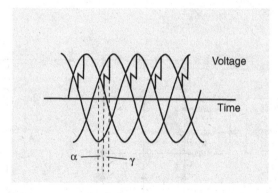

Figure 9.10 Waveforms of rectifier with instant of firing delayed by an angle α by means of gate control

$$V'_0 = \frac{1}{2\pi/3} \int_{\frac{\pi}{2}-\frac{\pi}{3}-\alpha}^{\frac{\pi}{2}+\frac{\pi}{3}+\alpha} \hat{V} \sin(\omega t) d(\omega t)$$

$$= \frac{3\hat{V}}{2\pi} \int_{-\frac{\pi}{3}-\alpha}^{\frac{\pi}{3}+\alpha} \cos(\omega t) d(\omega t) = \frac{3\hat{V}}{2\pi} 2 \sin\left(\frac{\pi}{3}\right) \cos \alpha$$

$$= \frac{3\sqrt{3}\hat{V}}{\pi} \cos \alpha$$

$$V'_0 = V_0 \cos \alpha \tag{9.3}$$

where V_0 is the maximum value of direct-output voltage as defined by equation (9.1).

9.4.1.3 Bridge Connection

The bridge arrangement shown in Figure 9.11 is the common implementation of CSC h.v.d.c schemes, mainly because the d.c. output voltage is doubled. There are always two valves conducting in series. The corresponding voltage waveforms are shown in Figure 9.12 along with the currents (assuming ideal rectifier operation).

The sequence of events in the bridge connection is as follows (see Figures 9.11 and 9.12). Assume that the transformer voltage V_A is most positive at the beginning of the sequence, then valve 1 conducts and the current flows through valve 1 and the load then returns through valve 6 as V_B is most negative. After this period, V_C becomes the most negative and current flows through valves 1 and 2. Next, valve 3 takes over from valve 1, the current still returning through valve 2. The complete sequence of valves conducting is therefore: 1 and 6; 1 and 2; 3 and 2; 3 and 4; 5 and 4; 5 and 6; 1 and 6. Control may be obtained in exactly the same manner as previously described.

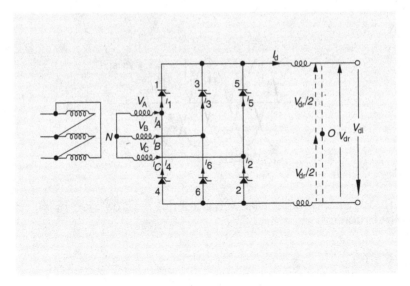

Figure 9.11 Bridge arrangement of valves

The direct-voltage output with the bridge can be calculated by defining a fictitious mid-point on the d.c. side (point O of Figure 9.11). The Bridge arrangement may then be redrawn as shown in Figure 9.13. This is the same as two three-phase rectifiers shown in Figure 9.8(a), but having an output voltage of $V_{dr}/2$ (here V_{dr} is the d.c.

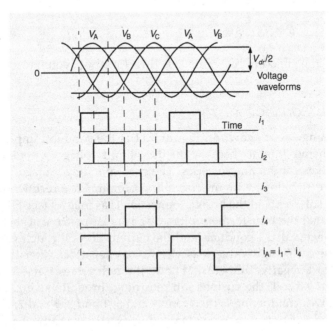

Figure 9.12 Idealized voltage and current waveforms for bridge arrangement

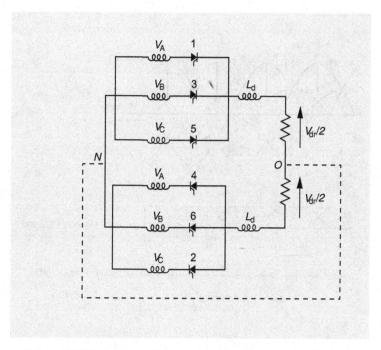

Figure 9.13 Equivalent circuit of the bridge arrangement

voltage of the bridge circuit). Hence, from equation (9.2), the mean d.c. voltage for the bridge rectifier is

$$V_0 = 2 \times \frac{3}{\sqrt{2}} \frac{V_L}{\pi} = \frac{3\sqrt{2}V_L}{\pi} = 1.35V_L \tag{9.4}$$

If the analysis used to obtain equation (9.3) is repeated for the bridge, it will be shown that $V'_0 = V_0 \cos \alpha$.

9.4.1.4 Current Relationships in the Bridge Circuit

The voltage waveforms with delay and commutation time accounted for are shown in Figure 9.14. Commutation from one valve to another can be explained by following the dark line in the positive part of Figure 9.14(a). Consider point P, where valve 3 is conducting and continues to conduct until point Q. At Q, valve 5 is triggered. Then the positive busbar (common cathode) voltage is increased from the phase (b) voltage to the average of phase (b) and (c) voltages (this is equal to half the inverse of the phase (a) voltage as shown in the dotted lines of Figure 9.14(a)). From R to S, both valves (3 and 5) conduct. At S, the commutation process finishes and only valve 5 conducts, then the positive busbar voltage is equal to the phase (c) voltage.

During the commutation process when two valves are conducting simultaneously, the two corresponding secondary phases of the supply transformer are

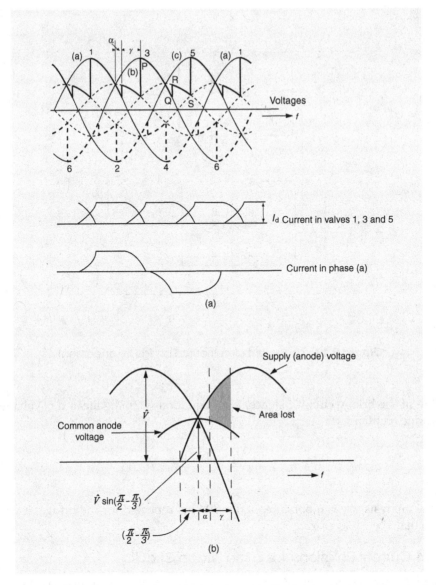

Figure 9.14 (a) Voltage and current waveforms in the bridge connection, including commutation (γ) and delay (α). Rectifier action, (b) Expanded waveforms showing voltage drop due to commutation

short-circuited and if the voltage drop across the valves is neglected the following analysis applies.

When two phases of the transformer each of leakage inductance L henries are effectively short-circuited, the short-circuit current (i_S) is governed by the equation,

$$2L\frac{di_S}{dt} = \hat{V}_L \sin \omega t$$

$\hat{V}_L \sin \omega t$ is the voltage across two phases that are short circuited by two conducting valves

$$\therefore i_S = -\frac{\hat{V}_L}{2L}\frac{\cos \omega t}{\omega} + A$$

where A is a constant of integration and $\hat{V}_L$ is the peak value of the line-to-line voltage. Now $\omega t = \alpha$ when $i_S = 0$

$$\therefore A = \frac{\hat{V}_L}{2\omega L}\cos \alpha$$

Also, when

$$\omega t = \alpha + \gamma \quad i_S = I_d$$

$$\therefore I_d = \frac{\hat{V}_L}{2\omega L}[\cos \alpha - \cos(\alpha + \gamma)]$$

$$= \frac{V_L}{\sqrt{2}\omega L}[\cos \alpha - \cos(\alpha + \gamma)]$$

where $V_L = $ r.m.s. line-to-line voltage and, as for the bridge circuit, $V_0 = \frac{3\sqrt{2}}{\pi}V_L$.
Hence

$$I_d = \frac{\pi V_0}{3\sqrt{2}} \cdot \frac{1}{\sqrt{2}X}[\cos \alpha - \cos(\alpha + \gamma)]$$

$$= \frac{\pi V_0}{6X} \cdot [\cos \alpha - \cos(\alpha + \gamma)]$$

(9.5)

The mean direct-output voltage, with gate delay angle α only considered, has been shown to be $V_0 \cos \alpha$. With both α and the commutation angle γ, the voltage with α only will be modified by the subtraction of a voltage equal to the mean of the area under the anode voltage curve lost due to commutation. Referring to Figure 9.14(b), the supply (anode) voltage is given by $\hat{V} \sin(\omega t)$ and the common anode voltage is given by $\hat{V} \sin\left(\frac{\pi}{2} - \frac{\pi}{3}\right)\cos(\omega t)$ (note that with respect to the supply voltage this is a cosine function with a peak value $\hat{V} \sin\left(\frac{\pi}{2} - \frac{\pi}{3}\right)$). Therefore the area between input voltage wave and common anode voltage is

$$\int_{\frac{\pi}{2}-\frac{\pi}{3}+\alpha}^{\frac{\pi}{2}-\frac{\pi}{3}+\alpha+\gamma} \hat{V} \sin(\omega t) d(\omega t) - \int_{\alpha}^{\alpha+\gamma} \hat{V} \sin\left(\frac{\pi}{2}-\frac{\pi}{3}\right) \cos(\omega t) d(\omega t)$$

$$= \hat{V}\left[-\cos\left(\frac{\pi}{2}-\frac{\pi}{3}+\alpha+\gamma\right) + \cos\left(\frac{\pi}{2}-\frac{\pi}{3}+\alpha\right) - \sin\left(\frac{\pi}{2}-\frac{\pi}{3}\right)\sin(\alpha+\gamma) + \sin\left(\frac{\pi}{2}-\frac{\pi}{3}\right)\sin\alpha\right]$$

$$= \hat{V}\left[-\sin\left(\frac{\pi}{3}-\alpha-\gamma\right) + \sin\left(\frac{\pi}{3}-\alpha\right) - \left(\cos\frac{\pi}{3}\right)\sin(\alpha+\gamma) + \left(\cos\frac{\pi}{3}\right)\sin\alpha\right]$$

$$= \hat{V}\left[\sin\frac{\pi}{3}\right][\cos\alpha - \cos(\alpha+\gamma)]$$

Voltage drop (mean value of lost area)

$$= \frac{\hat{V}\sin(\pi/3)}{2(\pi/3)}[\cos\alpha - \cos(\alpha+\gamma)]$$

$$= \frac{V_0}{2}[\cos\alpha - \cos(\alpha+\gamma)]$$

The direct-voltage output,

$$V_d = V_0 \cos\alpha - \frac{V_0}{2}[\cos\alpha - \cos(\alpha+\gamma)]$$

$$= \frac{V_0}{2}[\cos\alpha + \cos(\alpha+\gamma)] \tag{9.6}$$

Adding equations (9.5) and (9.6),

$$\frac{3X}{\pi}I_d + V_d = V_0 \cos\alpha$$

$$\therefore V_d = V_0 \cos\alpha - R_c I_d \tag{9.7}$$

where $R_c = \dfrac{3X}{\pi}$

Equation (9.7) may be represented by the equivalent circuit shown in Figure 9.15, the term $R_c I_d$ represents the voltage drop due to commutation and not a physical resistance drop. It should be remembered that V_0 is the theoretical maximum value of direct-output voltage and it is evident that V_d can be varied by changing V_0 (control of transformer secondary voltage by tap changing) and by changing α.

9.4.1.5 Power Factor

Applying Fourier analysis to the phase (a) current waveform shown in Figure 9.14 (a) and neglecting the effect of the commutation angle γ on the current waveform (Figure 9.12), the instantaneous value of phase (a) current is obtained as:

$$i_a = \frac{2\sqrt{3}}{\pi}I_d\left[\cos(\omega t) + \frac{1}{5}\cos(5\omega t) - \frac{1}{7}\cos(7\omega t) + \frac{1}{11}\cos(11\omega t) - \frac{1}{13}\cos(13\omega t) + \cdots\cdots\right]$$

$$\tag{9.8}$$

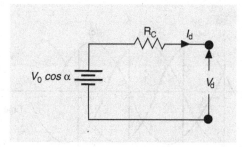

Figure 9.15 Equivalent circuit representing operation of a bridge rectifier. Reactance per phase $X\,(\Omega)$

The r.m.s value of the fundamental line current

$$I_L = \frac{2\sqrt{3}}{\pi}\frac{1}{\sqrt{2}}I_d = \frac{\sqrt{6}}{\pi}I_d \tag{9.9}$$

Neglecting losses and by equating a.c. input power to d.c. output power

$$\sqrt{3}V_L I_L \cos\phi = V_d I_d$$

Substituting for I_L and V_d from (9.6) and (9.9):

$$\sqrt{3}V_L \times \frac{\sqrt{6}}{\pi}I_d \cos\phi = \frac{V_0}{2}[\cos\alpha + \cos(\alpha + \gamma)] \times I_d$$

Since $V_0 = \dfrac{3\sqrt{2}V_L}{\pi}$, the power factor is given approximately (as this was derived by neglecting losses) by

$$\cos\phi = \frac{1}{2}[\cos\alpha + \cos(\alpha + \gamma)] \tag{9.10}$$

9.4.2 Inversion

With rectifier operation the output current I_d and output voltage V_d are such that power is absorbed by a load. For inverter operation it is required to transfer power from the direct-current to the alternating-current systems, and as current can flow only from anode to cathode (i.e., in the same direction as with rectification), the direction of the associated voltage must be reversed. An alternating-voltage system must exist on the primary side of the transformer, and gate control of the converters is essential.

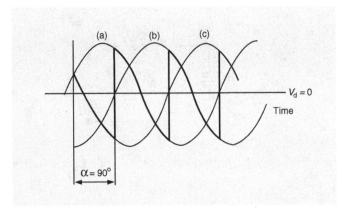

Figure 9.16 Waveforms with operation with $\alpha = 90°$, direct voltage zero. Transition from rectifier to inverter action

If the bridge rectifier is given progressively greater delay the output voltage decreases, becoming zero when α is 90°, as shown in Figure 9.16. With further delay the average direct voltage becomes negative and the applied direct voltage (from the rectifier) forces current through the valves against this negative or back voltage. The converter thus receives power and inverts. The inverter bridge is shown in Figure 9.17(a) and the voltage and current waveforms in Figure 9.17(b). Hence the change from rectifier to inverter action, and vice versa, is smoothly obtained by control of α. This may be seen by consulting Figures 9.9 ($\alpha = 0$), 9.16 ($\alpha = 90°$), and 9.17(b) ($\alpha > 90°$).

Consider Figure 9.17. With valve 3 conducting, valve 5 is triggered at time A and as the cathode is held negative to the anode by the applied direct voltage (V_d), current flows, limited only by the circuit impedance. When time B is reached, the voltage applied across anode-to-cathode is zero and the valve endeavours to cease conduction. The large d.c.-side inductance L_d, however, which has previously stored energy, now maintains the current constant ($e = -L_d(di/dt)$ and if $L_d \to \infty$, $di/dt \to 0$). Conduction in valve 5 continues until time C, when valve 1 is triggered. As the anode-to-cathode voltage for valve 1 is greater than for valve 5, valve 1 will conduct, but for a time valves 5 and 1 conduct together (commutation time), the current gradually being transferred from valve 5 to valve 1 until valve 5 is non-conducting at point D. The current changeover must be complete before (F) by a time (δ_0) equal to the recovery time of the valves. If triggering is delayed to point F, valve 5, which is still conducting, would be subject to a positively rising voltage and would continue to conduct into the positive half-cycle with breakdown of the inversion process. Hence, triggering must allow cessation of current flow before time F.

The angle (δ) between the extinction of valve 1 and the point F, where the anode voltages are equal, is called the extinction angle, that is sufficient time must be allowed for the gate to regain control. The minimum value of δ is δ_0.

For inverter operation it is usual to replace the delay angle α by $\beta = 180° - \alpha$, the angle of advance. Hence β is equal also to $(\gamma + \delta)$.

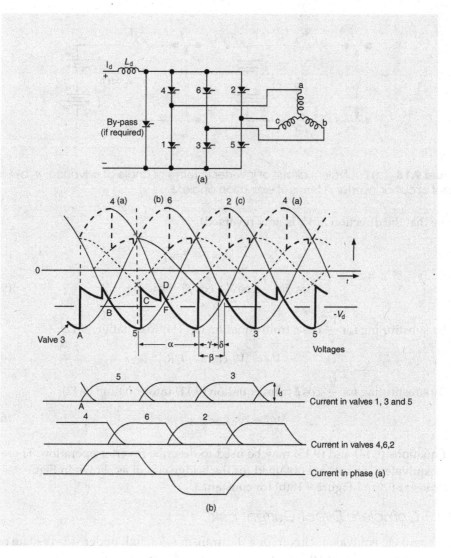

Figure 9.17 (a) Bridge connection–inverter operation, (b) Bridge connection–inverter voltage and current waveforms

Replacing α by $(180° - \beta)$ and γ by $(\beta - \delta)$ the following are obtained from equations (9.5), (9.6) and (9.10):

$$I_d = \frac{V_0}{2R_c} \cdot [\cos(180° - \beta) - \cos(180° - \delta)]$$

$$I_d = \frac{V_0}{2R_c} \cdot [\cos \delta - \cos \beta] \tag{9.11}$$

$$-V_d = \frac{V_0}{2}[\cos(180° - \beta) + \cos(180° - \delta)]$$

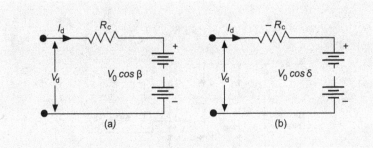

Figure 9.18 (a) Equivalent circuit of inverter in terms of angle of advance, β (b) Equivalent circuit of inverter in terms of extinction angle, δ

(note that the direction of V_d is now reversed)

$$V_d = \frac{V_0}{2}[\cos\delta + \cos\beta] \tag{9.12}$$

$$\cos\phi = \frac{1}{2}[\cos\delta + \cos\beta] \quad \text{leading} \tag{9.13}$$

By substituting for $\frac{V_0}{2}\cos\beta$ from equation (9.11) into equation (9.12):

$$V_d = [V_0\cos\delta - I_dR_c] \tag{9.14}$$

By substituting for $\frac{V_0}{2}\cos\delta$ from equation (9.11) into equation (9.12):

$$V_d = [V_0\cos\beta + I_dR_c] \tag{9.15}$$

Equations (9.14) and (9.15) may be used to describe inverter operation. Therefore two equivalent circuits are obtained for the bridge circuit as shown in Figure 9.18(a) for constant β and Figure 9.18(b) for constant δ.

9.4.3 Complete Direct-Current Link

The complete equivalent circuit for a d.c. transmission link under steady-state operation is shown in Figure 9.19. If both inverter and rectifier operate at constant delay angles the current transmitted

$$I_d = \frac{V_{dr} - V_{di}}{R_L}$$

From equations (9.7) and (9.15) for constant β operation:

$$I_d = \frac{[V_{0r}\cos\alpha - I_dR_{cr}] - [V_{0i}\cos\beta + I_dR_{ci}]}{R_L}$$

$$\therefore I_d = \frac{V_{0r}\cos\alpha - V_{0i}\cos\beta}{R_L + R_{cr} + R_{ci}} \tag{9.16}$$

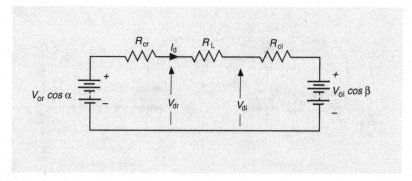

Figure 9.19 Equivalent circuit of complete link with operation with given delay angles for rectifier and for inverter

From equations (9.7) and (9.14) for constant δ operation:

$$I_d = \frac{[V_{0r}\cos\alpha - I_d R_{cr}] - [V_{0i}\cos\delta - I_d R_{ci}]}{R_L}$$

$$\therefore I_d = \frac{V_{0r}\cos\alpha - V_{0i}\cos\delta}{R_L + R_{cr} - R_{ci}} \tag{9.17}$$

where R_L is the loop resistance of the line or cable, and R_{cr} and R_{ci} are the effective commutation resistances of the rectifier and inverter, respectively.

The equations governing the operation of the inverter may be summarized as follows:

$$V_d = \frac{3\sqrt{2}V_L}{\pi}\cos\beta + \frac{3\omega L}{\pi}I_d \quad \text{constant } \beta \tag{9.18}$$

$$V_d = \frac{3\sqrt{2}V_L}{\pi}\cos\delta - \frac{3\omega L}{\pi}I_d \quad \text{constant } \delta \tag{9.19}$$

(Note that L is the leakage inductance per phase of the inverter transformer.

For the complete direct current link:

1. The magnitude of the direct current can be controlled by variation of α, β (or δ), V_{0r} and V_{0i} (the last two by tap-changing of the supply transformers).
2. V_{dr} becomes maximum and the power factor becomes closer to unity when $\alpha = 0°$ (see equation (9.10)). However α is maintained at a minimum value, α_{min}, to make sure that the converter valves have a small positive voltage during turn on.
3. Inverter control using constant delay angle has the disadvantage that if δ and hence β are too large, excessively high reactive-power demand results. It can be seen from Figure 9.17(b) that the inverter currents are then significantly out of phase with the anode voltages and hence a large requirement for reactive power

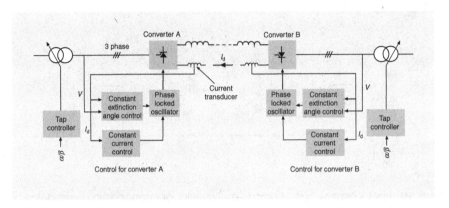

Figure 9.20 Schematic diagram of control of an h.v.d.c. system

is established. This also can be seen from equation (9.13) where power factor becomes small with large β and δ.

4. A reduction in the direct voltage (this could happen during grid faults) to the inverter results in an increase in the commutation angle γ, and if β is made large to cover this the reactive-power demand will again be excessive. Therefore it is more usual to operate the inverter with a constant δ, which is achieved by the use of suitable control systems.

9.4.4 Control of h.v.d.c. Link

A schematic diagram of the control systems is shown in Figure 9.20. The rectifier and inverter change roles as the required direction of power flow dictates, and it is necessary for each device to have dual-control systems. In Figure 9.21 the full characteristics of the two converters of a link are shown, with each converter operating as rectifier and inverter in turn. The thick line PQRST represents the operation of Converter A. Its operation can be described as follows:

1. From P to Q, Converter A acts as a rectifier and the optimum characteristic with a minimum α value, α_{min}, is shown.
2. As the current rating of the valves should not be exceeded, constant-current (CC) control is employed from Q to R. With constant-current control, α is increased, thus V_{dr} is reduced to maintain constant current in the d.c. link.
3. The output voltage-current characteristic crosses the $V_d = 0$ axis at point R, below which Converter A acts as an inverter. Constant current operation can still be maintained by decreasing the angle of advance, β.
4. At S, β reaches δ_0 and Converter A operates in constant excitation angle control.

A similar characteristic is shown for Converter B (VWXYZ), which commences as an inverter and with constant-current control (β increasing) eventually changes to

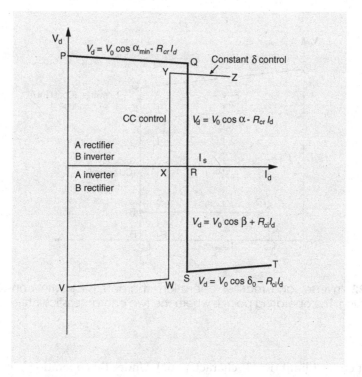

Figure 9.21 Voltage-current characteristics of converters with compounding

rectifier operation. It is seen that the current setting for Converter A is larger than that for Converter B.

Figure 9.22 shows the voltage current characteristics in the first quadrant where Converter A operates as a rectifier and Converter B operates as an inverter. In the rectifier the transmitted current is regulated in CC control by varying the delay angle and hence V_{dr}. This is described by QXR. The inverter characteristic is chosen such that I_{order} for the inverter is slightly lower than that of the rectifier. Therefore under normal operation the inverter operates in the constant extinction angle control and point X gives the operating condition. If, for example, the d.c. link power needs to be halved, I_{order} of the rectifier is halved and the new operating point becomes Z.

In case of an a.c. side remote fault where the d.c. voltage is depressed and the rectifier side PQ characteristics falls below the inverter constant excitation control characteristic, the inverter changes its role and tries to maintain the d.c. current flow in the same direction. The new operating point is Y and the current is maintained at value B thus the power transmitted is smaller than before. The rectifier now operates at a constant α (closer to α_{min}).

In case of an a.c. side fault closer to the rectifier, the d.c. voltage may be depressed to zero and in order to assist the d.c. link to recover from the fault, the original static characteristics of the rectifier shown in Figure 9.21 is modified using a voltage

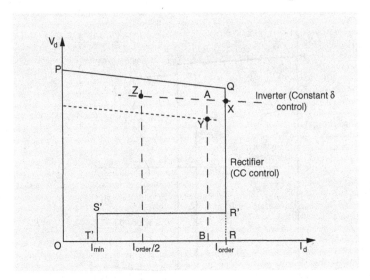

Figure 9.22 Inverter and rectifier operation characteristics with constant-current compounding. The operating point is where the two characteristics intersect

dependent current limit (R'S'T' characteristic). Under this operation α may increase to 90°.

The rectifier tap-changer is used to keep the delay angle close to α_{min} so that reactive power requirement is reduced. Further, when the current control is transferred to the inverter (point Q), the rectifier tap-changer tries to restore the original conditions. The value of the voltage margin is chosen to avoid frequent operation in the inverter control region.

To summarize, with normal operation the rectifier operates at constant current and the inverter at constant δ; under emergency conditions the rectifier is at zero firing-delay (that is $\alpha = 0$) and the inverter is at constant current.

9.4.5 Transmission Systems

The cheapest arrangement is a single conductor with a ground return. This, however, has various disadvantages. The ground-return current results in the corrosion of buried pipes, cable sheaths, and so on, due to electrolysis. With submarine cables the magnetic field set up may cause significant errors in ships' compass readings, especially when the cable runs north-south and unacceptable environmental impact. This system is shown in Figure 9.23(a). Two variations on two-conductor schemes are shown in Figure 9.23(b) and (c). The latter has the advantage that if the ground is used in emergencies, a double-circuit system is formed to provide some security of operation. Note that in the circuits of Figure 9.23(c), transformer connections provide twelve pulse operation, thereby reducing harmonics generated on the a.c. sides.

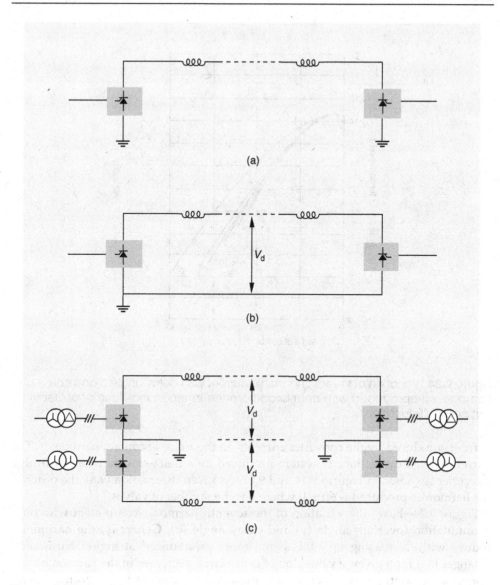

Figure 9.23 Possible conductor arrangements for d.c. transmissions: (a) ground return; (b) two conductors, return earthed at one end; (c) double-bridge arrangement

9.4.6 Harmonics

A knowledge of the harmonic components of voltage and current in a power system is necessary because of the possibility of resonance and also the enhanced interference with communication circuits. The direct-voltage output of a converter has a waveform containing a harmonic content, which results in current and voltage harmonics along the line. These are normally reduced by a smoothing inductor. The

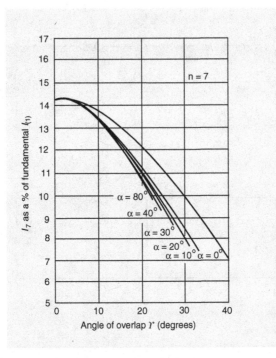

Figure 9.24 Variation of seventh harmonic current with delay angle a and commutation angle (Reproduced with permission from the International Journal of Electrical Engineering Education)

currents produced by the converter currents on the a.c. side contain harmonics. The current waveform in the a.c. system produced by a delta-star transformer bridge converter is shown in Figures 9.14 and 9.17. As given in equation (9.8), the order of the harmonics produced is $6n \pm 1$, where n is the number of valves.

Figure 9.24 shows the variation of the seventh harmonic component with both commutation (overlap) angle (γ) and delay angle (α). Generally, the harmonics reduce with decreasing in γ this being more pronounced at higher harmonics. Changes in α for a given γ value do not cause large decreases in the harmonic components, the largest change being for values between 0 and 10°. For normal operation, α is less than 10° and γ is perhaps of the order of 20°; hence the harmonics are small. During a severe fault, as discussed before, α may reach nearly 90°, γ is small, and the harmonics produced are large.

The harmonic voltages and currents produced in the a.c. system by the converter current waveform may be determined by representing the system components by their reactances at the particular harmonic frequency. Most of the system components have resonance frequencies between the fifth and eleventh harmonics.

It is usual to provide filters (L-C shunt resonance circuits) tuned to the harmonic frequencies. A typical installation is shown in detail in Figure 9.25. At the fundamental frequency the filters are capacitive and help to meet the reactive-power requirements of the converters.

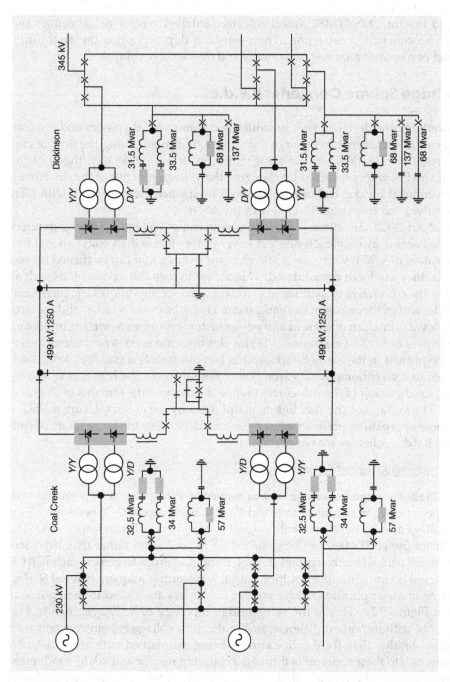

Figure 9.25 Single-line diagram of the main circuit of an h.v.d.c. scheme showing filter banks and shunt compensation. (Reproduced with permission from IEEE)

9.4.7 Variable Compensators

Rapid control of the reactive power compensation is achieved using TCR (thyristor-controlled reactor), MSR/MSC (mechanically switched reactor or capacitor) and TSC (thyristor-switched capacitor). Their selection depends upon the application and speed of response required for stability and overvoltage control.

9.5 Voltage Source Converter h.v.d.c.

The converters considered so far use naturally commutated thyristors and are connected to a d.c. circuit that has a large inductance thus maintaining the flow of current. Current always flows in the same direction (from anode to cathode of the thyristors) and power is reversed by changing the voltage polarity of the d.c. circuit. This conventional h.v.d.c. technology has been used since the mid-1950s with mercury arc valves and from around 1980 with thyristors.

Since about 2000, an alternative technology using voltage source converters (VSC), has become available although at lower power levels than conventional h.v. d.c. The valves of VSC h.v.d.c. use semiconductor devices that can be turned on and off, that is, they are force commutated. This ability to turn the valves off as well as on, allows the converters to synthesise a voltage wave of any frequency, phase and magnitude, within the rating of the equipment. Hence both the rectifier and inverter of a VSC h.v.d.c. link can operate at any power factor, exporting as well as importing reactive power from the a.c. systems. VSC h.v.d.c. does not need synchronous generators to be present in the a.c. networks and in fact can supply a passive (dead) load. In contrast to conventional h.v.d.c. the power flow through the link is reversed by changing the direction of the d.c. current while the d.c. voltage remains of the same polarity. The voltage of the d.c. link is maintained by capacitors. Using a VSC, a much closer approximation to a sine wave is created and so the large filters of conventional h.v.d.c. schemes are not needed.

9.5.1 Voltage Source Converter

Figure 9.26 shows a one-leg voltage source converter. When switch T_1 is on, the voltage at A becomes V_d with respect to N and $V_d/2$ with respect to O. When switch T_2 is on, the voltage at A becomes zero with respect to N and $-V_d/2$ with respect to O. The two switches use IGBTs that can be switched off as well as on rather than thyristors which turn off only when the current flowing through it drops to zero. If each IGBT is turned on and off for 10 ms, the resulting output waveform is a square wave at 50 Hz.

For three-phase applications, three one-leg converters are placed in each phase as shown in Figure 9.27. This is the same bridge topology of a CSC (shown in Figure 9.11) but with important differences. The d.c. link voltage is held constant with a capacitance rather than the d.c. link current being maintained with an inductance. The connection to the a.c. circuit is through a coupling reactor and so the fundamental operation of the VSC is of a voltage source behind a reactor in a manner similar to that of the synchronous generator described in Chapter 3.

The VSC is normally connected to a transformer to bring the output voltage to be equal to the a.c. system voltage. If the transformer is connected in star, then with

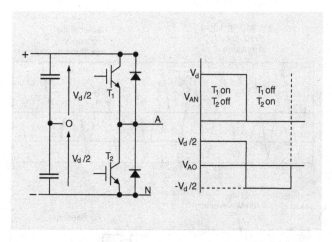

Figure 9.26 One-leg VSC

respect to the neutral point the voltage of the a-phase looks the same as V_{AO} of Figure 9.26. Other phase voltages will be phase shifted by 120°.

The simple bridge VSC of Figure 9.27 with each device switched once per 50 Hz cycle is not normally used due to the high harmonic content at the output. A square wave contains all the odd harmonics (excluding the triplen harmonics). As IGBTs can be switched many times during one a.c. cycle, the two switches of each phase are switched using a sine-triangular pulse width modulation (PWM) technique. The switching instances are determined by comparing a sinusoidal modulating signal with a triangular carrier signal (see Figure 9.28). A sinusoidal modulating signal representing the desired output voltage is compared with a high frequency triangular carrier signal. When the magnitude of the carrier is higher than the modulating signal, the upper switch is turned on. On the other hand, when the magnitude of the carrier is lower than the modulating signal, the lower switch is turned on.

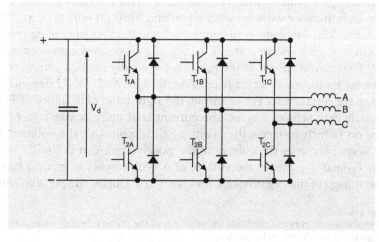

Figure 9.27 Bridge VSC

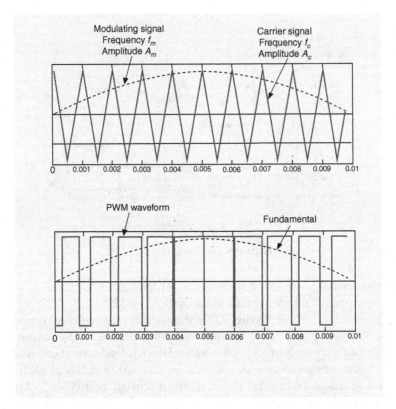

Figure 9.28 PWM waveform (only half a cycle is shown)

In high power applications, such as h.v.d.c., each switch in the bridge is made up of a large number of IGBTs connected in series to form a valve. It is important to operate every IGBT in a valve at the same time. This requires complex circuitry. Another disadvantage of this arrangement is that high frequency switching (around 2 kHz) of high current and voltage results in much higher losses than the equivalent CSC arrangement where each thyristor valve is switched on and turns off only once in each cycle.

An alternative VSC topology is to connect a large number of one-leg converters in series to form a multi-level converter. One leg of a three-level converter is shown in Figure 9.29. The ground shown in the figure is the solidly earthed neutral of the three phase converter transformer.[1] Point A provides 0, $V_d/2$ and $-V_d/2$ depending on the switches that are conducting. For example, the right hand side figure of Figure 9.29 shows the switching sequence when the current is at unity power factor. When T_{23} and T_{24} are on (which connects the positive rail to point A), the voltage at A with respect to ground becomes the voltage of the positive rail, that is $V_d/2$. On the other hand when T_{21} and T_{22} are on, the voltage at A with respect to ground becomes the same as the voltage of the negative rail, that is $-V_d/2$. During the positive half cycle of

---

[1] For ease of explanation it was assumed that the windings of the converter transformer connected to the VSC is star connected and the neutral is earthed.

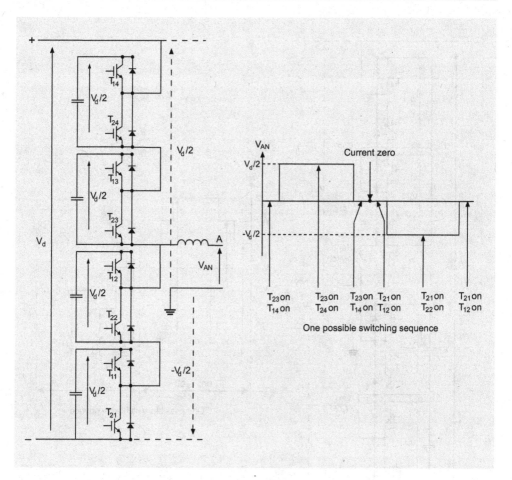

Figure 9.29 A three-level modular converter with the switching sequence for unity power factor operation

the current, if T_{23} and T_{14} are on, then the voltage at A with respect to ground becomes zero. During the negative half cycle of the current the voltage at A with respect to ground becomes zero, if T_{21} and T_{12} are on. However, when the power factor of the current is lagging or leading, the switching sequence becomes more complicated.

A five-level modified modular converter could be obtained with the same number of one-leg converters as Figure 9.29, but with two series switches as shown in Figure 9.30. Switch T_1 is on during the positive half cycle and upper two one-leg converters constructing the positive half cycle of the output. On the other hand, switch T_2 is on during the negative half cycle and lower two one-leg converters constructing the negative half cycle of the output.

Other multi-level configurations such as diode-clamped topology and the capacitor-clamped topology exist. However industrial practice is presently converging on modular topologies as this approach makes the implementation easier. A

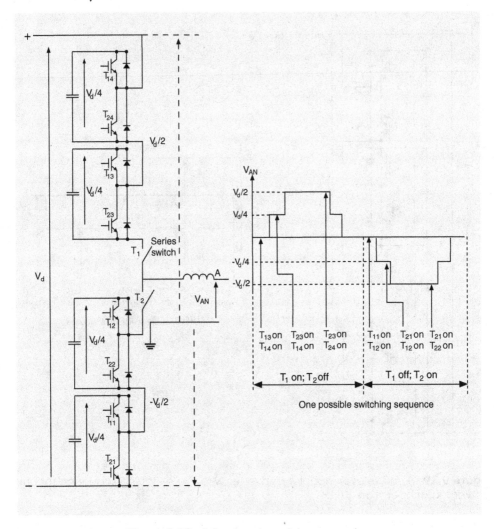

Figure 9.30 A five-level modular converter

multi-level converter used for h.v.d.c. application may have as many as 300 levels to make a waveform close to a sinusoid.

Compared to the bridge shown in Figure 9.27, the modular converter has lower losses as the valves in a bridge VSC operating with PWM turn on and off many times per cycle. In the modular VSC each valve switches only once per cycle. However modular converters require complex balancing circuits to balance the voltage in each d.c. capacitor.

9.5.2 Control of VSC h.v.d.c.

The VSC connected to the a.c. system can be considered as a voltage behind a reactance as shown in Figure 9.31. The reactance X represents the leakage reactance of

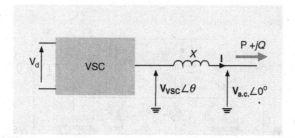

Figure 9.31 Equivalent circuit of VSC a.c. side

the transformer and any additional coupling reactance used between the transformer and VSC. The resistance is neglected. The VSC output voltage is $\mathbf{V_{VSC}} \angle \theta$ and the a.c. voltage is $\mathbf{V_{a.c.}} \angle 0°$.

Active and reactive power transfer to the grid is:

$$P + jQ = \mathbf{V_{a.c.}} \times \mathbf{I}^*$$

$$= \mathbf{V_{a.c.}} \angle 0° \times \left[\frac{\mathbf{V_{VSC}} \angle - \theta - \mathbf{V_{a.c.}} \angle 0°}{-jX} \right]$$

$$= j \left\{ \frac{\mathbf{V_{a.c.}} \mathbf{V_{VSC}} \angle - \theta - \mathbf{V}^2_{a.c.}}{X} \right\}$$

$$\therefore P = \frac{V_{a.c.} V_{VSC} \sin \theta}{X} \tag{9.20}$$

$$Q = \frac{V_{a.c.} [V_{VSC} \cos \theta - V_{a.c.}]}{X} \tag{9.21}$$

From equations (9.20) and (9.21), it is clear that by controlling both magnitude and angle of VSC output voltage, active and reactive power output of the VSC h.v.d.c. can be controlled.

The complete VSC h.v.d.c. scheme is shown in Figure 9.32. One converter controls active power flow as well as reactive power or a.c. voltage. The other converter controls the d.c. link voltage and reactive power or a.c. voltage. These control actions can be assigned to either of the two converters.

The controllers are normally implemented by transforming the voltage and current into a d-q rotating reference frame as shown in Figure 9.33 (d.c. voltage and reactive power control is shown). The d-q frame is locked, using a phase locked loop (PLL), on to the phasor of the a.c. system. The controller has two control loops. The outer control loop controls the DC voltage or power flow and reactive power. The inner current control loops regulate the d and q components of the currents.

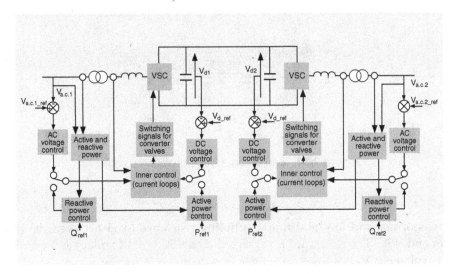

Figure 9.32 Complete control system for VSC based h.v.d.c

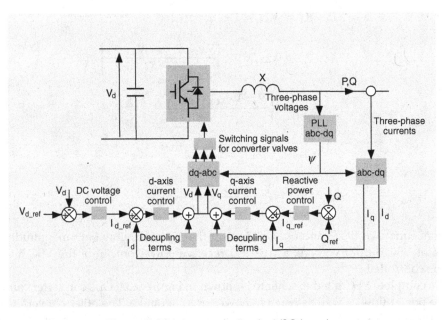

Figure 9.33 d-q controller for VSC h.v.d.c

Problems

9.1 A bridge-connected rectifier is fed from a 230 kV/120 kV transformer from the 230 kV supply. Calculate the direct voltage output when the commutation angle is 15° and the delay angle is (a) 0°; (b) 30°; (c) 60°.

(Answer: (a) 160 kV; (b) 127 kV; (c) 61.5 kV)

9.2 It is required to obtain a direct voltage of 100 kV from a bridge-connected rectifier operating with $\alpha = 30°$ and $\gamma = 15°$. Calculate the necessary line secondary voltage of the rectifier transformer, which is nominally rated at 345 kV/ 150 kV; calculate the tap ratio required.

(Answer: 94 kV line; 1.6)

9.3 If the rectifier in Problem 9.2 delivers 800 A d.c, calculate the effective reactance X (Ω) per phase.

(Answer: $X = 13.2\,\Omega$)

9.4 A d.c. link comprises a line of loop resistance $5\,\Omega$ and is connected to transformers giving secondary voltage of 120 kV at each end. The bridge-connected converters operate as follows:

Rectifier:	Inverter:
$\alpha = 10°$	$\delta_0 = 10°$; Allow 5° margin on δ_0 for δ
$X = 15\,\Omega$	$X = 15\,\Omega$

Calculate the direct current delivered if the inverter operates on constant δ control. If all parameters remain constant, except α, calculate the maximum direct current transmittable.

(Answer: 612; 1104 A)

9.5 The system in Problem 9.4 is operated with $\alpha = 15°$ and on constant β control. Calculate the direct current for $\gamma = 15°$.

(Answer: 482 A)

9.6 For a bridge arrangement, sketch the current waveforms in the valves and in the transformer windings and relate them, in time, to the anode voltages. Neglect delay and commutation times. Comment on the waveforms from the viewpoint of harmonics.

9.7 A direct current transmission link connects two a.c. systems via converters, the line voltages at the transformer-converter junctions being 100 kV and 90 kV. At the 100 kV end the converter operates with a delay angle of 10°, and at the 90 kV end the converter operates with a δ of 15°. The effective reactance per phase of each converter is $15\,\Omega$ and the loop resistance of the link is $10\,\Omega$. Determine the magnitude and direction of the power delivered if the inverter operates on constant δ control. Both converters consist of six valves in bridge connection. Calculate the percentage change required in the voltage of the transformer, which was originally at 90 kV, to produce a transmitted current of 800 A, other controls being unchanged. Comment on the reactive power requirements of the converters.

(Answer: 1.55 kA; 207 MW; 6.5%)

9.8   Draw a schematic diagram showing the main components of an h.v.d.c. link connecting two a.c. systems. Explain briefly the role of each component and how inversion into the receiving end a.c. system is achieved.

Discuss two of the main technical reasons for using h.v.d.c. in preference to a.c. transmission and list any disadvantages.

9.9   A VSC is fed from a 230kV/120 kV transformer from the 230 kV supply. Total series reactance between the VSC and the a.c. supply is 5 Ω (including the transformer leakage reactance). The output voltage of the VSC is 125 kV and leads the a.c. supply voltage by 5°. What is the active and reactive power delivered to the a.c. system?

(Answer: 262 MW; 109 MVAr)

9.10   A VSC based h.v.d.c. link is connected into a 33 kV a.c. system with a short circuit level of 150 MVA. The VSC can operate between 0.7–1.2 pu voltage and has a 21.8 Ω coupling reactor. Assume a voltage of 1 pu on the infinite busbar, and using a base of 150 MVA calculate:

a. The maximum active power that the d.c. link can inject into the a.c. system.
b. The angle of the VSC when operating at 10 MW exporting power.
c. The maximum reactive power (in MVAr) that the d.c. link can inject into the a.c. system.
d. The maximum reactive power (in MVAr) that the d.c. link can absorb from the a.c. system.

(Answer: 45 MW; 12.8°, 7.88 MVAr, 10.4 MVAr)

10

Overvoltages and Insulation Requirements

10.1 Introduction

The insulation requirements for lines, cables and substations are of critical importance in the design of power systems. When a transient event such as a lightning strike, a fault or a switching operation occurs, the network components can be subjected to very high stresses from the excessive currents and voltages that result. The internal insulation of individual items of plant (for example, generators and transformers) is designed to withstand the voltage transients that are anticipated at the location of the equipment. In addition, the insulation of the overall system, for example the number of insulators per string, and clearances of overhead lines, is designed to withstand these transient voltages. In some cases, additional devices such as over-running earth wires, lightning arrestors or arc gaps are provided to protect items of plant.

In earlier years, lightning largely determined the insulation requirements of both the transmission and distribution system. With the much higher transmission voltages now in use and with large cable networks, the voltage transients or surges due to switching, that is the opening and closing of circuit breakers, have become the major consideration.

A factor of major importance is the contamination of insulator surfaces caused by atmospheric pollution. This modifies the performance of insulation considerably, which becomes difficult to assess precisely. The presence of dirt or salt on the insulator discs or bushing surfaces results in these surfaces becoming slightly conducting, and hence flashover occurs.

A few terms frequently used in high-voltage technology need definition. They are as follows:

1. **Basic impulse insulation level or basic insulation level (BIL):** This is the reference level expressed as the impulse crest (peak) voltage with a standard wave of

Electric Power Systems, Fifth Edition. B.M. Weedy, B.J. Cory, N. Jenkins, J.B. Ekanayake and G. Strbac.
© 2012 John Wiley & Sons, Ltd. Published 2012 by John Wiley & Sons, Ltd.

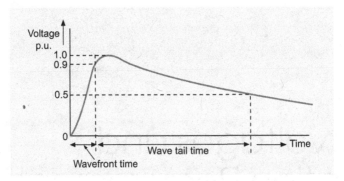

Figure 10.1 Basic impulse waveform; Shape of lightning and switching surges; the lightning impulse has a rise time of 1.2 µs and a fall time to half maximum value of 50 µs (hence 1.2/50 wave). Switching surges are much longer, the duration times varying with situation; a typical wave is 175/3000 µs

1.2 × 50 µs wave (see Figure 10.1). The insulation of power system apparatus, as demonstrated by suitable tests, shall be equal or greater than the BIL. The two standard tests are the power frequency and 1.2/50 impulse-wave voltage withstand tests. The withstand voltage is the level that the equipment will withstand for a given length of time or number of applications without disruptive discharge occurring, that is a failure of insulation resulting in a collapse of voltage and passage of current (sometimes termed 'sparkover' or 'flashover' when the discharge is on the external surface). Normally, several tests are performed and the number of flashovers noted. The BIL is usually expressed as a per unit of the peak (crest) value of the normal operating voltage to earth; for example for a maximum operating voltage of 362 kV,

$$1 \text{ p.u.} = \sqrt{2} \times \frac{362}{\sqrt{3}} = 300 \text{ kV}$$

so that a BIL of 2.7 p.u. = 810 kV.
2. **Critical flashover voltage (CFO):** This is the peak voltage with a 50% probability of flashover or disruptive discharge (sometimes denoted by V_{50}).
3. **Impulse ratio (for flashover or puncture of insulation):** This is the impulse peak voltage to cause flashover or puncture divided by the crest value of power-frequency voltage.

10.2 Generation of Overvoltages

10.2.1 Lightning Surges

A thundercloud is bipolar, often with positive charges at the top and negative at the bottom, usually separated by several kilometres. When the electric field strength exceeds the breakdown value a lightning discharge is initiated. The first discharge

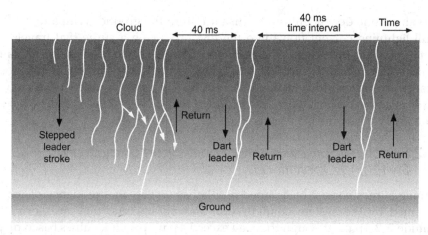

Figure 10.2 Sequence of strokes in a multiple lightning stroke

proceeds to the earth in steps (stepped leader stroke). When close to the earth a faster and luminous return stroke travels along the initial channel, and several such leader and return strokes constitute a flash. The ratio of negative to positive strokes is about 5:1 in temperate regions. The magnitude of the return stroke can be as high as 200 kA, although an average value is in the order of 20 kA.

Following the initial stroke, after a very short interval, a second stroke to earth occurs, usually in the ionized path formed by the original. Again, a return stroke follows. Usually, several such subsequent strokes (known as dart leaders) occur, the average being between three and four. The complete sequence is known as a multiple-stroke lightning flash and a representation of the strokes at different time intervals is shown in Figure 10.2. Normally, only the heavy current flowing over the first 50 μs is of importance and the current-time relationship has been shown to be of the form $i = i_{peak}(e^{-\alpha t} - e^{-\beta t})$. For a 1.2/50 μs wave, $\alpha = 1.46 \times 10^4$ and $\beta = 2.56 \times 10^6$.

When a stroke arrives on an overhead conductor, equal current surges of the above waveform are propagated in both directions away from the point of impact. The magnitude of each voltage surge set up is therefore $\frac{1}{2}Z_0 i_{peak}(e^{-\alpha t} - e^{-\beta t})$, where Z_0 is the conductor surge impedance. For a current peak of 20 kA and a Z_0 of 350 Ω the voltage surge will have a peak value of $(0.5 \times 350 \times 20 \times 10^3)$ that is 3500 kV.

When a ground or earth wire is positioned over the overhead line, a stroke arriving on a tower or on the wire itself sets up surges flowing in both directions along the wire. On reaching neighbouring towers the surge is partially reflected and transmitted further. This process continues over the length of the line as towers are encountered. If the towers are 300 m apart the travel time between towers and back to the original tower is $(2 \times 300)/(3 \times 10^8)$, that is 2 μs, where the speed of propagation is 3×10^8 m/s (the speed of light). The voltage distribution may be obtained by means of the Bewley lattice diagram, to be described in Section 10.6.

If an indirect stroke strikes the earth near a line, the induced current, which is normally of positive polarity, creates a voltage surge of the same waveshape which has an amplitude dependent on the distance from the ground. With a direct stroke the full lightning current flows into the line producing a surge that travels away from the point of impact in all directions. A direct stroke to a tower can cause a back flashover due to the voltage set up across the tower inductance and footing resistance by the rapidly changing lightning current (typically $10\,\text{kA}/\mu\text{s}$); this appears as an overvoltage between the top of the tower and the conductors (which are at a lower voltage).

10.2.2 Switching Surges–Interruption of Short Circuits and Switching Operations

Switching surges consist of damped oscillatory waves, the frequency of which is determined by the system configuration and parameters; they are normally of amplitude 2–2.8 p.u., although they can exceed 4 p.u. (per-unit values based on peak line-to-earth operating voltage).

A simple single phase network shown in Figure 10.3(a) is used to explain switching surges. If the transformer leakage inductance is L, the series resistance

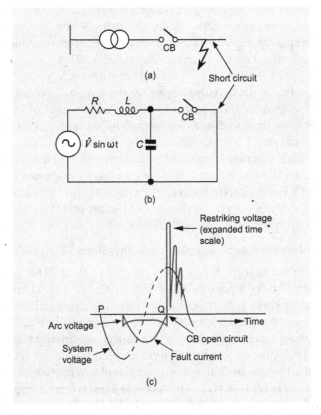

Figure 10.3 Restriking voltage set up on fault interruption, (a) System diagram, (b) Equivalent circuit, (c) Current and voltage waveforms

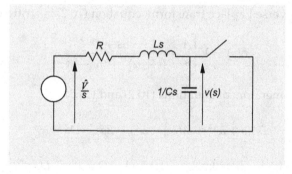

Figure 10.4 Laplace equivalent circuit with step applied

of the circuit is R and the transformer and connections capacitance to ground is C, then the equivalent circuit after a fault may be drawn as Figure 10.3(b). When a fault occurs the protection associated with the circuit breaker (CB) sends a control signal to open. At time instant P the contacts of the CB start to open but current continues to flow through the arc across the circuit breaker contacts. When the arc between the circuit-breaker contacts is extinguished (at time instant Q), the full system voltage (recovery voltage) suddenly appears across the R-L-C circuit and the open gap of the circuit breaker contacts as shown in Figure 10.3 (b). The resultant voltage appearing across the circuit is shown in Figure 10.3(c). It consists of a high-frequency component superimposed on the normal system voltage, the total being known as the restriking voltage. During the time for which this high-frequency term persists, the change in power frequency term is negligible. Therefore the equivalent circuit of Figure 10.3(b) may be analyzed by means of the Laplace transform assuming that a step input of $\hat{V}$ appears at Q. The Laplace equivalent is shown in Figure 10.4.

The voltage across the circuit breaker contacts, that is across C is given by:

$$v(s) = \frac{\hat{V}/s \times 1/Cs}{R + Ls + 1/Cs} = \frac{\hat{V}}{s(LCs^2 + RCs + 1)} \tag{10.1}$$

Defining $\omega_0 = \dfrac{1}{\sqrt{LC}}$ and $\alpha = \dfrac{R}{2L}$, equation (10.1) can be written as:

$$v(s) = \frac{\hat{V}\omega_0^2}{s(s^2 + 2\alpha s + \omega_0^2)}$$

$$= \frac{\hat{V}\omega_0^2}{s[(s+\alpha)^2 + (\omega_0^2 - \alpha^2)]} \tag{10.2}$$

To obtain the inverse Laplace transform, equation (10.2) is written in the form:

$$v(s) = \hat{V}\left[\frac{A}{s} - \frac{Bs + D}{(s + \alpha)^2 + (\omega_0^2 - \alpha^2)}\right]$$ (10.3)

Equating the numerators of equations (10.2) and (10.3):

$$A(s + \alpha)^2 + A(\omega_0^2 - \alpha^2) - Bs^2 - Ds = \omega_0^2$$

Therefore $A = 1$; $A - B = 0$; $\therefore B = 1$; $2\alpha A - D = 0$; $\therefore D = 2\alpha$. Now, equation (10.2) can be rewritten as:

$$v(s) = \hat{V}\left[\frac{1}{s} - \frac{s + 2\alpha}{(s + \alpha)^2 + (\omega_0^2 - \alpha^2)}\right]$$ (10.4)

By taking inverse Laplace transforms:

$$v(t) = \hat{V}\left[1 - e^{-\alpha t}\cos\left(\sqrt{\omega_0^2 - \alpha^2}\right)t - \frac{\alpha}{\omega_0}e^{-\alpha t}\sin\left(\sqrt{\omega_0^2 - \alpha^2}\right)t\right]$$ (10.5)

In practical systems $R \ll \sqrt{L/C}$ and thus $\alpha \ll \omega_0$. Therefore equation (10.5) can be simplified to:

$$v(t) = \hat{V}\left[1 - e^{-\alpha t}\cos \omega_0 t\right]$$ (10.6)

Equation (10.6) defines the restriking voltage in Figure 10.3(c) where ω_0 is the natural frequency of oscillation and α is the damping coefficient. The maximum value of the restriking voltage is $2\hat{V}$.

Example 10.1

Determine the relative attenuation occurring in five cycles in the overvoltage surge set up on a 66 kV cable fed through an air-blast circuit breaker when the breaker opens on a system short circuit. The network parameters are as follows:

$$R = 7.8\,\Omega \quad L = 6.4\,\text{mH} \quad C = 0.0495\,\mu\text{F}$$

Solution
From the network parameters:

$$\omega_0 = \frac{1}{\sqrt{LC}} = \frac{1}{\sqrt{6.4 \times 10^{-3} \times 0.0495 \times 10^{-6}}}$$
$$= 5.618 \times 10^4 \text{ rad/s}$$

that is transient frequency $= 5.618 \times 10^4 / 2\pi \, \text{Hz} = 8.93 \, \text{kHz}$.

$$\alpha = \frac{R}{2L} = \frac{7.8}{2 \times 6.4 \times 10^{-3}} = 609.4$$

Hence, time for five cycles $= \dfrac{5}{8.93 \times 10^3} = 0.56 \, \text{ms}$

The maximum voltage set up with the circuit breaker opening on a short circuit fault would be

$$\hat{V} = 2 \times \frac{66}{\sqrt{3}} \times \sqrt{2} = 107.78 \, \text{kV}$$

From equation (10.6), $v(t)$ at $t = 0.56 \, \text{ms}$ is:

$$v(t) = 107.78 \left[1 - e^{-609.4 \times 0.56 \times 10^{-3}} \cos(5.618 \times 10^4 \times 0.56 \times 10^{-3}) \right]$$
$$= 42.4 \, \text{kV}$$

The initial rate of rise of the surge is very important as this determines whether the contact gap, which is highly polluted with arc products, breaks down again after the initial open circuit occurs. From (10.5) if $\alpha = \omega_0$, that is $R = \dfrac{1}{2}\sqrt{\dfrac{L}{C}}$ and $v(t) = \hat{V}[1 - e^{-\alpha t}]$. Under this critically damped case, the severity of the transient is reduced. A critically damped surge can be obtained by connecting a resistor of $\dfrac{1}{2}\sqrt{\dfrac{L}{C}}$ across the contacts of the circuit breaker.

10.2.3 Switching Surges–Interruption of Capacitive Circuits

The interruption of capacitive circuits is shown in Figure 10.5. At the instant of arc interruption, the capacitor is left charged to value $\hat{V}$. If the capacitance, C, is large then the voltage across the capacitor decays slowly. The voltage across the circuit breaker is the difference between the system voltage and the voltage across the capacitor. At the negative crest of the system voltage, that is $-\hat{V}$, the gap voltage is $2\hat{V}$. If the gap breaks down, an oscillatory transient is set up (as previously discussed), which can increase the gap voltage to up to $3\hat{V}$, as shown in Figure 10.5.

10.2.4 Current Chopping

Current chopping arises with vacuum circuit breakers or with air-blast circuit breakers which operate on the same pressure and velocity for all values of interrupted current. Hence, on low-current interruption the breaker tends to open the circuit

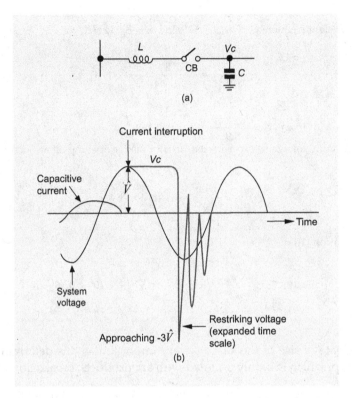

(a)

(b)

Figure 10.5 Voltage waveform when opening a capacitive circuit

before the current natural-zero, and the electromagnetic energy present is rapidly converted to electrostatic energy, that is.

$$\frac{1}{2}Li_0^2 = \frac{1}{2}Cv^2 \text{ and } v = i_0\sqrt{\frac{L}{C}} \tag{10.7}$$

The voltage waveform is shown in Figure 10.6.

The restriking voltage, v_r, can be obtained from equation (10.7) by including resistance and time:

$$v_r = i_0\sqrt{\frac{L}{C}}e^{-\alpha t}\sin \omega_0 t \tag{10.8}$$

where $\omega_0 = \dfrac{1}{\sqrt{LC}}$ and i_0 is the value of the current at the instant of chopping. High transient voltages may be set up on opening a highly inductive circuit such as a transformer on no-load.

10.2.5 Faults

Overvoltages may be produced by certain types of asymmetrical fault, mainly on systems with ungrounded neutrals. The voltages set up are of normal operating

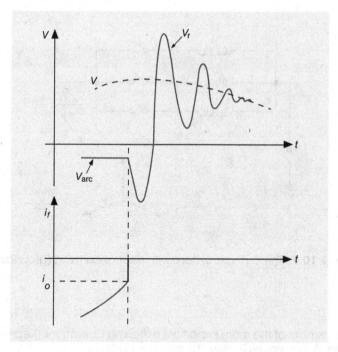

Figure 10.6 Voltage transient due to current chopping: i_f = fault current; v = system voltage, i_0 = current magnitude at chop, v_r = restriking voltage

frequency. Consider the circuit with a three-phase earth fault as shown in Figure 10.7. If the circuit is not grounded the voltage across the first gap to open is $1.5\,V_{phase}$. With the system grounded the gap voltage is limited to the phase voltage. This has been discussed more fully in Chapter 7.

10.2.6 Resonance

It is well known that in resonant circuits severe overvoltages occur and, depending on the resistance present; the voltage at resonance across the capacitance can be high. Although it is unlikely that resonance in a supply network can occur at normal supply frequencies, it is possible to have this condition at harmonic frequencies. Resonance is normally associated with the capacitance to earth of items of plant and is often brought about by an opened phase caused by a broken conductor or a fuse operating.

In circuits containing windings with iron cores, for example transformers, a condition due to the shape of the magnetization curve, known as ferroresonance, is possible. This can produce resonance with overvoltages and also sudden changes from one condition to another.

A summary of important switching operations is given in Table 10.1.

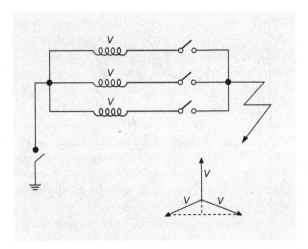

Figure 10.7 Three-phase system with neutral earthing (grounding)

Table 10.1 Summary of the more important switching operations (Reproduced with permission from Brown Boveri Review, December 1970)

Switching Operation	System	Voltage Across Contacts
1. Terminal short circuit		
2. Short line fault		
3. Two out-of-phase systems voltage depends on grounding conditions in systems		
4. Small inductive currents, current chopped (unloaded transformer)		
5. Interrupting capacitive currents capacitor banks, lines, and cables on no-load		See Figure 10.5

10.3 Protection Against Overvoltages

10.3.1 Modification of Transients

When considering the protection of a power system against overvoltages, the transients may be modified or even eliminated before reaching the substations. If this is not possible, the lines and substation equipment may be protected, by various means, from flashover or insulation damage.

By the use of over-running earth (ground) wires, phase conductors may be shielded from direct lightning strokes and the effects of induced surges from indirect strokes lessened. The shielding is not complete, except perhaps for a phase conductor immediately below the earth wire. The effective amount of shielding is often described by an angle ϕ, as shown in Figure 10.8; a value of 35° appears to agree with practical experience. Obviously, two earth wires horizontally separated provide much better shielding.

It has already been seen that the switching-in of resistance across circuit-breaker contacts reduces the high overvoltages produced on opening, especially on capacitive or low-current inductive circuits.

An aspect of vital importance, quite apart from the prevention of damage, is the maintenance of supply, especially as most flashovers on overhead lines cause no permanent damage and therefore a complete and lasting removal of the circuit from operation is not required. Transient faults may be removed by the use of autoreclosing circuit breakers. In distribution circuits and many transmission systems all three

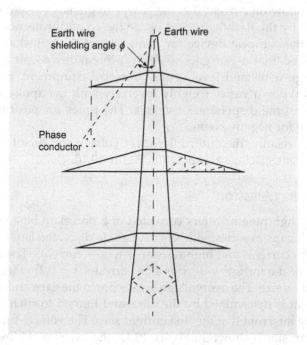

Figure 10.8 Single earth wire protection; shielding angle ϕ normally 35°

phases are operated together. Some countries use single phase auto-reclose to improve transient stability. Only the faulted phase is opened and re-closed while two phases remain connected.

It is uneconomic to attempt to modify or eliminate most overvoltages, and means are required to protect the various items of power systems. Surge divertors are connected in shunt across the equipment and divert the transient to earth; surge modifiers are connected to reduce the steepness of the wavefront-hence reduce its severity. A surge modifier may be produced by the connection of a shunt capacitor between the line and earth or an inductor in series with the line. Oscillatory effects may be reduced by the use of fast-acting voltage or current injections.

10.3.2 Surge Diverters

The basic requirements for diverters are that they should pass no current at normal voltage, interrupt the power frequency follow-on current after a flashover, and break down as quickly as possible after the abnormal voltage arrives.

10.3.2.1 Rod Gap

The simplest form of surge diverter is the rod gap connected across a bushing or insulator as shown in Figure 10.9. This may also take the form of rings (arcing rings) around the top and bottom of an insulator string. The breakdown voltage for a given gap is polarity dependent to some extent. It does not interrupt the post-flashover follow-on current (i.e. the power-frequency current which flows in the path created by the flashover), and hence the circuit protection must operate, but it is by far the cheapest device for plant protection against surges. It is usually recommended that a rod gap be set to a breakdown voltage at least 30% below the voltage-withstand level of the protected equipment. For a given gap, the time for breakdown varies roughly inversely with the applied voltage; there exists, however, some dispersion of values. The times for positive voltages are lower than those for negative ones.

Typical curves relating the critical flashover voltage and time to breakdown for rod gaps of different spacings are shown in Figure 10.10.

10.3.2.2 Lightning Arrestor

Early designs of lightning arresters consisted of a porcelain bushing containing a number of spark gaps in series with silicon carbide discs, the latter possessing low resistance to high currents and high resistance to low currents. The lightning arrestor obeys a law of the form $V = aI^k$ (for a SiC arrestor $k = 0.2$) where a depends on the material and its size. The overvoltage breaks down the gaps and then the power-frequency current is determined by the discs and limited to such a value that the gaps can quickly interrupt it at the first current zero. The voltage-time characteristic of a lightning arrester is shown in Figure 10.11; the gaps break down at S and the

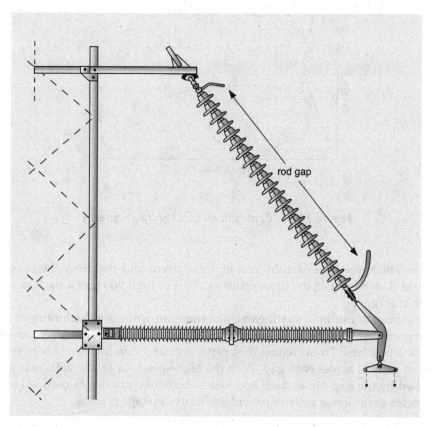

Figure 10.9 Insulating cross-arm for a double-circuit 420 kV line (Italian)

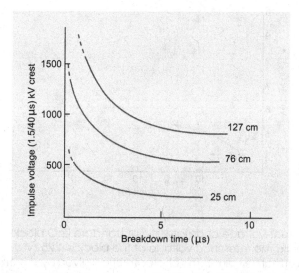

Figure 10.10 Breakdown characteristics of rod gaps

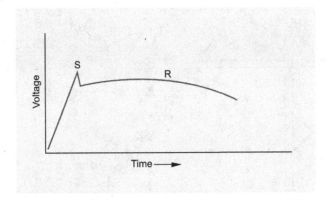

Figure 10.11 Characteristic of lightning arrester

characteristic after this is determined by the current and the discs; the maximum voltage at R should be in the same order as at S. For high voltages a stack of several such units is used.

Although with multiple spark gaps, diverters can withstand high rates of rise of recovery voltage (RRRV), the non-uniform voltage distribution between the gaps presents a problem. To overcome this, capacitors and non-linear resistors are connected in parallel across each gap. With the high-speed surge, the voltage is mainly controlled by the gap capacitance and hence capacitive grading is used. At power-frequencies a non-linear resistor provides effective voltage grading.

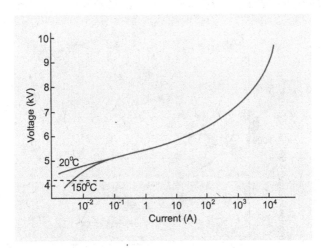

Figure 10.12 Current-voltage characteristic of standard ZnO blocks, 80 mm in diameter and 35 mm thick. The reference voltage of the block is 4.25 kV d.c. and its continuous rating is 3 kV (effective value); This characteristic is given by $V = aI^{0.03}$ (Reproduced with permission from the Electricity Association)

Developments in materials have produced an arrester which does not require gaps to reduce the current after operation to a low value. Zinc oxide material is employed in stacked cylindrical blocks encased in a ceramic weatherproof insulator. A typical current-voltage characteristic is shown in Figure 10.12, where it can be seen that with a designed steady-state voltage of 4.25 kV, the residual current is below 10^{-2} A, depending upon temperature. At an effective rating of 3 kV the current is negligible. Consequently, no action is required to extinguish the current after operation, as may be required with a gapped arrester.

Metal oxide arresters up to the highest transmission voltages are now employed almost to the exclusion of gapped devices.

10.4 Insulation Coordination

The equipment used in a power system comprises items having different breakdown or withstand voltages and different voltage-time characteristics. In order that all items of the system are adequately protected there is a need to consider the system as a whole and not items of plant in isolation, that is the insulation protection must be coordinated. To assist this process, standard insulation levels are recommended for power frequency flashovers, lightning impulses and switching impulses and these are summarized in Tables 10.2 (British substation practice) and 10.3 (U.S practice). The reduced basic insulation impulse levels for switching impulses are only specified for higher voltages. When there are more than one withstand voltage specified for a given nominal voltage, the standards relevant to different apparatus should be referred to in order to find the exact insulation level.

Coordination is made difficult by the different voltage-time characteristics of plant and protective devices, for example a gap may have an impulse ratio of 2 for a 20 μs front wave and 3 for a 5 μs wave. At the higher frequencies (shorter wavefronts) corona cannot form in time to relieve the stress concentration on the gap electrodes. With a lightning surge a higher voltage can be withstood because a discharge requires a certain discrete amount of energy as well as a minimum voltage, and the applied voltage increases until the energy reaches this value.

Table 10.2 Recommended BILs at various operating voltages following IEC 60071 (BS EN 60071-1:2006)

Nominal voltage (kV)	Standard lightning Impulse withstand (peak kV)	Power frequency withstand (peak kV)	Switching impulse withstand (peak kV)
11	60	28	
	75		
	90		
132	550	230	
	650	275	
400	1175	675	950
	1300		
	1425		

Table 10.3 Recommended BILs at various operating voltages following IEEE 1313.1-1996

Nominal voltage (kV)	Standard lightning Impulse withstand (peak kV)	Power frequency withstand (peak kV)	Switching impulse withstand (peak kV)
23	150	50	
115	350	140	
	450	185	
	550	230	
345	900		650
	975		750
	1050		825
	1175		900
	1300		975
			1050
500	1300		1175
	1425		1300
	1550		1425
	1675		1500
	1800		
765	1800		1300
	1925		1425
	2050		1550
			1675
			1800

Figure 10.13 shows the voltage-time characteristics for the system elements comprising a substation. The difficulty of coordination can be explained using the voltage time characteristics of the rod gap and the transformer. If they are subjected to a surge having a rate of rise less than the critical slope shown in Figure 10.13 (long rise time), then the rod gap breaks down before the transformer thus protecting it.

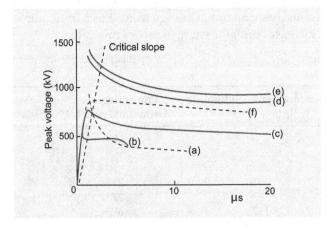

Figure 10.13 Insulation coordination in an H.V. substation. Voltage-time characteristics of plant. Characteristics of (a) rod gap (b) lightning arrester, (c) Transformer, (d) Line insulator string, (e) Busbar insulation, (f) Maximum surge applied waveform

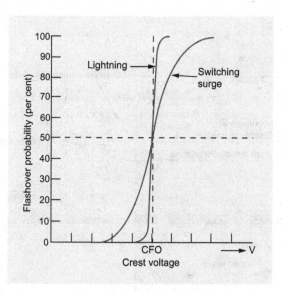

Figure 10.14 Flashover probability–peak surge voltage. Lightning flashover is close-grouped near the critical flashover voltage (CFO), whereas switching-surge probability is more widely dispersed and follows normal Gaussian distribution (Reproduced with permissions from Westinghouse Electrical Corporation)

On the other hand if the applied surge has a rate of rise greater than the critical slope (short rise time), the transformer breaks down before the rod gap. As can be seen, the arrester protects the transformer for all surges.

Up to an operating voltage of 345 kV the insulation level is determined by lightning, and the standard impulse tests are sufficient along with normal power frequency tests. Impulse tests are normally performed with a voltage wave shown in Figure 10.1; this is known as a 1.2/50 μs wave and typifies the lightning surge. At higher network operating voltages, the overvoltages resulting from switching are higher in magnitude and therefore they decide the insulation. The characteristics of air gaps and some solid insulation are different for switching surges than for the standard impulse waves, and closer coordination of insulation is required because of the lower attenuation of switching surges, although their amplification by reflection is less than with lightning. Research indicates that, for transformers, the switching impulse strength is of the order of 0.95 of the standard (lightning impulse) value while for oil-filled cables it is 0.7 to 0.8.

The design withstand level is selected by specifying the risk of flashover, for example for 550 kV towers a 0.13% probability has been used. At 345 kV, design is carried out by accepting a switching impulse level of 2.7 p.u. At 500 kV, however, a 2.7 p.u. switching impulse would require 40% more tower insulation than that governed by lightning. The tendency is therefore for the design switching impulse level to be forced lower with increasing system operating voltage and for control of the surges to be made by the more widespread use of resistance switching in the circuit breakers or use of surge diverters. For example, for the 500 kV network the level is

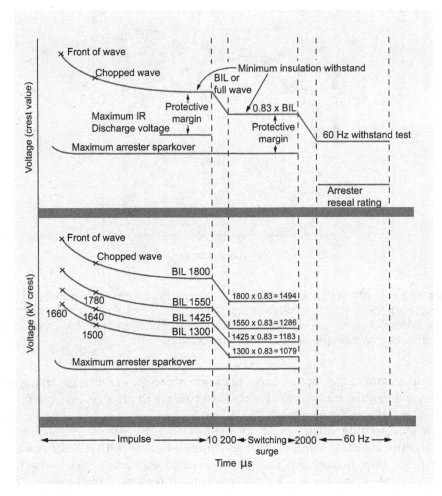

Figure 10.15 Coordination diagrams relating lightning-arrester characteristics to system requirements Top: Explanation of characteristics. Bottom: Typical values for 500 kV system (300 kV to ground) (Reproduced with permissions from Hubbell Power Systems Inc.)

2 p.u. and at 765 kV it is reduced to 1.7 p.u.; if further increases in system voltage occur, it is hoped to decrease the level to 1.5 p.u.

The problem of switching surges is illustrated in Figure 10.14, in which flashover probability is plotted against peak (crest) voltage; critical flashover voltage (CFO) is the peak voltage for a particular tower design for which there is a 50% probability of flashover. For lightning, the probability of flashover below the CFO is slight, but with switching surges having much longer fronts the probability is higher, with the curve following the normal Gaussian distribution; the tower clearances must be designed for a CFO much higher than the maximum transient expected.

An example of the application of lightning-arrester characteristics to system requirements is illustrated in Figure 10.15. In this figure, the variation of crest

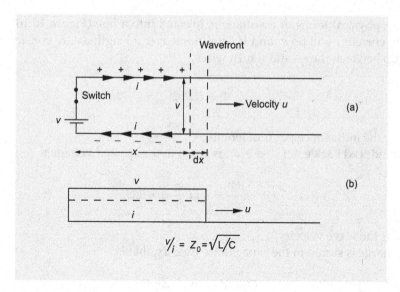

Figure 10.16 Distribution of charge and current as wave progresses along a previously unenergized line, (a) Physical arrangement, (b) Symbolic representation

voltage for lightning and switching surges with time is shown for an assumed BIL of the equipment to be protected. Over the impulse range (up to 10 μs) the fronts of wave and chopped-wave peaks are indicated. Over the switching surge region the insulation withstand strength is assumed to be 0.83 of the BIL (based on transformer requirements). The arrester maximum sparkover voltage is shown along with the maximum voltage set up in the arrester by the follow on current (10 or 20%). A protective margin of 15% between the equipment withstand strength and the maximum arrester sparkover is assumed.

10.5 Propagation of Surges

The basic differential equations for voltage and current in a distributed constant line are as follows:

$$\frac{\partial^2 v}{\partial x^2} = LC\frac{\partial^2 v}{\partial t^2} \text{ and } \frac{\partial^2 i}{\partial x^2} = LC\frac{\partial^2 i}{\partial t^2}$$

and these equations represent travelling waves. The solution for the voltage may be expressed in the form,

$$v = F_1\left(t - x\sqrt{LC}\right) + F_2\left(t + x\sqrt{LC}\right)$$

that is one wave travels in the positive direction of x and the other in the negative. Also, it may be shown that because $\partial v/\partial x = -L(\partial i/\partial t)$, the solution for current is

$$i = \sqrt{\frac{C}{L}}\left[F_1\left(t - x\sqrt{LC}\right) - F_2\left(t + x\sqrt{LC}\right)\right] \text{ noting that } \sqrt{\frac{C}{L}} = \frac{1}{Z_0}$$

In more physical terms, if a voltage is injected into a line (Figure 10.16) a corresponding current i will flow, and if conditions over a length dx are considered, the flux set up between the go and return wires,

$$\Phi = iLdx,$$

where L is the inductance per unit length.

As the induced back e.m.f., $-d\Phi/dt$, is equal to the applied voltage v

$$v = -\frac{d\Phi}{dt} = -Li\frac{dx}{dt} = -iLU \tag{10.9}$$

where U is the wave velocity

Also, charge is stored in the capacitance over dx, that is.

$$Q = idt = vCdx \text{ and}$$

$$i = vCU \tag{10.10}$$

From (10.9) and (10.10),

$$vi = viLCU^2 \text{ and } U = 1/\sqrt{LC}.$$

Substituting for U in (10.10),

$$i = vCU = \frac{vC}{\sqrt{LC}} = v\sqrt{\frac{C}{L}} = \frac{v}{Z_0}$$

where Z_0 is the characteristic or surge impedance.

For single-circuit three-phase overhead lines (conductors not bundled) Z_0 lies in the range 400–600 Ω. For overhead lines, $U = 3 \times 10^8$ m/s, that is the speed of light, and for cables

$$U = \frac{3 \times 10^8}{\sqrt{\varepsilon_r \mu_r}} \text{ m/s}$$

where ε_r is usually from 3 to 3.5, and $\mu_r = 1$.

From equations (10.9) and (10.10),

$$\frac{1}{2}Li^2 = \frac{1}{2}(iLU)\left(\frac{i}{U}\right)$$

$$= \frac{1}{2}v \times Cv = \frac{1}{2}Cv^2$$

The incident travelling waves of v_i and i_i, when they arrive at a junction or discontinuity, produce a reflected current i_r and a reflected voltage v_r which travel

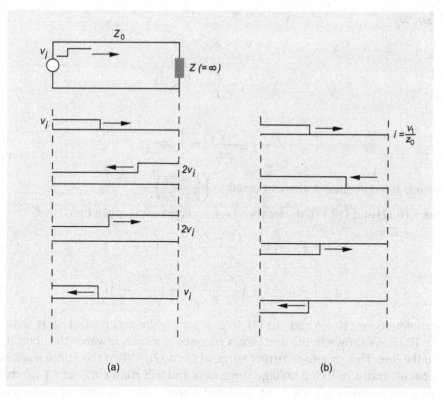

Figure 10.17 Application of voltage to unenergized loss-free line on open circuit at far end. (a) Distribution of voltage, (b) Distribution of current. Voltage source is an effective short circuit

back along the line. The incident and reflected components of voltage and current are governed by the surge impedance Z_0, so that

$$v_i = Z_0 i_i \text{ and } v_r = -Z_0 i_r$$

In the general case of a line of surge impedance Z_0 terminated in Z (Figure 10.17), the total voltage at Z is $v = v_i + v_r$ and the total current is $i = i_i + i_r$.
Also,

$$v_i + v_r = Z(i_i + i_r)$$
$$Z_0(i_i - i_r) = Z(i_i + i_r)$$

and hence

$$i_r = \left(\frac{Z_0 - Z}{Z_0 + Z}\right) i_i \tag{10.11}$$

Again,

$$v_i + v_r = Z(i_i + i_r)$$

$$= Z\left(\frac{v_i - v_r}{Z_0}\right)$$

or,

$$v_r = \left(\frac{Z - Z_0}{Z + Z_0}\right)v_i = \alpha v_i \tag{10.12}$$

where α is the reflection coefficient equal to $\left(\frac{Z - Z_0}{Z + Z_0}\right)$

From (10.12) and (10.11), the total voltage and current is given by:

$$v = v_i + v_r = v_i + \left(\frac{Z - Z_0}{Z + Z_0}\right)v_i = \left(\frac{2Z}{Z + Z_0}\right)v_i \tag{10.13}$$

$$i = i_i + i_r = i_i + \left(\frac{Z_0 - Z}{Z_0 + Z}\right)i_i = \left(\frac{2Z_0}{Z_0 + Z}\right)i_i \tag{10.14}$$

From equations (10.13) and (10.14), if $Z \to \infty$, $v = 2v_i$ and $i = 0$. This is shown in Figure 10.17. As shown in the first plot, a surge of v_i travels towards the open circuit end of the line. This creates a current surge of $i = v_i/Z_0$. When this surge reaches the open circuit end, a reflected voltage surge of v_i and a current surge of $-i$ are created and they start travelling back down the line. Therefore the total voltage surge is now $2v_i$ and the current is zero as shown in the second plot. At the voltage source, an effective short circuit, a voltage wave of $-v_i$ is reflected down the line. The reflected waves will travel back and forth along the line, setting up, in turn, further reflected waves at the ends, and this process will continue indefinitely unless the waves are attenuated because of resistance and corona.

From equations (10.13) and (10.14), if $Z = Z_0$ (matched line), $\alpha = 0$, that is there is no reflection. If $Z > Z_0$, then v_r is positive and i_r is negative, but if $Z < Z_0$, v_r is negative and i_r is positive.

Summarizing, at an open circuit the reflected voltage is equal to the incident voltage and this wave, along with a current $(-i_i)$, travels back along the line; note that at the open circuit the total current is zero. Conversely, at a short circuit the reflected voltage wave is $(-v_i)$ in magnitude and the current reflected is (i_i), giving a total voltage at the short circuit of zero and a total current of $2i_i$. For other termination arrangements, Thevenin's theorem may be applied to analyze the circuit. The voltage across the termination when it is open-circuited is seen to be $2v_i$ and the equivalent impedance looking in from the open-circuited termination is Z_0; the termination is then connected across the terminals of the Thevenin equivalent circuit (Figure 10.18).

Consider two lines of different surge impedance in series. It is necessary to determine the voltage across the junction between them (Figure 10.19). From (10.13):

$$v_{AB} = \left(\frac{2Z_1}{Z_1 + Z_0}\right)v_i = \beta v_i = \left(\frac{2v_i}{Z_1 + Z_0}\right)Z_1 \tag{10.15}$$

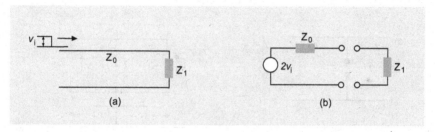

Figure 10.18 Analysis of travelling waves-use of Thevenin equivalent circuit, (a) System, (b) Equivalent circuit

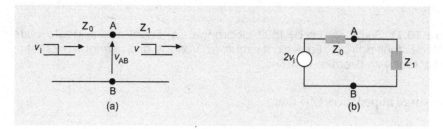

Figure 10.19 Analysis of conditions at the junction of two lines or cables of different surge impedance

The wave entering the line Z_1 is the refracted wave and β is the refraction coefficient, that is the proportion of the incident voltage proceeding along the second line (Z_1).

From (10.14), with $v_i = Z_0 i_i$

$$i = \left(\frac{2v_i}{Z_0 + Z}\right) \tag{10.16}$$

From equations (10.15) and (10.16), the equivalent circuit shown in Figure 10.19(b) can be obtained.

When several lines are joined to the line on which the surge originates (Figure 10.20), the treatment is similar, for example if there are three lines having

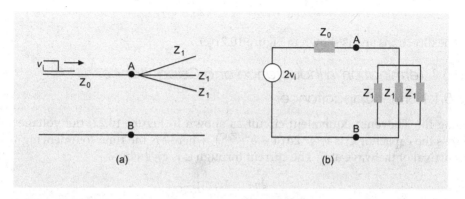

Figure 10.20 Junction of several lines, (a) System, (b) Equivalent circuit

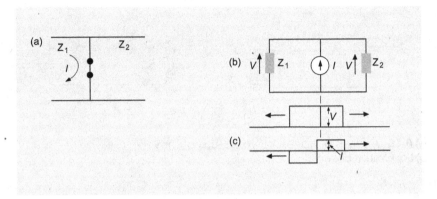

Figure 10.21 Surge set up by fault clearance, (a) Equal and opposite current (*I*) injected in fault path, (b) Equivalent circuit, (c) Voltage and current waves set up at a point of fault with direction of travel

equal surge impedances (Z_1) then

$$i = \left(\frac{2v_i}{Z_1/3 + Z_0}\right) \text{ and } v_{AB} = \left(\frac{2v_i}{Z_1/3 + Z_0}\right)\left(\frac{Z_1}{3}\right)$$

An important practical case is that of the clearance of a fault at the junction of two lines and the surges produced. The equivalent circuits are shown in Figure 10.21; the fault clearance is simulated by the insertion of an equal and opposite current (*I*) at the point of the fault. From the equivalent circuit, the magnitude of the resulting voltage surges (*v*)

$$= I\left(\frac{Z_1 Z_2}{Z_1 + Z_2}\right)$$

and the currents entering the lines are

$$I\left(\frac{Z_1}{Z_1 + Z_2}\right) \text{ and } I\left(\frac{Z_2}{Z_1 + Z_2}\right)$$

The directions are as shown in Figure 10.21(c).

10.5.1 Termination in Inductance and Capacitance

10.5.1.1 Shunt Capacitance

Using the Thevenin equivalent circuit as shown in Figure 10.22, the voltage rise across the capacitor C is $v_C = 2v_i(1 - e^{-t/Z_0 C})$, where *t* is the time commencing with the arrival of the wave at C. The current through C is given by,

$$i = \frac{dv_C}{dt} = \frac{2v_i}{Z_0}e^{-t/Z_0 C}$$

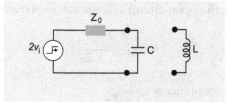

Figure 10.22 Termination of line (surge impedance Z_0) in a capacitor C or inductance L

The reflected wave,

$$v_r = v_C - v_i = 2v_i\left(1 - e^{-t/Z_0 C}\right) - v_i = v_i\left(1 - 2e^{-t/Z_0 C}\right) \tag{10.17}$$

When $t = 0$, $v_C = 0$ and $i = 2v_i/Z_0$ and when $t \to \infty$, $v_C = 2v_i$ and $i = 0$. That is, as to be expected, the capacitor acts initially as a short circuit and finally as an open circuit.

10.5.1.2 Shunt Inductance

Again, from the equivalent circuit shown in Figure 10.22 but with an inductor the voltage across the inductance is

$$v_L = 2v_i e^{-(Z_0/L)t}$$

and

$$v_r = v_L - v_i = v_i\left[2e^{-(Z_0/L)t} - 1\right] \tag{10.18}$$

Here, the inductance acts initially as an open circuit and finally as a short circuit.

10.5.1.3 Capacitance and Resistance in Parallel

Consider the practical system shown in Figure 10.23(a), where C is used to modify the surge.

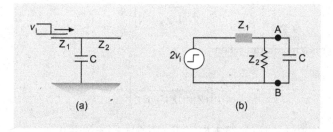

(a) (b)

Figure 10.23 Two lines surge impedances Z_1 and Z_2 grounded at their junction through a capacitor C. (a) System diagram, (b) Equivalent circuit

From Figure 10.23(b), the open-circuit voltage across AB without the capacitor:

$$= 2v_i \left(\frac{Z_2}{Z_1 + Z_2} \right)$$

Equivalent Thevenin resistance $= \dfrac{Z_1 Z_2}{Z_1 + Z_2}$

Voltage across AB is

$$v_{AB} = 2v_i \left(\frac{Z_2}{Z_1 + Z_2} \right) \left(1 - e^{-t(Z_1 + Z_2)/Z_1 Z_2 C} \right) \tag{10.19}$$

The reflected wave is given by $(v_{AB} - v_i)$.

Example 10.2

An overhead line of surge impedance 500 Ω is connected to a cable of surge impedance 50 Ω through a series resistor (Figure 10.24(a)). Determine the magnitude of the resistor such that it absorbs maximum energy from a surge originating on the overhead line and travelling into the cable.

Calculate:

a. the voltage and current transients reflected back into the line, and
b. those transmitted into the cable, in terms of the incident surge voltage;
c. the energies reflected back into the line and absorbed by the resistor.

Answer:

Let the incident voltage and current be v_i and i_i, respectively. From the equivalent circuit (Figure 10.24(b)),

$$v_B = \frac{2v_i \times Z_2}{Z_1 + Z_2 + R}$$

As $v_i = Z_1 i_i$,

$$i_B = \frac{2v_i}{Z_1 + Z_2 + R} = \frac{2Z_1 i_i}{Z_1 + Z_2 + R}$$

Power absorbed by the resistance

$$= \left[\frac{2Z_1 i_i}{Z_1 + Z_2 + R} \right]^2 R$$

This power is a maximum when

$$\frac{d}{dR} \left[\frac{R}{(Z_1 + Z_2 + R)^2} \right] = 0$$

$$(Z_1 + Z_2 + R)^2 - R \times 2(Z_1 + Z_2 + R) = 0$$

$$\therefore R = Z_1 + Z_2$$

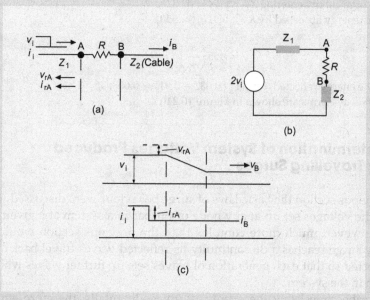

Figure 10.24 (a) System for Example 10.2. (b) Equivalent circuit, (c) Voltage and current surges

With this resistance the maximum energy is absorbed from the surge. Hence R should be $500 + 50 = 550 \, \Omega$.

a. the reflected voltage at A

$$v_{rA} = \frac{2v_i(Z_2 + R)}{Z_1 + Z_2 + R} - v_i = \frac{Z_2 + R - Z_1}{Z_1 + Z_2 + R} v_i$$

$$= \frac{50 + 550 - 500}{500 + 50 + 550} v_i = 0.091 \, v_i$$

And

$$i_{rA} = i_B - i_i = \left[\frac{Z_1 - Z_2 - R}{Z_1 + Z_2 + R} \right] i_i$$

$$= \left[\frac{500 - 50 - 550}{500 + 50 + 550} \right] i_i = -0.091 \, i_i$$

b. Voltage and current transmitted into the cable

$$i_B = \frac{2 \times 500 \times i_i}{500 + 50 + 550} = 0.91 i_i$$

$$v_B = 0.091 v_i$$

c. The surge energy entering $Z_2 = v_B i_B = 0.082 v_i i_i$
   The energy absorbed by $R = (0.91 i_i)^2 \times 550$

$$= 455 \left(\frac{v_i}{Z_1} \right) i_i = 0.91 v_i i_i$$

and the energy reflected $= v_i i_i (1 - 0.082 - 0.91) = 0.008 v_i i_i$
The waveforms are shown in Figure 10.24(c).

10.6 Determination of System Voltages Produced by Travelling Surges

In the previous section the basic laws of surge behaviour were discussed. The calculation of the voltages set up at any node or busbar in a system at a given instant in time is, however, much more complex than the previous section would suggest. When any surge reaches a discontinuity its reflected waves travel back and are, in turn, reflected so that each generation of waves sets up further waves which coexist with them in the system.

To describe completely the events at any node entails, therefore, an involved book-keeping exercise. Although many mathematical techniques are available and, in fact, used, the graphical method due to Bewley (1961) indicates clearly the physical changes occurring in time, and this method will be explained in some detail.

10.6.1 Bewley Lattice Diagram

This is a graphical method of determining the voltages at any point in a transmission system and is an effective way of illustrating the multiple reflections which take place. Two axes are established: a horizontal one scaled in distance along the system, and a vertical one scaled in time. Lines indicating the passage of surges are drawn such that their slopes give the times corresponding to distances travelled. At each point of change in impedance the reflected and transmitted waves are obtained by multiplying the incidence-wave magnitude by the appropriate reflection and refraction coefficients α and β. The method is best illustrated by Example 10.3.

Example 10.3

A loss-free system comprising a long overhead line (Z_1) in series with a cable (Z_2) will be considered. Typically, Z_1 is 500 Ω and Z_2 is 50 Ω.

Solution
Referring to Figure 10.25, the following coefficients apply:
   Line-to-cable reflection coefficient,

$$\alpha_1 = \frac{50 - 500}{50 + 500} = -0.818 \text{ (note order of values in numerator)}$$

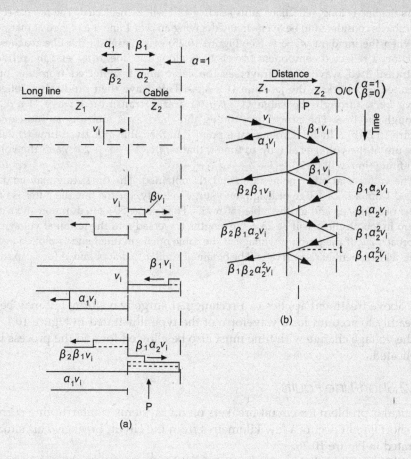

Figure 10.25 Bewley lattice diagram–analysis of long overhead line and cable in series, (a) Position of voltage surges at various instants over first complete cycle of events, that is up to second reflected wave travelling back along line, (b) Lattice diagram

Line-to-cable refraction coefficient,

$$\beta_1 = \frac{2 \times 50}{50 + 500} = 0.182$$

Cable-to-line reflection coefficient,

$$\alpha_2 = \frac{500 - 50}{500 + 50} = 0.818$$

Cable-to-line refraction coefficient,

$$\beta_1 = \frac{2 \times 500}{50 + 500} = 1.818$$

As the line is long, reflections at its sending end will be neglected. The remote end of the cable is considered to be open-circuited, giving an α of 1 and a β of zero at that point.

When the incident wave v_i (see Figure 10.25) originating in the line reaches the junction, a reflected component travels back along the line ($\alpha_1 v_i$), and the refracted or transmitted wave ($\beta_1 v_i$) traverses the cable and is reflected from the open-circuited end back to the junction ($1 \times \beta_1 v_i$). This wave then produces a reflected wave back through the cable ($1 \times \beta_1 \alpha_2 v_i$) and a transmitted wave ($1 \times \beta_1 \beta_2 v_i$) through the line. The process continues and the waves multiply as indicated in Figure 10.25(b). The total voltage at a point P in the cable at a given time (t) will be the sum of the voltages at P up to time t, that is $\beta_1 v_i(2 + 2\alpha_2 + 2\alpha_2^2)$, and the voltage at infinite time will be $2\beta_1 v_i(1 + \alpha_2 + \alpha_2^2 + \alpha_2^3 + \alpha_2^4 + \ldots)$.

The voltages at other points are similarly obtained. The time scale may be determined from a knowledge of length and surge velocity; for the line the latter is of the order of 300 m/µs and for the cable 150 m/µs. For a surge 50 µs in duration and a cable 300 m in length there will be 25 cable lengths traversed and the terminal voltage will approach $2v_i$. If the graph of voltage at the cable open-circuited end is plotted against time, an exponential rise curve will be obtained similar to that obtained for a capacitor.

The above treatment applies to a rectangular surge waveform, but may be modified readily to account for a waveform of the type illustrated in Figure 10.1. In this case the voltage change with time must also be allowed for and the process is more complicated.

10.6.2 Short-Line Faults

A particular problem for circuit breakers on h.v. systems is interrupting current if a solid short circuit occurs a few kilometres from the circuit breaker. This situation is illustrated in Figure 10.26.

It will be recalled that in Figure 10.3, fault interruption gives rise to a restriking transient on the supply side of the breaker. The sudden current interruption sends a step surge down the short line to the fault, which is reflected at the short-circuit point back to the breaker. In a manner similar to the Bewley lattice calculation of Figure 10.25, a surge is propagated back and forth in Z_0 as shown in Figure 10.26(b). The resulting combination of restriking voltage on the supply side and the continuously reflected surge on the line side of the breaker produces a steeply rising initial restriking voltage, greater than the restriking transient of Figure 10.3. Since this occurs just following current zero, it provides the most onerous condition of all for the circuit breaker to deal with, and must be included in the series of proving tests on any new design.

The first peak of the line-side component of the transient recovery voltage may be obtained from a knowledge of the fault current I_f and the value of Z_0. Assuming ideal interruption at current zero, the rate of rise of transient recovery voltage is

$$\frac{dv}{dt} = Z_0 \frac{di}{dt} = Z_0 \frac{\sqrt{2}I_f}{1/\omega} = \sqrt{2}\omega Z_0 I_f \times 10^{-6} \quad (V/\mu s)$$

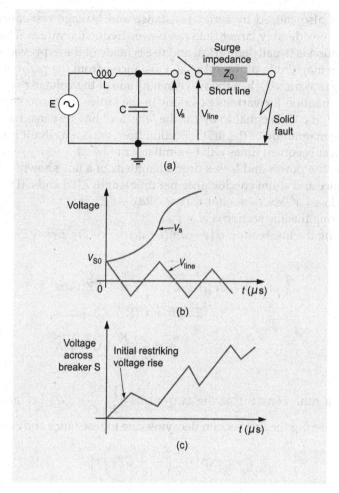

Figure 10.26 Short-line fault, (a) System diagram, (b) Supply-side and line-side transients, (c) Transient voltage across circuit breaker contacts

The amplitude of the first peak of reflected wave is

$$V = t\frac{dv}{dt} \quad (V)$$

where t is twice the transit time of the surge through Z_0 and ω is the system angular frequency. To this value should be added any change of v_s on the supply side, calculated as given by equation (10.6).

10.6.3 Effects of Line Loss

Attenuation of travelling waves is caused mainly by corona which reduces the steepness of the wavefronts considerably as the waves travel along the line.

Attenuation is also caused by series resistance and leakage resistance and these quantities are considerably larger than the power-frequency values. The determination of attenuation is usually empirical and use is made of the expression $v_x = v_i e^{-\gamma x}$, where v_x is the magnitude of the surge at a distance x from the point of origination. If a value for γ is assumed, then the wave magnitude of the voltage may be modified to include attenuation for various positions in the lattice diagram. For example, in Figure 10.25(b), if $e^{-\gamma x}$ is equal to α_L for the length of line traversed and α_C for the cable, then the magnitude of the first reflection from the open circuit is $\alpha_L v_i \alpha_C \beta_1$ and the voltages at subsequent times will be similarly modified.

Considering the power and losses over a length dx of a line shown in Figure 10.27 where resistance and shunt conductance per unit length $R(\Omega)$ and $G(\Omega^{-1})$.

The power loss $= i^2 R dx + (v - dv)^2 G dx \approx i^2 R dx + i^2 Z_0^2 G dx$

Power entering the line section $= vi = i^2 Z_0$

Power leaving the line section $= (v - dv)(i - di) = (i - di)^2 Z_0 \approx i^2 Z_0 - 2i Z_0 di$

Therefore,

$$vi - (v - dv)(i - di) = i^2 R dx + i^2 Z_0^2 G dx$$

$$2i Z_0 di = i^2 (R + Z_0^2 G) dx$$

$$\frac{di}{i} = \frac{1}{2}\left(\frac{R}{Z_0} + G Z_0\right) dx$$

This gives a time constant of the current as $\left[\frac{1}{2}\left(\frac{R}{Z_0} + G Z_0\right)x\right]$ and a surge of amplitude i_i entering the line section decaying due to resistance and conductance,

$$i = i_i \exp\left[-\frac{1}{2}\left(\frac{R}{Z_0} + G Z_0\right)x\right] \tag{10.20}$$

Also, it may be shown that

$$v = v_i \exp\left[-\frac{1}{2}\left(\frac{R}{Z_0} + G Z_0\right)x\right] \tag{10.21}$$

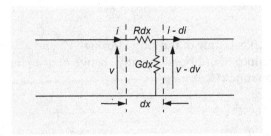

Figure 10.27 Lossy line section (L and C are not shown)

and the power at x is

$$vi = v_i i_i \exp\left[-\left(\frac{R}{Z_0} + GZ_0\right)x\right]$$ (10.22)

If R and G are realistically assessed (including corona effect), attenuation may be included in the travelling-wave analysis.

10.6.4 Digital Methods

The lattice diagram becomes very cumbersome for large systems and digital methods are usually applied. Digital methods may use strictly mathematical methods, that is the solution of the differential equations or the use of Fourier or Laplace transforms. These methods are capable of high accuracy, but require large amounts of data and long computation times. The general principles of the graphical approach described above may also be used to develop a computer program which is very applicable to large systems. Such a method will now be discussed in more detail.

A rectangular wave of infinite duration is used. The theory developed in the previous sections is applicable. The major role of the program is to scan the nodes of the system at each time interval and compute the voltages. In Figure 10.28 a particular system (single-phase representation) is shown and will be used to illustrate the method.

The relevant data describing the system are given in tabular form (see Table 10.4). Branches are listed both ways in ascending order of the first nodal number and are referred to under the name BRANCH. The time taken for a wave to travel along a branch is recorded in terms of a positive integer (referred to as PERIOD) which converts the basic time unit into actual travel time. Reflection coefficients are stored and referred to as REFLECT and the corresponding refraction coefficients are obtained, that is $(1 + \alpha_{ij})$. Elements of time as multiples of the basic integer are also shown in Table 10.4.

The method is illustrated by examining the system after the arrival of the rectangular wave at node 3 at Time (0). This voltage (magnitude 1 p.u.) is entered in the BRANCH (3, 2), TIME (0) element of the BRANCH-TIME matrix. On arrival at node

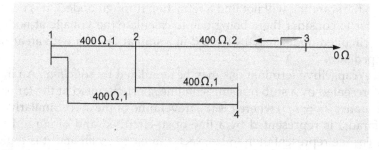

Figure 10.28 Application of digital method. Each line is labelled with surge impedance and surge travel time (multiples of basic unit), for example 400 Ω 1

Table 10.4 Data for lattice calculations

BRANCH i		1	1	2	2	2	3	4	4
j		2	4	1	3	4	2	1	2
PERIOD		1	1	1	2	1	2	1	1
REFLEC (α_{ij})		$-1/3$	0	0	-1	0	$-1/3$	0	$-1/3$
$(1 + \alpha_{ij})$		2/3	1	1	0	1	2/3	1	2/3
TIME	0						1		
	1								
	2			2/3	$-1/3$	2/3			
	3	0	2/3						
	4								

2 at time equal to zero plus *PERIOD* (3, 2), two waves are generated, on *BRANCH* (2, 1) and *BRANCH* (2, 4), both of magnitude $1(1 + \alpha_{32})$, that is $2/3$. A reflected wave is also generated on *BRANCH* (2, 3), of magnitude $1 \times \alpha_{32}$, that is $-1/3$. These voltages are entered in the appropriate *BRANCH* in the *TIME* (2) row of Table 10.4. On reaching node 1, *TIME* (3), a refracted wave of magnitude $2/3(1 + \alpha_{12})$, that is $2/3$, is generated on *BRANCH* (1, 4) and a reflected wave $2/3 \times \alpha_{12}$, that is 0, is generated on *BRANCH* (1, 2). This process is continued until a specified time is reached. All transmitted waves for a given node are placed in a separate *NODE-TIME* array; a transmitted wave is considered only once even though it could be entered into several *BRANCH-TIME* elements. Current waves are obtained by dividing the voltage by the surge impedance of the particular branch. The flow diagram for the digital solution is shown in Figure 10.29.

It is necessary for the programs to cater for semi-infinite lines (that is lines so long that waves reflected from the remote end may be neglected) and also inductive/capacitive terminations. Semi-infinite lines require the use of artificial nodes labelled (say) 0. For example, in Figure 10.30, node 2 is open-circuited and this is accounted for by introducing a line of infinite surge impedance between nodes 2 and 0, and hence, if

$$Z \to \infty \quad \alpha_{12} = \frac{Z - Z_0}{Z + Z_0} = 1$$

Although the scanning will not find a refraction through node 2, it is necessary for the computer to consider there being one to calculate the voltage at node 2. Lines with short-circuited nodes may be treated in a similar fashion with an artificial line of zero impedance.

Inductive/capacitive terminations may be simulated by *stub lines*. An inductance L (H) is represented by a stub transmission line short-circuited at the far end and of surge impedance $Z_L = L/t$, where t is the travel time of the stub. Similarly, a capacitance C (farads) is represented by a line open-circuited and of surge impedance $Z_C = t/C$. For the representation to be exact, t must be small and it is found necessary for Z_L to be of the order of 10 times and Z_C to be one-tenth of the combined surge impedance of the other lines connected to the node.

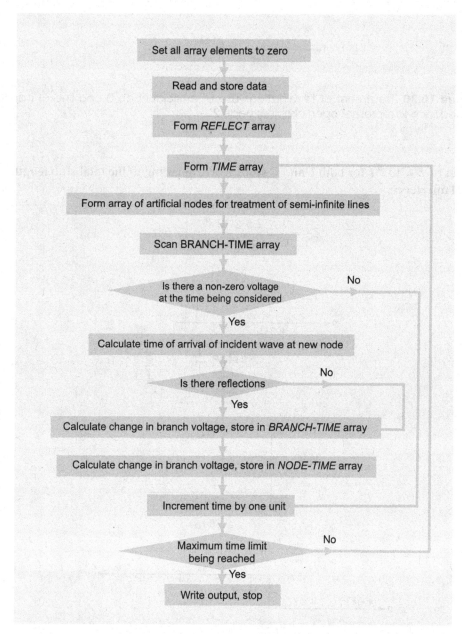

Figure 10.29 Flow diagram of digital method for travelling-wave analysis

For example, consider the termination shown in Figure 10.31(a). The equivalent stub-line circuit is shown in Figure 10.31(b). Stub travel times are chosen to be short compared with a quarter cycle of the natural frequency, that is.

$$\frac{2\pi\sqrt{LC}}{4} = \frac{2\pi}{4}\sqrt{0.01 \times 4 \times 10^{-8}} = 0.315 \times 10^{-4}\ \text{s}$$

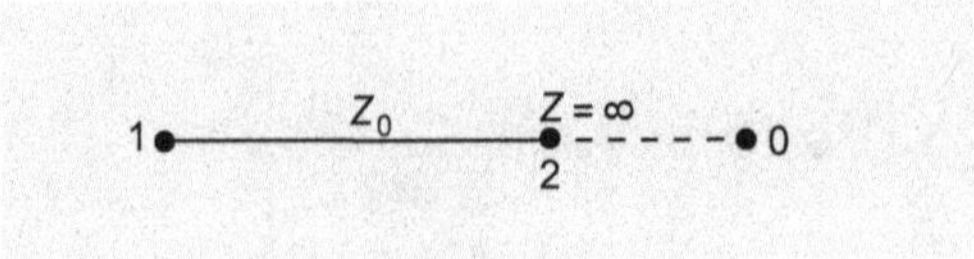

Figure 10.30 Treatment of terminations–use of artificial node 0 and line of infinite impedance to represent open circuit at node 2

Let $t = 5 \times 10^{-6}$ s for both L and C stubs (corresponding to the total stub length of 1524 m). Hence,

$$Z_C = \frac{t}{C} = \frac{2.5 \times 10^{-6}}{4 \times 10^{-8}} = 67.5\,\Omega$$

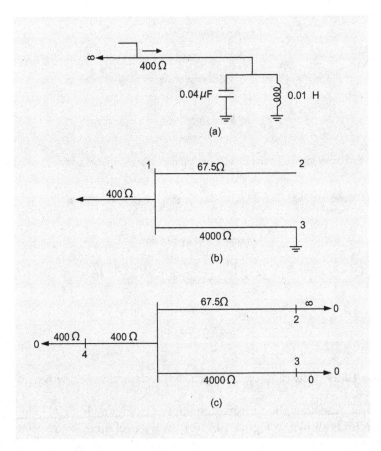

Figure 10.31 Representation of line terminated by L-C circuit by means of stublines, (a) Original system, (b) Equivalent stub lines, (c) Use of artificial nodes to represent open- and short-circuited ends of stub lines

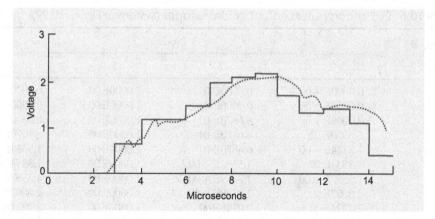

Figure 10.32 Voltage-time relationship at nodes 1 and 4 of system in Figure 10.31, —— computer solution; ⋯⋯, transient analyzer (Reproduced with permission from IEEE)

and

$$Z_C = \frac{L}{t} = \frac{0.01}{2.5 \times 10^{-6}} = 4000 \, \Omega$$

The configuration in a form acceptable for the computer program is shown in Figure 10.31(c). Refinements to the program to incorporate attenuation, wave shapes, and non-linear resistors may be made without changing its basic form.

A typical application is the analysis of the nodal voltages for the system shown in Figure 10.29. This system has previously been analyzed by a similar method by Barthold and Carter (1961) and good agreement found. The print out of nodal voltages for the first 20 μs is shown in Table 10.5, and in Figure 10.33 the voltage plot for nodes 1 and 4 is compared with a transient analyzer solution obtained by Barthold and Carter.

10.6.5 Three-Phase Analysis

The single-phase analysis of a system as presented in this chapter neglects the mutual effects which exist between the three phases of a line, transformer, and so on. The transient voltages due to energization may be further increased by this mutual coupling and also by the three contacts of a circuit breaker not closing at the same instant.

10.7 Electromagnetic Transient Program (EMTP)

The digital method previously described is very limited in scope. A much more powerful method has been developed by the Bonneville Power Administration and

Table 10.5 Digital computer printout: node voltages (system of Figure 10.29)

Time (µs)	Node			
	1	2	3	4
0	0.0000E + 00	0.0000E-01	1.0000E 00	0.0000E + 00
1	0.0000E + 00	0.0000E-01	1.0000E 00	0.0000E + 00
2	0.0000E + 00	6.6670E-01	1.0000E 00	0.0000E + 00
3	6.6670E-01	6.6670E-01	1.0000E 00	6.6670E-01
4	1.3334E + 00	6.6670E-01	1.0000E 00	1.3334E + 00
5	1.3334E 00	1.5557E + 00	1.0000E 00	1.3334E + 00
6	1.5557E + 00	1.7779E + 00	1.0000E 00	1.5557E + 00
7	2.0002E 00	1.7779E + 00	1.0000E 00	2.0002E 00
8	2.2224E 00	2.0743E 00	1.0000E 00	2.2224E 00
9	2.2965E 00	1.7779E + 00	1.0000E 00	2.2965E 00
10	1.8520E + 00	1.8520E + 00	1.0000E 00	1.8520E + 00
11	1.4075E + 00	1.9508E + 00	1.0000E 00	1.4075E + 00
12	1.5062E 00	1.0617E + 00	1.0000E 00	1.5062E + 00
13	1.1604E + 00	7.6534E-01	1.0000E 00	1.1604E + 00
14	4.1944E-01	8.2297E-01	1.0000E 00	4.1955E-01
15	8.2086E-02	2.6307E-01	1.0000E 00	8.2086E-02
16	−7.4396E-02	1.6435E-01	1.0000E 00	−7.4396E-02
17	7.8720E-03	1.0732E-02	1.0000E 00	7.8720E-03
18	9.3000E-02	−2.5556E-01	1.0000E 00	9.3000E-02
19	−1.7043E-01	4.1410E-01	1.0000E 00	−1.7043E-01
20	1.5067E-01	6.2634E-01	1.0000E 00	1.5067E-01

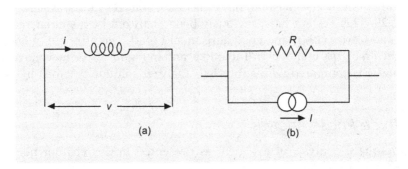

Figure 10.33 (a) Lumped inductance, (b) Equivalent circuit

is known as EMTP. This is widely used, especially in the USA and now has a number of implementations.

It is assumed that the variables of interest are known at the previous time step $t − \Delta t$, where Δt is the time step. The value of Δt must be small enough to give reasonable accuracy with a finite difference method.

10.7.1 Lumped Element Modelling

Consider an inductance L (Figure 10.33(a)),

$$v = L\frac{di}{dt} \tag{10.23}$$

Equation (10.23) can be rearranged to form:

$$i(t) = i(t - \Delta t) + \int_{t-\Delta t}^{t} v\, dt \tag{10.24}$$

Applying the trapezoidal rule gives:

$$i(t) = i(t - \Delta t) + \frac{\Delta t}{2L}[v(t) + v(t - \Delta t)]$$

$$= v(t)\frac{\Delta t}{2L} + i(t - \Delta t) + v(t - \Delta t)\frac{\Delta t}{2L} \tag{10.25}$$

$$= \frac{v(t)}{R} + I$$

where

$$R = \frac{2L}{\Delta t} \text{ and } I = i(t - \Delta t) + v(t - \Delta t)\frac{\Delta t}{2L}$$

Here, R is constant and I varies with time. The equivalent circuit is shown in Figure 10.33(b).

A similar treatment applies to capacitance (C) (see Figure 10.34(a)). Here,

$$i = C\frac{dv}{dt} \quad \therefore v = \int \frac{i}{C}\, dt$$

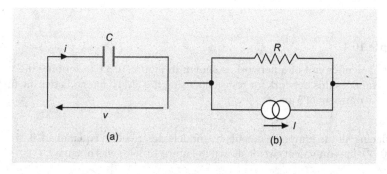

Figure 10.34 Equivalent circuits for transient analysis

Integrating and applying the trapezoidal rule gives

$$v(t) = v(t - \Delta t) + \frac{1}{C}\int_{t-\Delta t}^{t} i dt$$

$$= v(t - \Delta t) + \frac{\Delta t}{2C}(i(t) + i(t - \Delta t))$$

Rearranging gives:

$$i(t) = \frac{2C}{\Delta t}v(t) - i(t - \Delta t) - \frac{2C}{\Delta t}v(t - \Delta t)$$

$$= \frac{v(t)}{R} + I$$

(10.26)

Where

$$R = \frac{\Delta t}{2C} \text{ and } I = -i(t - \Delta t) - v(t - \Delta t)\frac{2C}{\Delta t}$$

giving the equivalent circuit of Figure 10.34(b).
    Resistance is represented directly by,

$$i(t) = \frac{v(t)}{R}$$

(10.27)

The procedure for solving a network using EMTP is as follows:

1. Replace all the lumped elements by models described in equations (10.25), (10.26) and (10.27).
2. From initial conditions, determine $i(t - \Delta t) = i(0)$ and $v(t - \Delta t) = v(0)$.
3. Solve for $i(t)$ and $v(t)$. Increment time step by Δt and calculate new values of i and v, and so on. The analysis is carried out by use of the nodal admittance matrix and Gaussian elimination. If mutual coupling between elements exists then the representation becomes very complex.

Example 10.4

The equivalent circuit of a network is shown in Figure 10.35. Determine the network which simulates this network for transients using the EMTP method after the first time step of the transient of 5 µs.

Solution
By replacing all the lumped elements by models described in equations (10.25), (10.26) and (10.27), the equivalent circuit shown in Figure 10.36(a) was obtained.

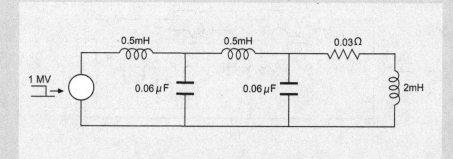

Figure 10.35 Figure for Example 10.4

At $t = 0$, note current sources are zero and equivalent circuit is reduced to Figure 10.36(b).

By applying the mesh method to Figure 10.36(b):

$$
\begin{bmatrix} 10^6 \\ 0 \\ 0 \end{bmatrix} = \begin{bmatrix} 241.7 & -41.7 & 0 \\ -41.7 & 241.7 & -41.7 \\ 0 & -41.7 & 800 \end{bmatrix} \begin{bmatrix} i_1 \\ i_2 \\ i_3 \end{bmatrix}
$$

Invert

$$
\begin{bmatrix} i_1 \\ i_2 \\ i_3 \end{bmatrix} = \begin{bmatrix} 0.0043 & 0.0043 & 0.0007 \\ 3.8 \times 10^{-5} & 0.0007 & 0.0043 \\ 0.0002 & 3.8 \times 10^{-5} & 0.0002 \end{bmatrix} \begin{bmatrix} 10^6 \\ 0 \\ 0 \end{bmatrix}
$$

$i_1 = 4300$ A, $i_2 = -38$ A and $i_3 = -200$
From Figure 10.36(a) and (b):

$I_{L1} = 4300$ A, $I_{L2} = -38$ A and $I_{L3} = -200$ A
$I_{C1} = i_1 - i_2 = 4338$ A and $I_{C2} = i_2 - i_3 = 238$ A

The equivalent circuit after $5\,\mu s$ is shown in Figure 10.36(c). The process is then repeated for the next $5\,\mu s$ step using Figure 10.36(c) as the starting condition.

10.7.2 Switching

The various representations are shown in Figure 10.37. A switching operation changes the topology of the network and hence the $[\mathbf{Y}]$ matrix. If the $[\mathbf{Y}]$ matrix is formed with all the switches open, then the closure of a switch is obtained by the adding together of the two rows and columns of $[\mathbf{Y}]$ and the associated rows of $[i]$.

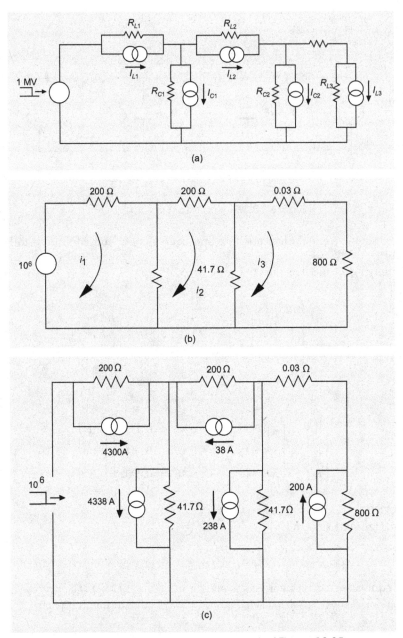

Figure 10.36 The equivalent circuit of Figure 10.35

Another area where switching is used is to account for non-linear $\psi - i$ characteristics of transformers, reactors, and so on. The representation is shown in Figure 10.38(a) and (b), in which

$$R = \frac{2b_k}{\Delta t}$$

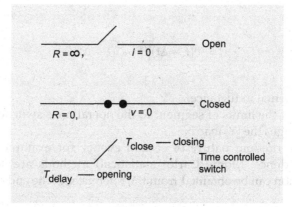

Figure 10.37 Representation of switch

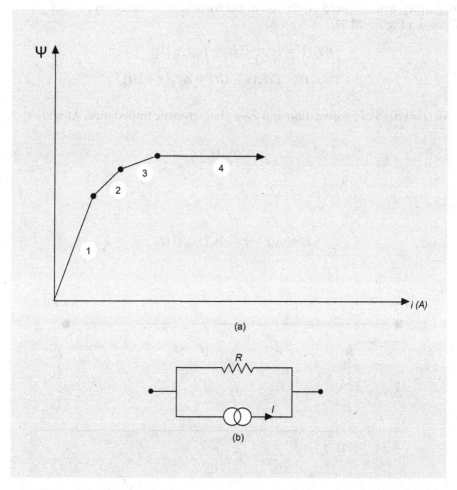

Figure 10.38 (a) Non-linear characteristic; (b) circuit representation

and

$$I = v(t - \Delta t)/R + i(t - \Delta t) \qquad (10.28)$$

where b_k = incremental inductance.

If ψ is outside of the limits of segment K, the operation is switched to either $k - 1$ or $k + 1$. This changes the $[\mathbf{Y}]$ matrix.

Because of the random nature of certain events, for example switching time or lightning incidence, Monte Carlo (statistical) methods are sometimes used. Further information can be obtained from the references at the end of this book.

10.7.3 Travelling-Wave Approach

Lines and cables would require a large number of n circuits for accurate representation. An alternative would be the use of the travelling-wave theory.

Consider Figure 10.39,

$$i(x, t) = f_1(x - Ut) + f_2(x + Ut)$$

$$v(x, t) = Z_0 f_1(x - Ut) + Z_0 f_2(x + Ut)$$

where U = speed of propagation and Z_0 = characteristic impedance. At node k,

$$i_k(t) = \frac{v_k(t)}{Z_0} + I_k$$

where

$$I_k = i_k(t - \tau) - v_m(t - \tau)/Z_0$$

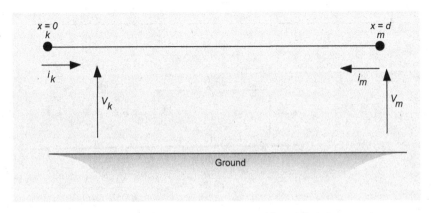

Figure 10.39 Transmission line with earth return

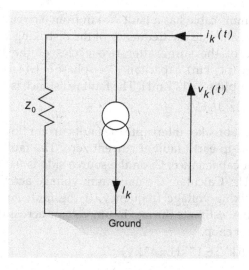

Figure 10.40 Equivalent circuit for single-phase line

and

$$\tau = \frac{d}{U} = \text{travel time}$$

The equivalent circuit is shown in Figure 10.40. An advantage of this method is that the two ends of the line are decoupled. The value of I_k depends on the current and voltage at the other end of the line τ seconds' previously, for example if $\tau = 0.36\,\text{ms}$ and $\Delta t = 100\,\mu s$, storage of four previous times is required.

Detailed models for synchronous machines and h.v.d.c. converter systems are given in the references.

Problems

10.1 A 345 kV, 60 Hz system has a fault current of 40 kA. The capacitance of a bus-bar to which a circuit breaker is connected is 25 000 pF. Calculate the surge impedance of the busbar and the frequency of the restriking (recovery) voltage on opening.

(Answer: 727 Ω, 8760 Hz)

10.2 A highly capacitive circuit of capacitance per phase 100 μF is disconnected by circuit breaker, the source inductance being 1 mH. The breaker gap breaks down when the voltage across it reaches twice the system peak line-to-neutral voltage of 38 kV. Calculate the current flowing with the breakdown, and its frequency, and compare it with the normal charging current of the circuit.

(Answer: 24 kA, 503 Hz; note $I = 2V_p/Z_0$)

10.3   A 10 kV, 64.5 mm^2 cable has a fault 9.6 km from a circuit breaker on the supply side of it. Calculate the frequency of the restriking voltage and the maximum voltage of the surge after two cycles of the transient. The cable parameters are (per km), capacitance per phase $= 1.14\,\mu$F, resistance $= 5.37\,\Omega$, inductance per phase $= 1.72$ mH. The fault resistance is $6\,\Omega$.

(Answer: 374 Hz; 16 kV)

10.4   A 132 kV circuit breaker interrupts the fault current flowing into a symmetrical three-phase-to-earth fault at current zero. The fault infeed is 2500 MVA and the shunt capacitance, C, on the source side is 0.03 μF. The system frequency is 50 Hz. Calculate the maximum voltage across the circuit breaker and the restriking-voltage frequency. If the fault current is prematurely chopped at 50 A, estimate the maximum voltage across the circuit breaker on the first current chop.

(Answer: 215.5 kV; 6.17 kHz; 43 kV)

10.5   An overhead line of surge impedance 500 Ω is connected to a cable of surge impedance 50 Ω. Determine the energy reflected back to the line as a percentage of incident energy.

(Answer: reflected energy $=- 0.67 \times$ incident surge energy)

10.6   A cable of inductance 0.188 mH per phase and capacitance per phase of 0.4 μF is connected to a line of inductance of 0.94 mH per phase and capacitance 0.0075 μF per phase. All quantities are per km. A surge of 1 p.u. magnitude travels along the cable towards the line. Determine the voltage set up at the junction of the line and cable.

(Answer: 1.88 p.u)

10.7   A long overhead line has a surge impedance of 500 Ω and an effective resistance at the frequency of the surge of 7 Ω/km. If a surge of magnitude 500 kV enters the line at a certain point, calculate the magnitude of this surge after it has traversed 100 km and calculate the resistive power loss of the wave over this distance. The wave velocity is 3×10^5 km/s.

(Answer: 250 kV; 375 MW)

10.8   A rectangular surge of 1 p.u. magnitude strikes an earth (ground) wire at the centre of its span between two towers of effective resistance to ground of 200 and 50 Ω. The ground wire has a surge impedance of 500 Ω. Determine the voltages transmitted beyond the towers to the earth wires outside the span.

(Answer: $0.44v_i$ from 200 Ω tower and $0.17v_i$ from 50 Ω tower)

10.9   A system consists of the following elements in series: a long line of surge impedance 500 ft, a cable (Z_0 of 50 Ω), a short line (Z_0 of 500 Ω), a cable (Z_0 of 50 Ω), a long line (Z_0 of 500 Ω). A surge takes 1 μs to traverse each cable (they are of equal length) and 0.5 μs to traverse the short line connecting the cables. The short line is half the length of each cable. Determine, by means of a

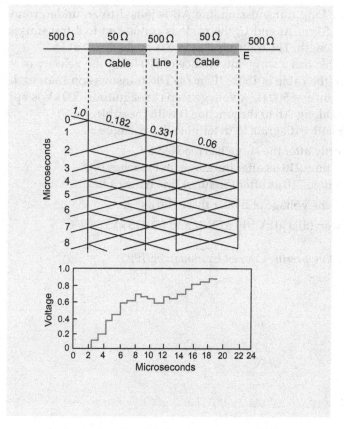

Figure 10.41 Solution of Problem 10.9

lattice diagram, the p.u. voltage of the junction of the cable and the long line if the surge originates in the remote long line.

(Answer: see Figure 10.41)

10.10 A 3 p.h., 50 Hz, 11 kV star-connected generator, with its star point earthed, is connected via a circuit breaker to a busbar. There is no load connected to the busbar. The capacitance to earth on the generator side terminals of the circuit breaker is 0.007 μF per phase. A three-phase-to-earth short circuit occurs at the busbar with a symmetrical subtransient fault current of 5000 A. The fault is then cleared by the circuit breaker. Assume interruption at current zero. (a) Sketch the voltage across the circuit breaker terminals of the first phase to clear. (b) Neglecting damping, calculate the peak value of the transient recovery voltage of this phase. (c) Determine the time to this peak voltage and hence the average rate of rise of recovery voltage.

(Answer: (b) 17.96 kV; (c) 16.7 μs, 1.075 kV/μs)

(From Engineering Council Examination, 1996)

10.11 A very long transmission line AB is joined to an underground cable BC of
length 5 km. At end C, the cable is connected to a transmission line CD of
15 km length. The transmission line is open-circuit at D.

The cable has a surge impedance of 50 Ω and the velocity of wave propagation in the cable is 150×10^6 m/s. The transmission lines each have a surge
impedance of 500 Ω. A voltage step of magnitude 500 kV is applied at A and
travels along AB to the junction B with the cable.

Use a lattice diagram to determine the voltage at:

a. D shortly after the surge has reached D;
b. D at a time 210 μs after the surge first reaches B;
c. B at a time 210 μs after the surge first reaches B.

Sketch the voltage at B over these 210 μs.

(Answer: (a) 330 kV, (b) 823 kV and (c) 413 kV)

(From Engineering Council Examination, 1995)

11

Substations and Protection

11.1 Introduction

Chapter 7 described techniques for the analysis of various types of faults which occur in a power system. The major uses of the results of fault analysis are for the specification of switchgear and the setting of protective relays, both being housed in substations. However, the design of all items in an electrical plant is influenced by knowledge of fault conditions, for example, the bursting forces experienced by underground cables.

Circuit-breaker ratings are determined from the fault currents that may flow at their particular locations and are expressed in kA (fault current flow) or MVA (short circuit level). The maximum circuit-breaker rating is of the order of 50 000–60 000 MVA, and this is achieved by the use of several interrupter heads in series per phase. Not only has the circuit breaker to extinguish the fault-current arc, with the substation connections it has also to withstand the considerable forces set up by short-circuit currents, which can be very high.

Knowledge of the currents resulting from various types of fault at a location is essential for the effective specification and setting of system protection. Faults on a power system result in high currents with possible loss of sychronism of generators and must be removed very quickly. Automatic means, therefore, are required to detect abnormal currents and voltages and, when detected, to open the appropriate circuit breakers. It is the object of protection to accomplish this. In a large interconnected network, considerable design knowledge and skill is required to remove the faulty part from the network and leave the healthy remainder working intact.

There are many varieties of automatic protective systems, ranging from simple over-current relays to sophisticated systems transmitting high-frequency signals along the power lines. The simplest but extremely effective form of protection is the over-current relay, which closes contacts and hence energizes the circuit-breaker opening mechanisms when currents due to faults pass through the equipment.

Electric Power Systems, Fifth Edition. B.M. Weedy, B.J. Cory, N. Jenkins, J.B. Ekanayake and G. Strbac.
© 2012 John Wiley & Sons, Ltd. Published 2012 by John Wiley & Sons, Ltd.

The protection used in a network can be looked upon as a form of insurance in which a percentage of the total capital cost (about 5%) is used to safeguard apparatus and ensure continued operation of the power system when faults occur. In a highly industrialized society the maintenance of an uninterrupted supply to consumers is of paramount importance and the adequate provision of protection systems is essential. It is important the protection system *discriminates* between those items of plant that are faulted and must be removed from the system and those that are sound and should remain in service.

Summarizing, protection and the automatic tripping (opening) of associated circuit breakers has two main functions: (1) to isolate faulty equipment so that the remainder of the system can continue to operate successfully; and (2) to limit damage to equipment caused by overheating and mechanical forces .

11.2 Switchgear

For maintenance to be carried out on high voltage plant, it must be isolated from the rest of the network and hence switches must be provided on each side. Different designs of these switches have very different capabilities and costs. Circuit breakers are able to interrupt fault current but some isolators can interrupt load current while others can only be operated off-circuit that is with no voltage applied. Some isolators are able to close on to faults safely but not interrupt fault current. In addition earth (ground) switches are used to connect to earth the isolated equipment which is to be worked on. Switchgear must always be operated within its capability as otherwise very considerable danger will arise.

Owing to the high cost of circuit breakers, much thought is given, in practice, to obtaining the largest degree of flexibility in connecting circuits with the minimum number of circuit breakers. Popular arrangements of switches are shown in Figure 11.1 and a typical feeder bay layout of a transmission substation in Figure 11.2.

High-voltage circuit breakers, 11 kV and above, take five basic forms:

- oil immersed or bulk oil
- air blast
- small oil volume
- sulphur hexafluoride (SF_6)
- vacuum.

In addition, air circuit breakers with long contact gaps and large arc-splitter plates have been used at up to 11 kV, but they are bulky and expensive compared with the five types listed above.

All high voltage circuit breakers operate by their contacts separating and creating an arc that is extinguished at a subsequent current-zero. The voltage across the separated contacts is then established in the medium of the interrupter (oil, air, SF_6 or vacuum).

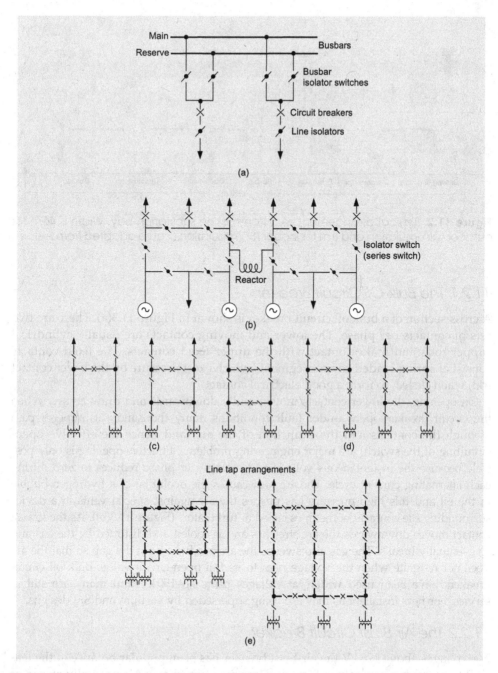

Figure 11.1 Possible switchgear arrangements for transmission and generation substations, (a) Double-busbar selection arrangements, (b) Double-ring busbars with connecting reactor, busbars can be isolated for maintenance, but circuits cannot be transferred from one side of the reactor to the other, (c) Open-mesh switching stations, transformers not switched, (d) Open mesh switching stations, transformers switched, (e) Closed-mesh switching stations. (Isolators are sometimes called series switches.) Arrangements can be indoors or outdoors

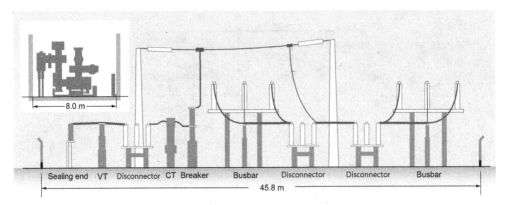

Figure 11.2 Typical arrangement of a transmission substation bay. Width is 46 m for outdoor (air) insulation and 8 m for indoor (SF$_6$) insulation (Figure adapted from Alstom)

11.2.1 The Bulk-Oil Circuit Breaker

A cross-section of a bulk oil circuit breaker is shown in Figure 11.3(a). There are two sets of contacts per phase. The lower and moving contacts are usually cylindrical copper rods and make contact with the upper fixed contacts. The fixed contacts consist of spring-loaded copper segments which exert pressure on the lower contact rod, when closed, to form a good electrical contact.

On opening, the lower contacts move rapidly downwards and draw an arc. When the circuit breaker opens under fault conditions many thousands of amperes pass through the contacts and the extinction of the arc (and hence the effective open-circuiting of the switch) are major engineering problems. Effective opening is only possible because the instantaneous voltage and current per phase reduces to zero during each alternating current cycle. The arc heat causes the evolution of a hydrogen bubble in the oil and this high-pressure gas pushes the arc against special vents in a device surrounding the contacts, sometimes called a 'turbulator' (Figure 11.3(b)). As the lowest contact moves downwards the arc stretches and is cooled and distorted by the gas and so eventually breaks. The gas also sweeps the arc products from the gap so that the arc does not re-ignite when the voltage rises to its full open-circuit value. Bulk oil circuit breakers have been used widely at voltages from 6.6–150 kV and many are still in service. For new installations they are being superseded by vacuum and SF$_6$ designs.

11.2.2 The Air-Blast Circuit Breaker

For voltages above 120 kV the air-blast breaker has been popular because of the feasibility of having several contact gaps in series per phase. Air normally stored at 1.38 MN/m^2 is released and directed at the arc at high velocities, thus extinguishing it. The air also actuates the mechanism of the movable contact. Figure 11.4 shows a 132 kV air-blast breaker with two breaks (interrupters) per phase and its associated isolator or series switch. Schematic diagrams of two types of air-blast head are shown in Figure 11.5 while the arrangement adopted for very high voltages, with grading resistors, is shown in Figure 11.6.

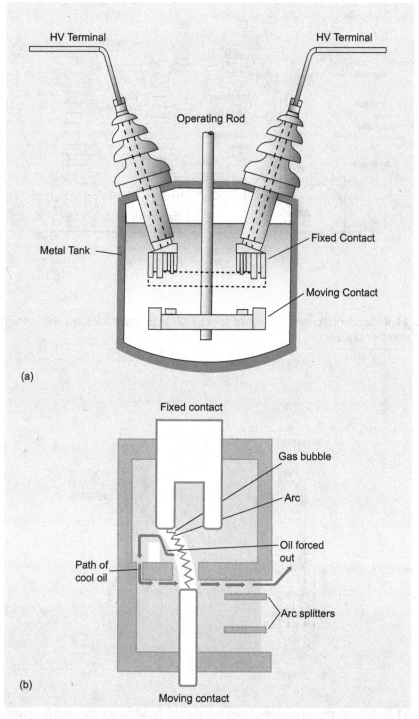

Figure 11.3 (a) Cross-section of a bulk-oil circuit breaker. Three phases are placed in one tank (Figure adapted from Alstom). (b) Cross-jet explosion pot for arc extinction in bulk-oil circuit breakers (Figure adapted from Alstom)

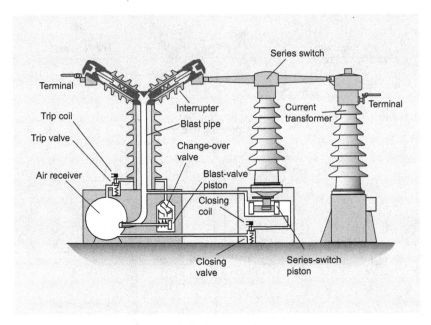

Figure 11.4 Schematic arrangement for a 132 kV air-blast circuit breaker with two interrupters per phase

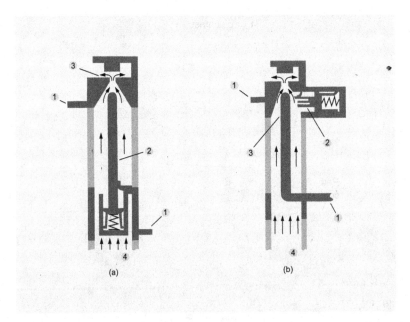

Figure 11.5 Schematic diagrams of two types of air-blast head, (a) Axial flow with axially moving contact, (b) Axial flow with side-moving contact. 1 = Terminal, 2 = moving contact, 3 = fixed contact, and 4 = blast pipe (Figure adapted from Siemens)

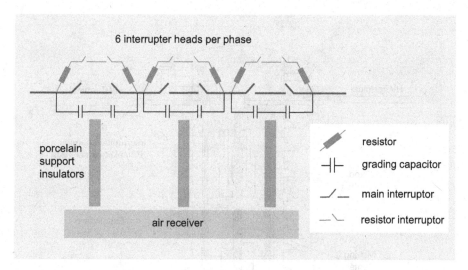

Figure 11.6 275 kV air-blast circuit breaker

Although air-blast circuit breakers have been developed and installed up to the highest voltages they have now been largely superseded for new installations by the SF_6 gas circuit breaker

11.2.3 Small- or Low-Oil-Volume Circuit Breakers

Here, the quenching mechanisms are enclosed in vertical porcelain insulation compartments and the arc is extinguished by a jet of oil issuing from the moving (lower contact) as it opens (Figure 11.7). The volume of oil is much smaller than in the bulk-oil type, thereby reducing hazards from explosion and fire. Although many oil circuit breakers are in use, particularly at distribution voltages, modern practice is to use vacuum or SF_6 to avoid the presence of flammable liquids for circuit interruption purposes.

11.2.4 Sulphur Hexafluoride (SF_6) Gas

The advantages of using sulphur hexafluoride (SF_6) as an insulating and interrupting medium in circuit breakers arise from its high electric strength and outstanding arc-quenching characteristics. SF_6 circuit breakers are much smaller than air blast circuit breakers of the same rating, the electric strength of SF_6 at atmospheric pressure being roughly equal to that of air at the pressure of 10 atm. Temperatures in the order of 30 000 K are likely to be experienced in arcs in SF_6 and these are, of course, well above the dissociation temperature of the gas (about 2000 K); however, nearly all the decomposition products are electronegative so that the electric strength of the gas recovers quickly after the arc has been extinguished. Filters are provided to render the decomposition products harmless and only a small amount of fluorine reacts with metallic parts of the breaker. A sectional view of an SF_6 switchgear unit is shown in Figure 11.8.

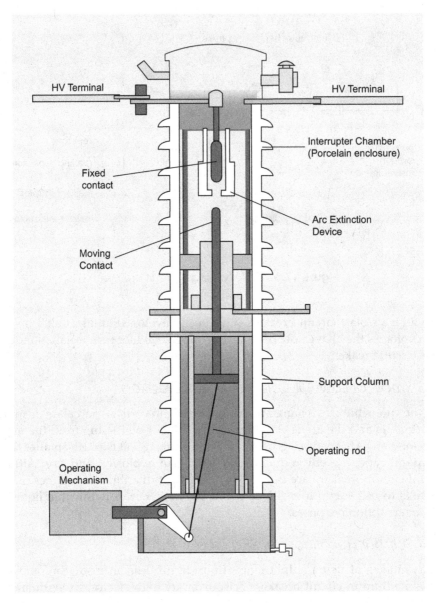

Figure 11.7 Minimum oil circuit breaker (Figure adapted from Alstom)

The switchgear also contains all necessary measuring and other facilities as follows: SF_6-insulated toroidal current transformers and voltage transformers, cable terminations, gas storage cylinders, cable isolators, grounding switches, bus isolators, and busbar system. Such substations are of immense value in urban areas because of their greatly reduced size compared with air blast switchgear. SF_6 circuit breakers rated at 45 GVA are available and designs for 1300 kV have been produced.

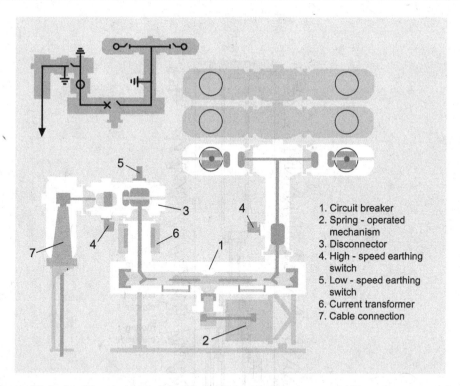

1. Circuit breaker
2. Spring - operated mechanism
3. Disconnector
4. High - speed earthing switch
5. Low - speed earthing switch
6. Current transformer
7. Cable connection

Figure 11.8 Sectional view of an SF_6 insulated switchgear unit (Figure adapted from Alstom)

SF_6 switchgear is now widely used at lower voltages from 6.6 to 132 kV for distribution systems. A typical interrupter for outdoor use, shown in Figure 11.9, is enclosed in a sealed porcelain cylinder with SF_6 under about 5 atm pressure. The movement of the contact forces gas into the opening contact by a 'puffer' action, thereby forcing extinction. SF_6 interrupters are very compact and robust; they require very little maintenance when mounted in metal cabinets to form switchboards.

SF_6 is an extremely powerful greenhouse gas and if released into the atmosphere has an effect on global warming some 20 000 times more powerful than an equivalent amount of CO_2. Hence during switchgear maintenance every effort is made to recover the SF_6 and avoid venting it to atmosphere

11.2.5 Vacuum Interrupter

A pair of contacts opening in vacuum draws an arc which burns in the vaporized contact material. Consequently, the contact material and its arcing root shape are crucial to the design of a commercial interrupter. A typical design is illustrated in Figure 11.10.

The main advantages of a vacuum interrupter are: (1) the very small damage normally caused to the contacts on operation, so that a life of 30 years can be expected

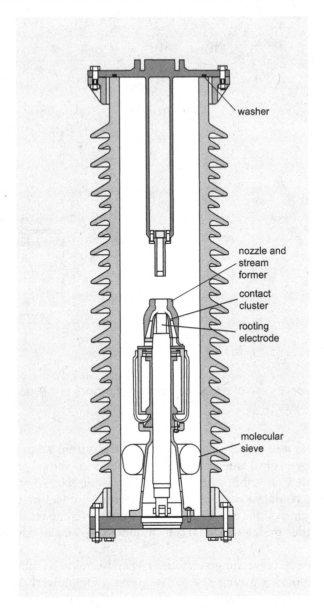

Figure 11.9 An SF_6 single-pressure puffer-type interrupter (Figure adapted from Siemens)

without maintenance; (2) the small mechanical energy required for tripping; and (3) the low noise caused on operation. The nature of the vacuum arc depends on the current; at low currents the arc is diffuse and can readily be interrupted, but at high currents the arc tends to be constricted. Electrode contour geometries have been produced to give diffuse arcs with current densities of 10^6–10^8 A/cm^2; electron velocities of 10^8 cm/s are experienced in the arc and ion velocities of 10^6 cm/s.

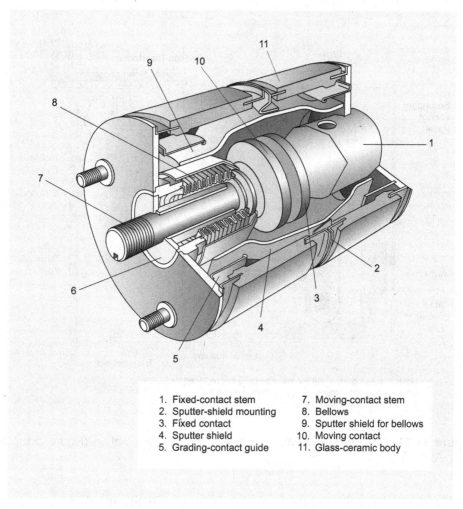

1. Fixed-contact stem	7. Moving-contact stem
2. Sputter-shield mounting	8. Bellows
3. Fixed contact	9. Sputter shield for bellows
4. Sputter shield	10. Moving contact
5. Grading-contact guide	11. Glass-ceramic body

Figure 11.10 Constructional features of an 11 kV vacuum interrupter) (Figure adapted from Siemens)

Considerable progress has been made in increasing the current-breaking capacity of vacuum interrupters, but not their operating voltage. Today, they are available with ratings of 500 MVA at 30 kV and are extensively used in 11, 20, and 25 kV switchboards. A three-phase vacuum breaker with horizontally mounted interrupters is shown in Figure 11.11 with SF_6 insulation surrounding the interrupters.

11.2.6 Summary of Circuit-Breaker Requirements

A circuit breaker must fulfil the following conditions:

1. Open and close in the shortest possible time under any network condition.
2. Conduct rated current without exceeding rated design temperature.

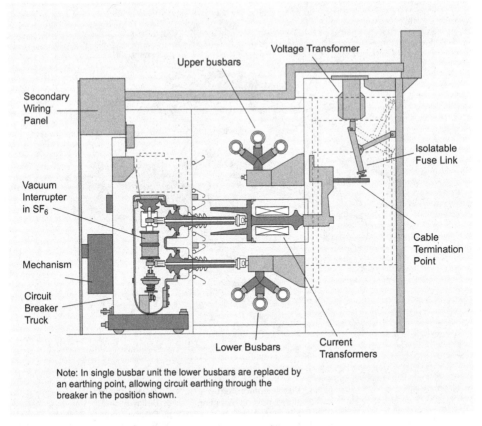

Note: In single busbar unit the lower busbars are replaced by
an earthing point, allowing circuit earthing through the
breaker in the position shown.

Figure 11.11 Double-busbar MV switchgear, horizontal isolation (Figure adapted from Alstom)

3. Withstand, thermally and mechanically, any short-circuit currents.
4. Maintain its voltage to earth and across the open contacts under both clean and polluted conditions.
5. Not create any large overvoltage during opening and closing.
6. Be easily maintained.
7. Be not too expensive.

Although air has now been largely overtaken by SF_6 as an interrupting medium at high voltages, it has given good performance over the years as it was easily able to achieve 2-cycle (~40 ms) interruption after receipt of a tripping signal to the operating coil. The increasing need, at voltages up to 725 kV, to reduce interruption times to $1\frac{1}{2}$ cycles to maintain stability and to reduce fault damage has meant that SF_6 switchgear has been required. A 60 kA rating has been possible with resistive switching to avoid unnecessary overvoltages, and a much lower noise level on operation has further confirmed SF_6 as the preferred medium. Being factory-sealed and fully enclosed has considerably reduced radio-interference and audible noise due to corona discharge.

Increasing performance and low maintenance has led to the development of on-line monitoring of the health of switchgear and the associated connections and components (known as 'condition monitoring'). This is particularly required for SF_6 enclosed units as even small dust particles can cause breakdown. Fast protection gear is essential to match these developments in switchgear.

11.3 Qualities Required of Protection

The effectiveness of protective gear can be described as:

1. **Selectivity or discrimination** – its effectiveness in isolating only the faulty part of the system.
2. **Stability** – the property of remaining inoperative with faults occurring outside the protected zone (called external faults).
3. **Speed of operation** – this property is more obvious. The longer the fault current continues to flow, the greater the damage to equipment. Of great importance is the necessity to open faulty sections of network before the connected synchronous generators lose synchronism with the rest of the system. A typical fault clearance time in h.v. systems is 80 ms, and this requires very high-speed relaying as well as very fast operation of the circuit breakers.
4. **Sensitivity** – this is the level of magnitude of fault current at which operation occurs, which may be expressed in current in the actual network (primary current) or as a percentage of the current-transformer secondary current.
5. **Economic considerations** – in distribution systems the economic aspect almost overrides the technical one, owing to the large number of feeders and transformers, provided that basic safety requirements are met. In transmission systems the technical aspects are more important. The protection is relatively expensive, but so is the system or equipment protected, and security of supply is vital. In transmission systems two separate protective systems are used, one main (or primary) and one back-up.
6. **Reliability** – this property is self-evident. A major cause of circuit outages is mal-operation of the protection itself. On average, in the British system (not including faults on generators), nearly 10% of outages are due to mal-operation of protection systems.

11.3.1 Protective Zones and Back-Up Protection

Each main protective scheme protects a defined area or zone of the power system. These zones are designed to overlap so that the entire power system is protected with fast acting main protection. Overlapping zones of protection are shown in Figure 11.12.

Back-up protection, as the name implies, is a completely separate arrangement which operates to remove the faulty part should the main protection fail to operate. The back-up system should be as independent of the main protection as possible, possessing its own current transformers and relays. Often, only the circuit-breaker tripping and voltage transformers are common.

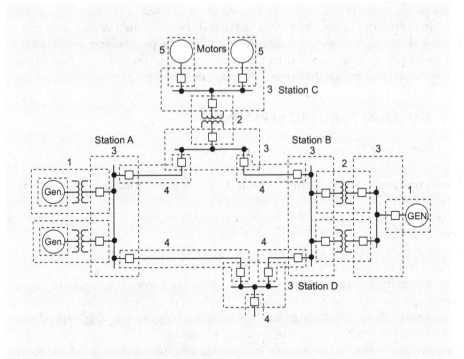

Figure 11.12 Line diagram of a typical system and the overlapping zones of protection

In distribution the application of back-up protection is not as widespread as in transmission systems; often it is sufficient to apply it at strategic points only or overlap the main protection further. Remote back-up is slow and usually disconnects more of the supply system than is necessary to remove the faulty part.

11.4 Components of Protective Schemes

11.4.1 Current Transformers (CTs)

In order to obtain currents which can be used in control circuits and that are proportional to the system (primary) currents, current transformers are used. Often the primary conductor itself, for example a busbar, forms a single primary turn (bar primary). Whereas instrument current transformers have to remain accurate only up to slight over-currents, protection current transformers must retain proportionality up to at least 20 times normal full load. The nominal secondary current rating of current transformers is now usually 1 A, but 5 A has been used in the past.

A major problem can exist when two current transformers are used which should retain identical characteristics up to the highest fault current, for example in pilot wire schemes. Because of saturation in the silicon steel used and the possible existence of a direct component in the fault current, the exact matching of such current transformers is difficult.

11.4.1.1 Linear Couplers

The problems associated with current transformers have resulted in the development of devices called linear couplers, which serve the same purpose but, having air cores, remain linear at the highest currents. These are also known as Rogowski coils and are particularly suited to digital protection schemes.

11.4.2 Voltage (or Potential) Transformers (VTs or PTs)

These provide a voltage which is much lower than the system voltage, the nominal secondary voltage being 110 V. There are two basic types: the wound (electromagnetic), virtually a small power transformer, and the capacitor type. In the latter a tapping is made on a capacitor bushing (usually of the order of 12 kV) and the voltage from this tapping is stepped down by a small electromagnetic voltage transformer. The arrangement is shown in Figure 11.13; the reactor (X) and the

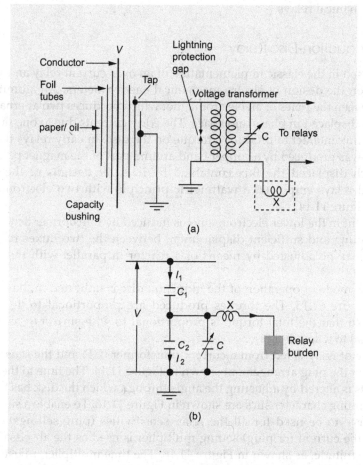

Figure 11.13 Capacitor voltage transformer, (a) Circuit arrangement, (b) Equivalent circuit–burden — impedance of transformer and load referred to primary winding

capacitor (C) constitute a tuned circuit which corrects the phase-angle error of the secondary voltage. In h.v. systems, the capacitor divider is a separate unit mounted within an insulator. It can also be used as a line coupler for high frequency signalling (power line carrier, see Section 11.10).

11.4.3 Relays

A relay is a device which, when supplied with appropriately scaled quantities, indicates an abnormal or fault condition on the power system. When the relay contacts close, the associated circuit-breaker trip-circuits are energized and the breaker opens, isolating the faulty part of the power network. Historically electromagnetic and semiconductor relays were installed and are still in use on the system. Modern practice is to install digital (numerical) protection. Although now almost all new relays use micro-processors and the measured quantities are manipulated digitally, the underlying techniques are often those developed for electro-mechanical relays.

11.4.3.1 Induction-Disc Relay

This was used in the classic implementation of an over-current relay and was based originally on the design of electro-mechanical energy meters. An aluminium disc rotates between the poles of an electromagnet which produces two alternating magnetic fields displaced in phase and space. The eddy currents due to one flux and the remaining flux interact to produce a torque on the disc. In early relays the flux displacement was produced by a copper band around part of the magnet pole (shading ring) which displaced the flux contained by it. Later designs of these electro-mechanical relays employed a wattmetric principle with two electromagnets, as shown in Figure 11.14.

The current in the lower electromagnet is induced by transformer action from the upper winding and sufficient displacement between the two fluxes results. This, however, may be adjusted by means of a reactor in parallel with the secondary winding.

The basic mode of operation of the induction disc is indicated in the phasor diagram of Figure 11.15. The torques produced are proportional to $\Phi_2 i_1 \sin \alpha$ and $\Phi_1 i_2 \sin \alpha$, so that the total torque is proportional to $\Phi_1 \Phi_2 \sin \alpha$ or $i_1 i_2 \sin \alpha$ as Φ_1 is proportional to i_1 and Φ_2 to i_2.

This type of relay is fed from a current transformer (CT) and the sensitivity may be varied by the plug arrangement shown in Figure 11.14. The time to the closing of the contacts is altered by adjusting the angle through which the disk has to rotate.

The operating characteristics are shown in Figure 11.16. To enable a single characteristic curve to be used for all the relay sensitivities (plug settings) a quantity known as the current (or plug) setting multiplier is used as the abscissa instead of current magnitude, as shown in Figure 11.16. The time multiplier adjusts the angle through which the disk rotates and so translates the curve vertically.

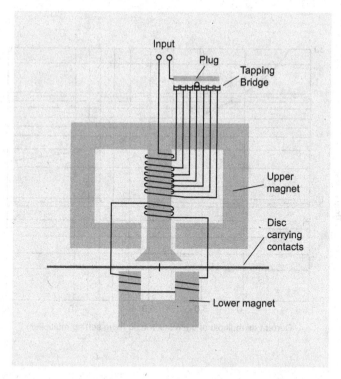

Figure 11.14 Induction-disc relay

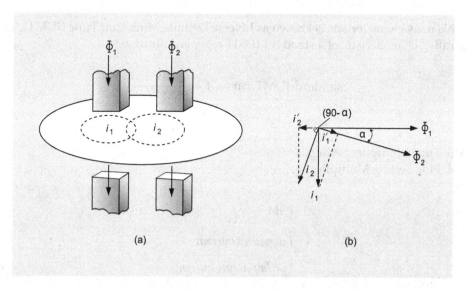

Figure 11.15 Operation of disc-type electromagnetic relay, (a) Fluxes, (b) Phasor diagram. i_1 and i_2 are induced currents in disc

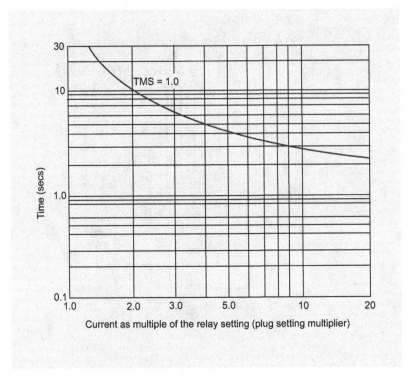

Figure 11.16 Time-current characteristics of a typical induction disc in terms of the plug-setting multiplier. TMS = time multiplier setting

This relay characteristic is known as Inverse Definite Minimum Time (IDMT). The operating characteristic of a standard IDMT relay is defined as:

$$\text{Standard IDMT curve: } t = \frac{0.14 \times \text{TMS}}{\text{PSM}^{0.02} - 1} \tag{11.1}$$

TMS: Time Multiplier Setting
PSM: Plug Setting Multiplier

$$\text{PSM} = {}^{I}\!/_{I_S}$$

I measured current

I_S *relay setting current*

To illustrate the use of this curve (usually shown on the relay casing) the following example is given.

Example 11.1

Determine the time of operation of a 1 A, 3 s overcurrent relay having a Plug Setting of 125% and a Time Multiplier of 0.6. The supplying CT is rated 400 : 1 A and the fault current is 4000 A.

Solution
The relay coil current for the fault $= (4000/400) \times 1 = 10$ A. The nominal relay coil current is $1 \times (125/100) = 1.25$ A. Therefore the relay fault current as a multiple of the Plug Setting $= (10/1.25) = 8$ (Plug Setting Multiplier). From the relay curve (Figure 11.16), the time of operation is 3.3 s for a time setting of 1. The time multiplier (TM) controls the time of operation by changing the angle through which the disc moves to close the contacts. The actual operating time $= 3.3 \times 0.6 = 2.0$ s. This can be obtained directly from equation (11.1) as:

$$t = \frac{0.14 \times \text{TMS}}{\text{PSM}^{0.02} - 1} = \frac{0.14 * 0.6}{8^{0.02} - 1} = 1.98 \text{ s}$$

Induction-disc relays may be made responsive to real power flow by feeding the upper magnet winding in Figure 11.14 from a voltage via a potential transformer and the lower winding from the corresponding current. As the upper coil will consist of a large number of turns, the current in it lags the applied voltage by 90°, whereas in the lower (small number of turns) coil they are almost in phase. Hence, Φ_1 is proportional to **V**, and Φ_2 is proportional to **I**, and torque is proportional to $\Phi_1\Phi_2 \sin \alpha$, that is to $VI \sin (90 - \alpha)$, or $VI \cos \alpha$ (where α is the angle between **V** and **I**).

The direction of the torque depends on the power direction and hence the relay is directional. A power relay may be used in conjunction with a current operated relay to provide a directional overcurrent property.

11.4.3.2 Balanced Beam

The basic form of this relay is shown in Figure 11.17. The armatures at the ends of the beam are attracted by electromagnets which are operated by the appropriate parameters, usually voltage and current. A slight mechanical bias is incorporated to keep the contacts open, except when operation is required.

The pulls on the armatures by the electromagnets are equal to K_1V^2 and K_2I^2, where K_1 and K_2 are constants, and for operation (that is, contacts to close), that is.

$$K_1V^2 > K_2I^2$$

then

$$\frac{V}{I} < \sqrt{\frac{K_2}{K_1}} \text{ or } Z < \sqrt{\frac{K_2}{K_1}}$$

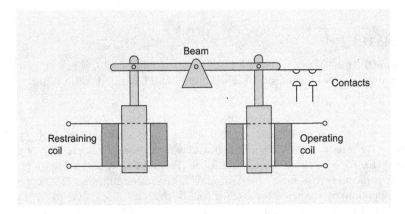

Figure 11.17 Schematic diagram of balanced-beam relay

This shows that the relay operates when the impedance it 'sees' is less than a pre-determined value. The characteristic of this impedance relay, when drawn on an R and jX axes, is a circle as shown in Figure 11.18.

11.4.3.3 Distance Relays

The balanced-beam relay, because it measures the impedance of the protected line, effectively measures distance to the fault. Two other relay forms also may be used for this purpose:

1. The reactance relay which operates when $\dfrac{V}{I}\sin\phi <$ constant, having the straight line characteristic shown in Figure 11.18.
2. The mho or admittance relay, the characteristic of which is also shown in Figure 11.18.

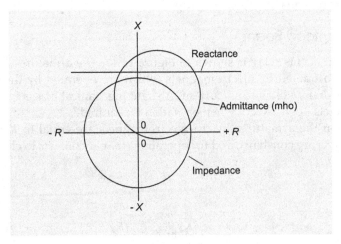

Figure 11.18 Characteristics of impedance, reactance, and admittance (mho) relays shown on the R-X diagram

These relays operate with any impedance phasor lying inside the characteristic circle or below the reactance line. In practice modern distance relays calculate the operating characteristics using digital micro-processors and some manufacturers offer the ability to create more complex shapes (for example, trapezoids).

11.4.3.4 Solid-State Relays

Electromechanical relays are vulnerable to corrosion, shock vibration and contact bounce and welding. They require regular maintenance by skilled personnel and over their life are expensive. Hence they were replaced initially with solid state electronic relays that fulfilled similar functions using analogue and simple digital circuits.

These relays are extremely fast in operation, having no moving parts and are very reliable. Detection involving phase angles and current and voltage magnitudes are made with appropriate electronic circuits. Most of the required current-time characteristics may be readily obtained. Inverse-characteristic, overcurrent and earth-fault relays have a minimum time lag and the operating time is inversely related to some power of the input (for example, current). In practical static relays it is advantageous to choose a circuit which can accommodate a wide range of alternative inverse time characteristics, precise minimum operating levels, and definite minimum times.

11.4.3.5 Digital (Numerical) Relaying

With the advent of integrated circuits and the microprocessor, digital (numerical) protection devices are now the norm. Having monitored currents and voltages through primary transducers (CTs and VTs), these analogue quantities are sampled and converted into digital form for numerical manipulation, analysis, display and recording. This process provides a flexible and very reliable relaying function, thereby enabling the same basic hardware units to be used for almost any kind of relaying scheme. With the continuous reduction in digital-circuit costs and increases in their functionality, considerable cost-benefit improvement ensues. It is now often the case that the cost of the relay housing, connections and EMC protection dominates the hardware but, as is usual in digital systems, the software development and proving procedure are the most expensive items in the overall scheme.

Since digital relays can store data, they are particularly suited to post-fault analysis and can even be used in a self-adaptive mode, which is impossible with traditional electro-mechanical or analogue electronic devices. Additionally, numerical relays are capable of self-monitoring and communication with hierarchical controllers. By these means, not only can fast and selective fault clearance be obtained, but also fault location can be flagged to mobile repair crews. Minor (non-vital) protection-system faults can also be indicated for maintenance attention. With the incorporation of a satellite-timing signal receiver using Global Positioning Satellite (GPS) to give a 1 μs synchronized signal, faults on overhead circuits can be located to within 300 m.

The basic hardware elements of a digital relay are indicated in Figure 11.19. The signals coming from CTs and VTs (only a single input is shown even through measurements of each phase are connected to the relay) are first connected to a surge protection circuit so as to prevent any surges entering into the relay. As these signals

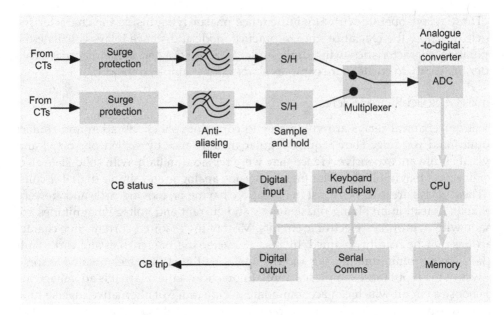

Figure 11.19 Basic components of a digital relay

contain dc offset, harmonics and noise they are first filtered using anti-aliasing filters. Then the filtered analogue signals are converted to digital signals by the Analogue to Digital Converter (ADC). Since the ADC takes a finite conversion time, sample and hold circuits are used to retain the signals coming from CTs and VTs while the multiplexer passes one signal to the ADC. Digitised signals are then processed by the relay algorithms held in the memory. The memory has a volatile component (RAM) where some processing data is stored and non-volatile components (EPROM and ROM) where relay algorithms, and event records are kept.

In numerical relays (also known as IEDs, Intelligent Electronic Devices, as they perform more than protection functions) the hardware is similar for different relays. Depending on the application different relay algorithms are stored in the EPROM. For example, if the relay is an overcurrent relay, the relay algorithm reads settings such as the type of characteristic (standard inverse, very inverse, and so on), the time and current settings. Digitized current measurements from the ADC are compared with the current setting to check whether the fault current is greater than the setting. If the fault current exceeds the setting then a trip signal is generated with an appropriate delay. For the advanced student, many good references to numerical relaying, as well as information from the major equipment manufacturers, are now available (see Further Reading at the end of this book).

11.5 Protection Systems

The application of the various relays and other equipment to form adequate schemes of protection forms a large and complex subject. Also, the various schemes may depend on the methods of individual manufacturers. The intention here is to present a survey of general practice and to outline the principles of the methods used. Some schemes are discriminative to fault location and involve several parameters, for example time, direction, current, distance, current balance, and phase comparison. Others discriminate according to the type of fault, for example negative-sequence relays for unbalanced faults in generators, and some use a combination of location and type of fault.

A convenient classification is the division of the systems into unit and non-unit types. Unit protection signifies that an item of equipment or zone is being uniquely protected independently of the adjoining parts of the system. Non-unit schemes are those in which several relays and associated equipment are used to provide protection covering more than one zone. Non-unit schemes represent the most widely used and cheapest forms of protection and these will be discussed first.

11.5.1 Over-Current Protection Schemes

This basic method is widely used in distribution networks and as a back-up in transmission systems. It is applied to generators, transformers and feeders. The arrangement of the components is shown in Figure 11.20. In the past the relay normally

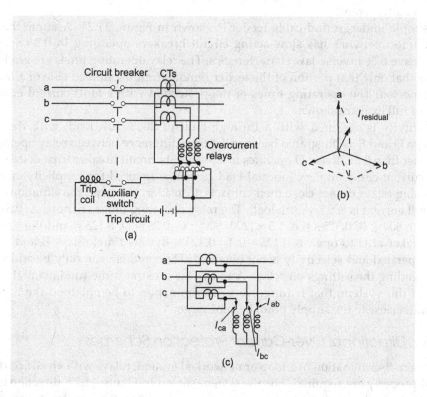

Figure 11.20 Circuit diagram of simple overcurrent protection scheme, (a) CTs in star connection, (b) Phasor diagram of relay currents, star connection, (c) CTs in delta connection

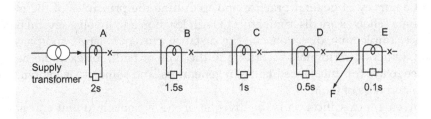

Figure 11.21 Application of overcurrent relays to feeder protection

employed was the induction-disc type (Figure 11.14), but numerical over-current relays are now used.

On an underground cable feeder, fault currents only reduce slightly as the position of the fault moves further from the in-feed point of supply. The impedance of the feeder is small compared with that of the source. Therefore it is necessary to use different operating times to provide discrimination. Grading of relays across transformers, which introduce a large impedance in the circuit relies more on the variations in fault current.

A simple underground cable feeder is shown in Figure 11.21. Assume that the distribution network has slow-acting circuit breakers operating in 0.3 s and the relays have true inverse-law characteristics. The relay operating times are graded to ensure that only that portion of the feeder remote from the in-feed side of a fault is disconnected. The operating times of the protection with a fault current equal to 200% of full load are shown.

Selectivity is obtained with a through-fault of 200% full load, with the fault between D and E as illustrated because the time difference between relay operations is greater than 0.3 s. Relay D operates in 0.5 s and its circuit breaker trips in 0.8 s. The fault current ceases to flow (normal-load current is ignored for simplicity) and the remaining relays do not close their contacts. Consider, however, the situation when the fault current is 800% of full load. The relay operating times are now: A, 0.5 s [i.e. $2 \times (200/800)$]; B, 0.375 s [i.e. $1.5 \times (200/800)$]; C, 0.25 s; D, 0.125 s; and the time for the breaker at D to open is $0.125 + 0.3 = 0.425$ s. By this time, relays B and C will have operated and selectivity is not obtained. This problem can only be addressed by extending the settings on relays A to C. This illustrates the fundamental drawback of this system, that is for correct discrimination to be obtained the times of operation close to the supply point become large.

11.5.2 Directional Over-Current Protection Schemes

To obtain discrimination in a loop or networked system, relays with an added directional property are required. For the system shown in Figure 11.22, directional and non-directional over-current relays have time lags for a given fault current as shown. Current feeds into fault at the location indicated from both directions, and the first relay to operate is at B (0.6 s). The fault is now fed along route ACB only,

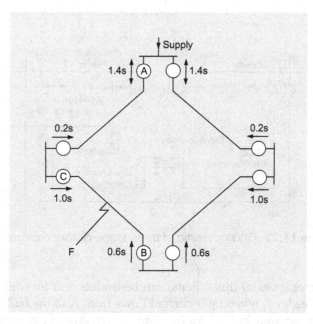

Figure 11.22 Application of directional overcurrent relays to a loop network. ↔ Relay responsive to current flow in both directions; → relay responsive to current flow in direction of arrow

and next the relay at C (1 s) operates and completely isolates the fault from the system. Assuming a circuit-breaker clearance time of 0.3 s, complete selectivity is obtained at any fault position. Note, however, that directional relays require a voltage input, non-directional over-current relays do not.

11.6 Distance Protection

The shortcomings of over-current relays graded by current and time led to the widespread use of distance protection. The distance between a relay and the fault is proportional to the ratio (voltage/current) measured by the distance relay. Relays responsive to impedance, admittance (mho), or reactance may be used.

Although a variety of time-distance characteristics are available for providing correct selectivity, the most popular one is the stepped characteristic shown in Figure 11.23. Here, A, B, C, and D are distance relays with directional properties and A and C only measure distance when the fault current flows in the indicated direction. Relay A trips its associated breaker instantaneously if a fault occurs within the first 80% of the length of feeder 1. For faults in the remaining 20% of feeder 1 and the initial 30% of feeder 2 (called the stage 2 zone), relay A initiates tripping after a short time delay. A further delay in relay A is introduced for faults further along feeder 2 (stage 3 zone). Relays B and D have similar characteristics when the fault current flows in the opposite direction.

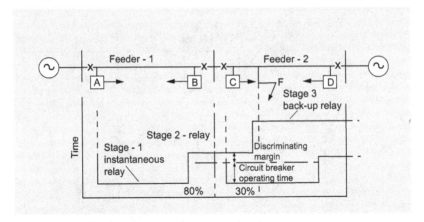

Figure 11.23 Characteristic of three-stage distance protection

The selective properties of this scheme can be understood by considering a fault such as at F in feeder 2, when fault current flows from A to the fault. For this fault, relay A starts to operate, but before the tripping circuit can be completed, relay C trips its circuit breaker and the fault is cleared. Relay A then resets and feeder 1 remains in service. The time margin of selectivity provided is indicated by the vertical intercept between the two characteristics for relays A and C at the position F, less the circuit-breaker operating time.

It will be noted that the stage 1 zones are arranged to extend over only 80% of a feeder from each end. The main reason for this is because practical distance relays and their associated equipment have errors, and a margin of safety has to be allowed if incorrect tripping for faults which occur just inside the next feeder is to be avoided. Similarly, the stage 2 zone is extended well into the next feeder to ensure definite protection for that part of the feeder not covered by stage 1. The object of the stage 3 zone is to provide general back-up protection for the rest of the adjacent feeders.

The characteristics shown in Figure 11.23 require three basic features: namely, response to direction, response to impedance and timing. These features need not necessarily be provided by three separate relay elements, but they are fundamental to all distance protective systems. As far as the directional and measuring relays are concerned, the number required in any scheme is governed by the consideration that three-phase, phase-to-phase, phase-to-earth, and two-phase-to-earth faults must be catered for. For the relays to measure the same distance for all types of faults, the applied voltages and currents must be different. With electro-mechanical relays it was common practice, therefore, to provide two separate sets of relays, one set for phase faults and the other for earth faults, and either of these caters for three-phase faults and double-earth faults. Each set of relays was, in practice, usually further divided into three, since phase faults may concern any pair of phases, and, similarly, any phase can be faulted to earth. With digital relays, a pre-selection of relaying quantities allows just one processor to deal with all types of fault.

11.7 Unit Protection Schemes

With the ever-increasing complexity of modern power systems the methods of protection so far described may not be adequate to afford proper discrimination, especially when the fault current flows in parallel paths. In unit schemes, protection is limited to one distinct part or element of the system that is disconnected if any internal fault occurs. The protected part should be tripped for an internal fault, but remain connected with the passage of current flowing into an external fault.

11.7.1 Differential Relaying

At the extremities of the zone to be protected, the currents are continuously compared and balanced by suitable relays. Provided that the currents flowing into and out of the zone are equal in magnitude and phase, no relay operation will occur. If, however, an internal fault (inside the protected zone) occurs, this balance will be disturbed (see Figure 11.24) and the relay will operate. The current transformers at

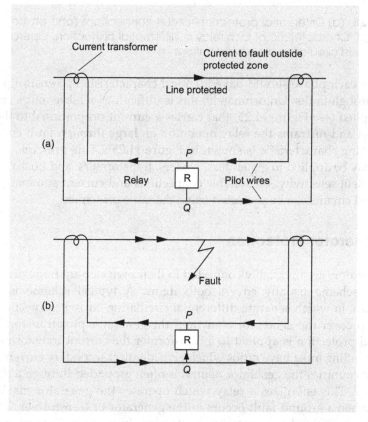

Figure 11.24 Circulating current, differential protection (one phase only shown), (a) Current distribution with through-fault–no current in relay, (b) Fault on line, unequal currents from current transformers and current flows in relay coil. Relay contacts close and trip circuit breakers at each end of the line

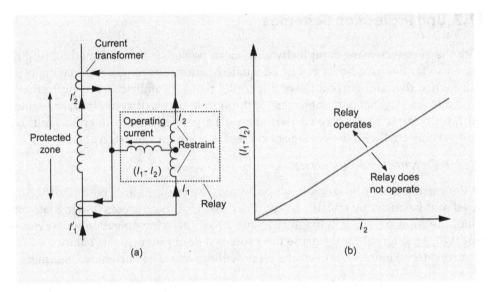

Figure 11.25 (a) Differential protection–circuit connections (one phase only)–relay with bias, (b) Characteristic of bias relay in differential protection. Operating current plotted against circulating or restraint current

the ends of each phase should have identical characteristics to ensure perfect balance on through-faults. Unfortunately, this is difficult to achieve and a restraint, or bias, is applied (see Figure 11.25) that carries a current proportional to the full system current and restrains the relay operation on large through-fault currents. The corresponding characteristic is shown in Figure 11.25b. This principle (circulating current) may be applied to generators, feeders, transformers, and busbars, and provides excellent selectivity. By suitable connections and current summation, teed or multi-ended circuits can be protected using the same principles.

11.8 Generator Protection

Large generators are invariably connected to their own step-up transformer and the protective scheme usually covers both items. A typical scheme is shown in Figure 11.26, in which separate differential circulating-current protection schemes are used to cover the generator alone and the generator plus transformer. When differential protection is applied to a transformer the current transformer on each side of a winding must have ratios which give identical secondary currents.

In many countries the generator neutral is often grounded through a distribution transformer. This energizes a relay which operates the generator main and field breakers when a ground fault occurs in the generator or transformer. The ground fault is usually limited to about 10 A by the distribution transformer or a resistor, although an inductor has some advantages. The field circuit of the generator must be opened when the differential protection operates in order to avoid the machine feeding the fault.

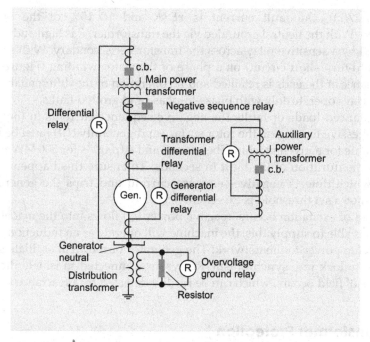

Figure 11.26 Protection scheme for a generator and unit transformer

The relays of the differential protection on the stator windings (see Figure 11.27) are set to operate at about 10–15% of the circulating current produced by full-load current in order to avoid current-transformer errors. If the phase e.m.f. generated by the winding is E, the minimum current for a ground fault at the star-point end, and hence with the whole winding in circuit, is E/R, where R is the neutral effective resistance. For a fault at a fraction x along the winding from the neutral

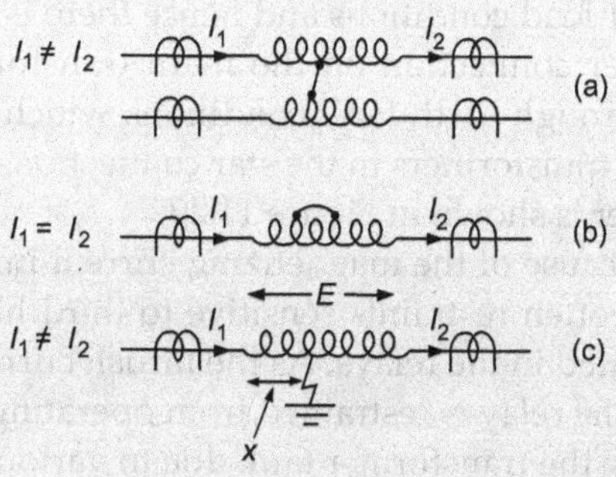

Figure 11.27 Generator winding faults and differential protection, (a) Phase-to phase fault, (b) Interturn fault, (c) Phase-to-earth fault

(Figure 11.27(c)), the fault current is xE/R and 10–15% of the winding is unprotected. With the neutral grounded via the transformer, R is high and earth faults are detected by a sensitive relay across the transformer secondary. With an interturn fault (turn-to-turn short circuit) on a phase of the stator winding (Figure 11.27(b)), current balance at the ends is retained and no operation of the differential relay takes place, the relays operate only with phase-to-phase and ground faults.

On unbalanced loads or faults the negative-sequence currents in the generator produce excessive heating on the rotor surface and generally ($I_2^2 t$) must be limited to a certain value for a given machine (between 3 and 4 (p.u.)^2 s for 500 MW machines), where t is the duration of the fault in seconds. To ensure this happens, a relay is installed which detects negative-sequence current and trips the generator main breakers when a set threshold is exceeded.

When loss of excitation occurs, reactive power (Q) flows into the machine, and if the system is able to supply this, the machine will operate as an induction generator, still supplying power to the network. The generator output will oscillate slightly as it attempts to lock into synchronism. Relays are connected to isolate the machine when a loss of field occurs, which can be readily detected by a reactance relay.

11.9 Transformer Protection

A typical protection scheme is shown in Figure 11.28(a), in which the differential circulating-current arrangement is used. The specification and arrangement of the current transformers is complicated by the main transformer connections and ratio. Current-magnitude differences are corrected by adjusting the turns ratio of the current transformers to account for the voltage ratio at the transformer terminals. In a differential scheme the phase of the secondary currents in the pilot wires must also be accounted for with star-delta transformers. In Figure 11.28(a) the primary-side current transformers are connected in delta and the secondary ones in star. The corresponding currents are shown in Figure 11.28(b) and it is seen that the final (pilot) currents entering the connections between the current transformers are in phase for balanced-load conditions and hence there is no relay operation. The delta current-transformer connection on the main transformer star-winding also ensures stability with through earth-fault conditions which would not be obtained with both sets of current transformers in the star connection. The distribution of currents in a Y-Δ transformer is shown in Figure 11.29.

Troubles may arise because of the magnetizing current inrush when energization operates the relays, and often restraints sensitive to third-harmonic components of the current are incorporated in the relays. As the inrush current has a relatively high third-harmonic content the relay is restrained from operating.

Faults occurring inside the transformer tank due to various causes give rise to the generation of gas from the insulating oil or liquid. This may be used as a means of fault detection by the installation of a gas/oil-operated relay in the pipe between the tank and conservator. The relay normally comprises hinged floats and is known as the Buchholz relay (see Figure 11.30). With a small fault, bubbles rising to flow into

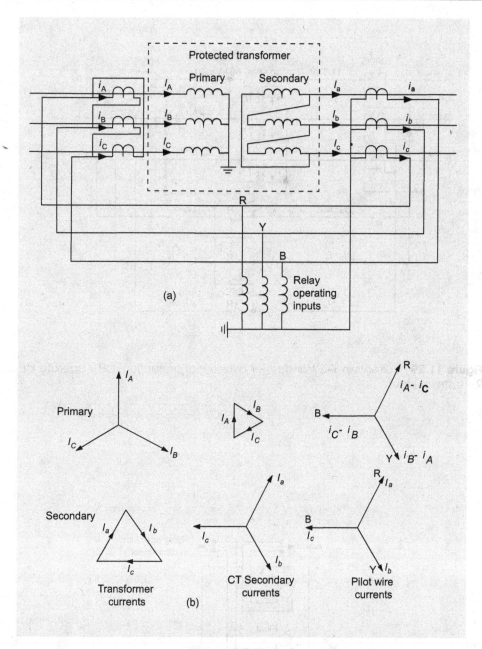

Figure 11.28 Differential protection applied to Y-Δ transformers, (a) Current transformer connections, (b) Phasor diagrams of currents in current transformers

the conservator are trapped in the relay chamber, disturbing the float which closes contacts and operates an alarm. On the other hand, a serious fault causes a violent movement of oil which moves the floats, making other contacts which trip the main circuit breakers.

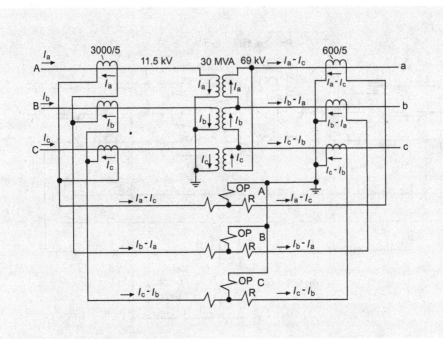

Figure 11.29 Currents in Y-Δ transformer differential protection. OP = operate input; R = restraint input

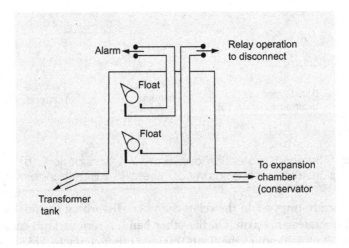

Figure 11.30 Schematic diagram of Buchholz relay arrangement

11.10 Feeder Protection

11.10.1 Differential Pilot Wire

The differential system already described can be applied to feeder protection. The current transformers situated at the ends of the feeder are connected by insulated wires known as pilot wires. In Figure 11.24, P and Q must be at the electrical midpoints of the pilot wires, and often resistors are added to obtain a geographically convenient midpoint. By reversing the current-transformer connections (Figure 11.31) the current-transformer e.m.f.s oppose and no current flows in the wires on normal or through-fault conditions. This is known as the opposed voltage method. As, under these conditions, there are no back ampere-turns in the current-transformer secondary windings, on heavy through-faults the flux is high and saturation occurs. Also, the voltages across the pilot wires may be high under this condition and unbalance may occur due to capacitance currents between the pilot wires. To avoid this, sheathed pilots are used.

In some differential protection schemes it is necessary to transmit the secondary currents of the current transformers considerable distances in order to compare them with currents elsewhere. To avoid the use of wires from each of the three CTs in a three-phase system, a summation transformer is used which gives a single phase output, the magnitude of which depends on the nature of the fault. The arrangement is shown in Figure 11.32, in which the ratios of the turns are indicated. On balanced through-faults there is no current in the winding between c and n. The phase (a) current energizes the 1 p.u. turns between a and b and the phasor sum of Ia and Ib flows in the 1 p.u. turns between b and c. The arrangement gives a much greater sensitivity to earth faults than to phase faults. When used in phase-comparison systems, however, the actual value of output current is not important and the transformer usually saturates on high-fault currents, so protecting the secondary circuits against high voltages.

Pilot wires may be installed underground or strung on towers. In the latter method, care must be taken to cater for the induced voltages from the power-line conductors. Sometimes it is more economical to rent wires from the telephone companies, although special precautions to limit pilot voltages are then required.

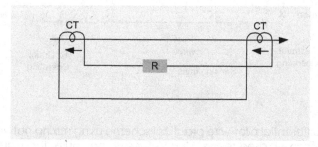

Figure 11.31 Pilot-wire differential feeder protection – opposed voltage connections

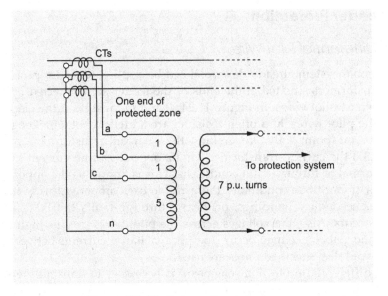

Figure 11.32 Summation transformer

A typical scheme using circulating current is shown in Figure 11.33, in which a mixing or summation device is used. With an internal fault at F2 the current entering end A will be in phase with the current entering end B, as in H.V. networks the feeder will inevitably be part of a loop network and an internal fault will be fed from both ends. V_A and V_B become additive, causing a circulating current to flow, which causes relay operation. Thus, this scheme could be looked on as a phase-comparison method. If the pilot wires become short-circuited, current will flow and the relays can give a false trip. In view of this, the state of the wires is constantly monitored by the passage of a small d.c. current.

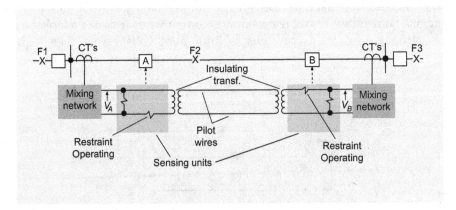

Figure 11.33 Differential pilot-wire practical scheme using mixing network (or summation transformer) and biased relays. $V_A = V_B$ for external faults, for example at F1 and F3; $V_A \neq V_B$ for internal faults, for example at F2

11.10.2 Carrier-Current Protection

Because of pilot capacitance, pilot wire relaying is limited to line lengths below 40 km (30 miles). Above this, distance protection may be used, although for discrimination of the same order as that obtained with pilot wires, carrier current equipment may be used. In carrier-current schemes a high-frequency signal in the band 80–500 kHz and of low power level (1 or 2 W) is transmitted via the power-line conductors from each end of the line to the other. It is not convenient to superimpose signals proportioned to the magnitude of the line primary current, and usually the phases of the currents entering and leaving the protected zone are compared. Alternatively, directional and distance relays are used to start the transmission of a carrier signal to prevent the tripping of circuit breakers at the line ends on through faults or external faults. On internal faults other directional and distance relays stop the transmission of the carrier signal, the protection operates, and the breakers trip. A further application, known as transferred tripping, uses the carrier signal to transmit tripping commands from one end of the line to the other. The tripping command signal may take account of, say, the operation or nonoperation of a relay at the other end (permissive intertripping) or the signal may give a direct positive instruction to trip alone (intertripping).

Carrier-current equipment is complex and expensive. The high-frequency signal is injected on to the power line by coupling capacitors and may be coupled either to one phase conductor (phase-to-earth) or between two conductors (phase-to-phase), the latter being technically better but more expensive. A schematic diagram of a phase-comparison carrier-current system is shown in Figure 11.34. The wave or line trap is tuned to the carrier frequency and presents a high impedance to it, but a low

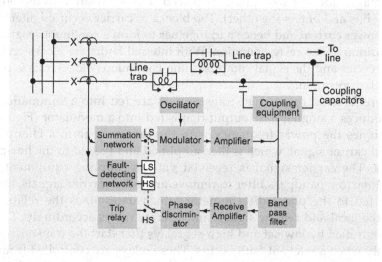

Figure 11.34 Block diagram measuring and control equipment for carrier-current phase-comparison scheme. LS = low-set relay; HS = high-set relay

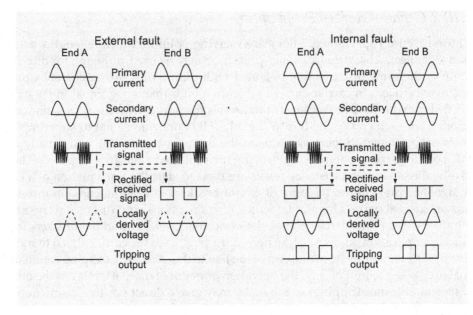

Figure 11.35 Waveforms of transmitted signals in carrier-type line protection

impedance to power-frequency currents; it thus confines the carrier to the protected line. Information regarding the phase angles of the currents entering and leaving the line is transmitted from the ends by modulation of the carrier by the power current, that is by blocks of carrier signal corresponding to half-cycles of power current (Figure 11.35). With through faults or external faults the currents at the line ends are equal in magnitude but 180° phase-displaced (that is, relative to the busbars, it leaves one bus and enters the other). The blocks of carrier occur on alternate half-cycles of power current and hence add together to form a continuous signal, which is the condition for no relay operation. With internal faults the blocks occur in the same half-cycles and the signal comprises non-continuous blocks; this is processed to cause relay operation.

The currents from the current transformers are fed into a summation device which produces a single-phase output that is fed into a modulator (Figure 11.34). This combines the power frequency with the carrier to form a chopped 100% modulated carrier signal, which is then amplified and passed to the line-coupling capacitors. The carrier signal is received via the coupling equipment, passed through a narrow-bandpass filter to remove any other carrier signals, amplified, and then fed to the phase discriminator which determines the relative phase between the local and remote signals and operate relays accordingly. The equipment is controlled by low-set and high-set relays that start the transmission of the carrier only when a relevant fault occurs. These relays are controlled from a starting network. Although expensive, this form of protection is very popular on overhead transmission lines.

11.10.3 Voice-Frequency Signalling

Increasingly, as the telephone network expands with the use of fibre-optic cabling and high-frequency multiplexing of communication channels, it is no longer possible to obtain a continuous metallic connection between the ends of unit-protection schemes. Consequently, differential protection, utilizing voice frequencies (600–4000 Hz), has been developed. The pilot wires (after the insulating transformers) feed into a voltage-to-frequency (v.f.) converter with send and receive channels. The summated or derived 50 (or 60) Hz relaying signal is frequency-modulated onto the channel carrier and demodulated at the far end. The signal is then compared with the local signal and if a discrepancy is detected then relay operation occurs. It is important to ensure that both signal magnitude and phase are faithfully reproduced at each end after demodulation, independently of the channel characteristics which can change or distort during adverse transmission conditions. The usual methods built into the v.f. channel relays to ensure reliability utilize an automatic control of signal level and a regular measurement (say, every 100 ms) of channel delay so that phase correction can be applied to the demodulated signal.

Problems

11.1   A 132 kV supply feeds a line of reactance 13 Ω which is connected to a 100 MVA 132/33 kV transformer of 0.1 p.u. reactance as shown in Figure 11.36. The transformer feeds a 33 kV line of reactance 6 Ω which, in turn, is connected to an 80 MVA, 33/11 kV transformer of 0.1 p.u. reactance. This transformer supplies an 11 kV substation from which a local 11 kV feeder of 3 Ω reactance is supplied. This feeder energizes a protective overcurrent relay through 100/1 A current transformers. The relay has a true inverse-time characteristic and operates in 10 s with a coil current of 10 A.

If a three-phase fault occurs at the load end of the 11 kV feeder, calculate the fault current and time of operation of the relay.

(Answer: 1575 A; 6.35 s)

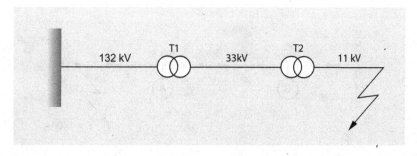

Figure 11.36 Circuit for Problem 11.1

11.2 A ring-main system consists of a number of substations designated A, B, C, D, and E, connected by transmission lines having the following impedances per phase (Ω): AB $(1.5+j2)$; BC $(1.5+j2)$; CD $(1+j1.5)$; DE $(3+j4)$; EA $(1+j1)$.

The system is fed at A at 33 kV from a source of negligible impedance. At each substation, except A, the circuit breakers are controlled by relays fed from 1500/5 A current transformers. At A, the current transformer ratio is 4000/5. The characteristics of the relays are as follows:

Current (A)	7	9	11	15	20
Operating times of relays at A, D' and C	3.1	1.95	1.37	0.97	0.78
Operating times at relays at B and E	4	2.55	1.8	1.27	1.02

Examine the sequence of operation of the protective gear for a three-phase symmetrical fault at the midpoint of line CD.

Assume that the primary current of the current transformer at A is the total fault current to the ring and that each circuit breaker opens 0.3 s after the closing of the trip-coil circuit. Comment on the disadvantages of this system.

11.3 The following currents were recorded under fault conditions in a three-phase system:

$$I_A = 1500 \angle 45° \text{ A},$$
$$I_B = 2500 \angle 150° \text{ A},$$
$$I_C = 1000 \angle 300° \text{ A}.$$

If the phase sequence is A-B-C, calculate the values of the positive, negative and zero phase-sequence components for each line.

(Answer: $I_0 = -200 + j480$ A, $I_{A1} = 20 - j480$, $I_{A2} = 1240 + j1060$)

11.4 Determine the time of operation of 1 A standard IDMT over-current relay having a Plug Setting (PS) of 125% and a Time Multiplier setting (TMS) of 0.6. The CT ratio is 400 : 1 and the fault current 4000 A.

(Answer: 2 sec)

11.5 The radial circuit shown in Figure 11.37 employs two IDMT relays of 5 A. The plug settings of the relays are 125% and time multiplier of relay A is 0.05 sec.

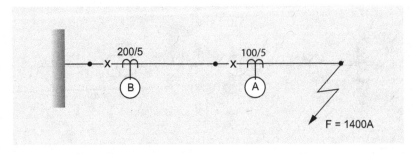

Figure 11.37 Circuit for Problem 11.5

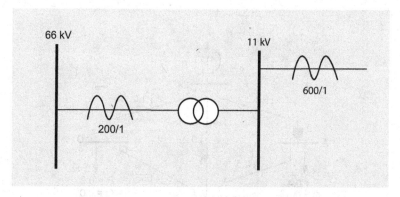

Figure 11.38 Circuit for Problem 11.6

Find the time multiplier setting of relay B to coordinate two relays for a fault of 1400 A. Assume a grading margin of 0.4 sec.

(Answer: TMS 0.15)

11.6 A 66 kV busbar having a short circuit level of 800 MVA is connected to a 15 MVA 66/11 kV transformer having a leakage reactance of 10% on its rating as shown in Figure 11.38.

   i. Write down the three-phase short circuit current (in kA) for a fault on the 66 kV terminals of the transformer.

   ii. Calculate the three-phase short circuit current (in kA) for a fault on the 11 kV feeder.

   iii. The 11 kV relay has a Current Setting (Plug Setting) of 100% and Time Multiplier of 0.5. Use the IDMT characteristic to calculate the operating time for a three-phase fault on the 11 kV feeder.

   iv. The 66 kV relay has a Current Setting (Plug Setting) of 125%. Choose a Time Multiplier to give a grading margin of 0.4 seconds for a fault on the 11 kV feeder.

   v. What is the operating time for a three-phase fault on the 66 kV winding of the transformer?

   vi. In practice, what form of transformer protection would operate first for a fault on the 66 kV winding?

(Answers: 7 kA, 6.6 kA, 1.4 sec, 0.5, 1 sec)

11.7 A three-phase, 200 kVA, 33/11 kV transformer is connected as delta-star. The CTs on the 11 kV side have turns ratio of 800/5. What should be the CT ratio on the HV side?

(Answer: 150/5)

11.8 A three zone distance protection relay is located at busbar A as shown in Figure 11.39 (square). The VT ratio is 132 kV/220 V and CT ratio is 1000/1. Find the impedance setting for zone 1 and zone 2 protection of the relay

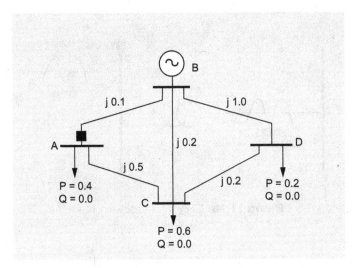

Figure 11.39 Circuit for Problem 11.8

assuming that zone 1 covers 80% of line section AB and zone 2 covers 100% of line section AB plus 30% of the shortest adjacent line.

(All quantities are in p.u. on 100 MVA, 132 kV base)

(Answers: $j0.135$, $j0.27$ p.u.)

12

Fundamentals of the Economics of Operation and Planning of Electricity Systems

Figure 12.1 shows the structure of the traditional electricity system with its four main sectors: generation, bulk transmission, distribution and demand.

In most of the industrialized countries electricity systems have been designed since the 1950s to support economic growth and benefit from developments in generation technology. The conventional power system is characterized by small numbers of very large generators, mainly coal, oil, hydro and nuclear, and more recently gas fuelled generation. Typical power station ratings would be from a few hundred MWs to several thousand MWs. These stations are connected to a very high voltage transmission network (operating typically at 275/220 kV and 400 kV). The role of the transmission system is to provide bulk transport of electricity from these large stations to demand centres, that is cities. The electricity is then taken by the distribution networks through a number of voltage transformations (typically Extra High Voltage (132/110 kV, 33/35 kV), High Voltage (20/11/10 kV)) and Low Voltage (400/230 V). These networks provide the final delivery of electricity to consumers. The flow is unidirectional from higher to lower voltage levels.

In a typical consumer bill in most European countries, generation costs (investment and operation) make up about 60% while transmission and distribution network costs are in the order of 10% and 30% respectively. Annual electricity losses in transmission and distribution networks in the majority of industrialized countries are around 1–3% and 4–7% respectively.

Electric Power Systems, Fifth Edition. B.M. Weedy, B.J. Cory, N. Jenkins, J.B. Ekanayake and G. Strbac.
© 2012 John Wiley & Sons, Ltd. Published 2012 by John Wiley & Sons, Ltd.

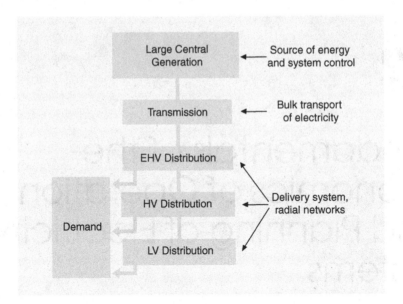

Figure 12.1 Schematic diagram of the conventional power system with four main sectors: generation, bulk transmission, distribution and demand

Key drivers of system operation and planning are economy and security. More recently these concerns have been supplemented by sustainability considerations aimed at limiting the CO_2 emissions that accompany the use of fossil fuel in power stations.

12.1 Economic Operation of Generation Systems

The input-output characteristic of a fossil fuel generating set is of great importance when economic operation is considered. Typical characteristics of a fossil fuelled generator are shown in Figure 12.2. The graph of the heat rate input (GJ/h) against output (MW) is known as the Willans line. For large turbines with a single valve, and for gas turbines, the heat rate is approximately a no-load offset with a reasonably straight line over the operating range, Figure 12.3(a). Most steam turbines in Britain are of this type. With multivalve turbines (as used in the USA) the Willans line curves upwards more markedly Figure 12.2(b).

The incremental heat rate is defined as the slope of the input output curve at any given output. The value taken for the incremental heat rate of a generating set is sometimes complicated because if only one or two shifts are being operated (there are normally three shifts per day) heat has to be expended in banking boilers when the generator is not required to produce output.

Instead of plotting heat rate or incremental heat rate against power output for the turbine-generator, the incremental fuel cost may be used. This is advantageous

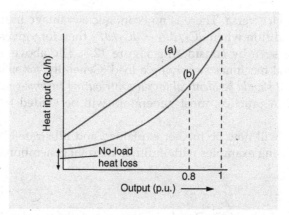

Figure 12.2 Input-output characteristic of a turbogenerator set. (a) single valve steam turbine or gas turbine, (b) steam turbine with multiple valves

when allocating load to generators for optimum economy as it incorporates differences in the fuel costs of the various generating stations. Usually, the graph of incremental fuel cost against power output can be approximated by a straight line (Figure 12.3). Consider two turbine-generator sets having the following different incremental fuel costs, dC_1/dP_1 and dC_2/dP_2, where C_1 is the cost of the fuel input to unit number 1 for a power output of P_1 and, similarly, C_2 and P_2 relate to unit number 2. It is required to load the generators to meet a given requirement in the most economic manner. Obviously the load on the machine with the higher dC/dP will be reduced by increasing the load taken by the machine with the lower dC/dP. This transfer will be beneficial until the values of dC/dP for both sets are equal, after which the machine with the previously higher dC/dP now becomes the one with the

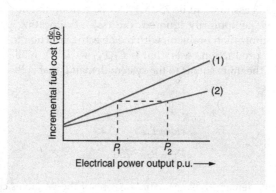

Figure 12.3 Idealized graphs of incremental fuel cost against output for two machines sharing a load equal to P_1 and P_2

lower value, and vice versa. There is no economic advantage in further transfer of load, and the condition when $dC_1/dP_1 = dC_2/dP_2$ therefore gives optimum economy; this can be seen by considering Figure 12.3. The above argument can be extended to several machines supplying a load. Generally, *for optimum economy the incremental fuel cost should be identical for all contributing turbine-generator sets* on free governor action. In practice, most generators will be loaded to their maximum output.

These concepts will now be further explained and illustrated on economic despatch problems, using examples with multiple thermal generation units.

Example 12.1

Generating units fired with fossil fuels are characterized by their input–output curves that define the amount of fuel required to produce a given and constant electrical power output for one hour. Consider two coal-fired steam units with minimum stable generation of 100 MW and 230 MW (that is, the minimum power that can be produced continuously) and maximum outputs of 480 MW and 660 MW. On the basis of measurements taken at the plants, the input–output curves of these units are estimated as

$$H_1(P_1) = 110 + 8.2P_1 + 0.02P_1^2 \text{ [MJ/h]} \tag{12.1}$$

$$H_2(P_2) = 170 + 9.6P_2 + 0.01P_2^2 \text{ [MJ/h]} \tag{12.2}$$

The hourly cost of operating these units is obtained by multiplying the input–output curve by the cost of fuel (expressed in £/MJ). Assuming that the cost of coal is 1.3 £/MJ, the cost curve of these generating units is given by the following expressions:

$$C_1(P_1) = 143 + 10.66P_1 + 0.026 + P_1^2 \text{ [£/h]} \tag{12.3}$$

$$C_2(P_2) = 221 + 12.48P_2 + 0.013 + P_2^2 \text{ [£/h]} \tag{12.4}$$

System demand is assumed to be 750 MW. If the minimum and maximum production constraints are temporarily ignored, the task of allocating production can be described as an optimization problem, with the objective function to be minimized representing the total production costs, that is $C_1(P_1) + C_2(P_2)$, while ensuring that the joint production of the units equals to the system demand, that is $P_1 + P_2 = 750$ MW:

$$\min_{P_1, P_2} C_1(P_1) + C_2(P_2)$$

subject to $\tag{12.5}$

$$P_1 + P_2 = 750$$

Using the supply-demand balance equation, one of the variables can be eliminated, so that the number of variables reduces to one. At the optimum, the derivative of the objective function should be equal to zero:

$$\min_{P_1} C_1(P_1) + C_2(750 - P_1) \Rightarrow \frac{d}{dP_1}[C_1(P_1) + C_2(750 - P_1)] = 0$$

$$\Rightarrow \frac{d}{dP_1}\left[143 + 10.66P_1 + 0.026P_1^2 + 221 + 12.48 \cdot (750 - P_1) + 0.013 \cdot (750 - P_1)^2\right] = 0$$

$$(12.6)$$

Finding the derivative of the cost function and then solving the resulting linear equation gives $P_1 = 273.33$ MW. Given that the joint production of the units must equal system demand, that is $P_1 + P_2 = 750$ MW, the production of the second unit is $P_2 = 476.67$ MW.

We now also observe that both units operate within the constraints on their outputs, that is within minimum and maximum generation, and hence the solution obtained represents the true optimum.

A key feature of this solution, is that the marginal cost of each of the units, at the optimal levels of production are equal:

$$\frac{dC_1(P_1)}{dP_1} = 10.66 + 0.052P_1 \bigg|_{p_1=273.33} = 24.87 \; [\text{£/MWh}]$$

$$(12.7)$$

$$\frac{dC_2(P_2)}{dP_2} = 12.48 + 0.026P_2 \bigg|_{p_2=476.67} = 24.87 \; [\text{£/MWh}]$$

The marginal cost of 24.87 £/MWh is the cost of increasing production from the present demand of 750 MW by one additional MW.

Alternatively, the original problem (12.5) can be expressed more formally using the corresponding Lagrange function:

$$\min_{P_1, P_2, \lambda} C_1(P_1) + C_2(P_2) + \lambda \cdot [750 - (P_1 + P_2)] \qquad (12.8)$$

Note that in this formulation the term $\lambda \cdot [750 - (P_1 + P_2)]$ is added to the objective function of Equation (12.5) to form Equation (12.8). However, the value of this term is zero and hence the value of the objective functions Equations (12.5) and (12.8) are the same, as the generation-load balance constraint will be satisfied (that is, $P_1 + P_2 = 750$ MW).

Optimality conditions require that the first partial derivatives of the objective function with respect to all three variables (P_1, P_2 and λ) are zero:

$$\frac{d}{dP_1}\{C_1(P_1) + C_2(P_2) + \lambda \cdot [750 - (P_1 + P_2)]\} = 0$$

$$\frac{d}{dP_2}\{C_1(P_1) + C_2(P_2) + \lambda \cdot [750 - (P_1 + P_2)]\} = 0 \qquad (12.9)$$

$$\frac{d}{d\lambda}\{C_1(P_1) + C_2(P_2) + \lambda \cdot [750 - (P_1 + P_2)]\} = 0$$

This leads to:

$$10.66 + 0.052P_1 = \lambda$$

$$12.48 + 0.026P_2 = \lambda \qquad (12.10)$$

$$P_1 + P_2 = 750$$

Equations (12.10) are three linear equations with three unknowns, that is P_1, P_2 and λ. Solving these gives:

$$P_1 = 273.33 \, \text{MW}$$

$$P_2 = 476.67 \, \text{MW} \qquad (12.11)$$

$$\lambda = 24.87 [\text{£/MWh}]$$

Note that the first two expressions in (12.10) represent the *marginal cost* of the two units respectively and that the optimality condition requires that these are *equal*.[1]

Parameter λ is a Lagrange multiplier associated with the production-demand balance equation and represents numerically the *change* in the total production costs that would correspond to an *incremental* change in demand around the operating point of 750 MW. This parameter hence represents *system marginal cost* at a demand of 750 MW.

This solution is also presented in Figure 12.4.

Finally, the total cost of supplying this demand over an hour is

$$C_T = C_1(P_1) + C_2(P_2) = 14\,122.77 \, [\text{£}] \qquad (12.12)$$

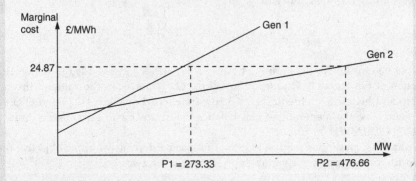

Figure 12.4 Marginal cost curves of two generators with output and the system marginal cost indicated for demand of 750 MW

---

[1] This will be the case when the units operate within their operating constraints but not at the limits of these constraints.

Which gives the average production cost

$$a_C = \frac{C_T}{P_1 + P_2} = 18.83 \, [\pounds/MWh] \qquad (12.13)$$

In this particular case the marginal production cost is higher than the average production cost. If demand is to be charged at the marginal cost (which corresponds to the economic optimum), there will be some profit that the generators will make from supplying this load.

Example 12.2

Four generators are available to supply a power system peak load of 472.5 MW. The cost of power $C_i(P_i)$ from each generator, and maximum output:

$$\begin{aligned}
C_1(P_1) &= 200 + 15P_1 + 0.20P_1^2 \, [\pounds/h] \text{ Max. output 100 MW} \\
C_2(P_2) &= 300 + 17P_2 + 0.10P_2^2 \, [\pounds/h] \text{ Max. output 120 MW} \\
C_3(P_3) &= 150 + 12P_3 + 0.15P_3^2 \, [\pounds/h] \text{ Max. output 160 MW} \\
C_4(P_4) &= 500 + 2P_4 + 0.07P_4^2 \, [\pounds/h] \text{ Max. output 200 MW}
\end{aligned} \qquad (12.14)$$

The spinning reserve is to be 10% of peak load.

As in the previous example, the objective is to calculate the optimal loading of each generator, system marginal cost, and the cost of operating the system at peak.

We will start with deriving the marginal cost for each of the generators. These are given (in £/MWh) by:

$$\frac{dC_1}{dP_1} = 15 + 0.40P_1 \quad P_1 \leq 100 \, \text{MW}$$

$$\frac{dC_2}{dP_2} = 17 + 0.20P_2 \quad P_2 \leq 120 \, \text{MW}$$

$$\frac{dC_3}{dP_3} = 12 + 0.30P_3 \quad P_3 \leq 160 \, \text{MW} \qquad (12.15)$$

$$\frac{dC_4}{dP_4} = 2 + 0.14P_4 \quad P_4 \leq 200 \, \text{MW}$$

The marginal cost curves are plotted in Figure 12.5 for each of the generators up to their maximum output. From the curves, at 30 £/MWh the outputs are:

$$\begin{aligned}
P_1 &= 23 \, \text{MW} \\
P_2 &= 64 \, \text{MW} \\
P_3 &= 60 \, \text{MW} \\
P_4 &= 200 \, \text{MW} \\
\text{Total} &= 347 \, \text{MW}
\end{aligned} \qquad (12.16)$$

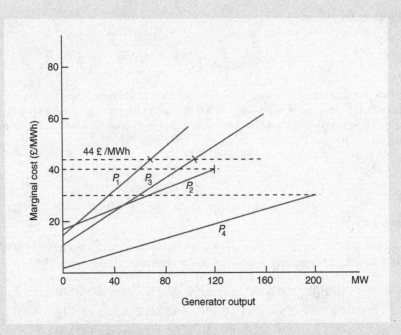

Figure 12.5 Marginal cost curves for four generators

Hence, P_4 runs at the full output and the remaining generators must sum to 272.5 MW.

At 40£/MWh the outputs sum to 446 MW. At 41£/MW P_2 reaches its maximum of 120 MW, and $P_1 = 40$ MW and $P_3 = 98$ MW (total 458 MW).

This leaves an additional 14.5 MW to be found from P1, and P3.

Adjusting the marginal cost to 44£/MWh provides 44 MW from P_1 and 108 MW from P_3, giving a total of 472 MW, which is considered to be near enough when using this graphical method. The spinning reserve on these generators is 108 MW, more than enough to cover 10% of 472.5 MW.

The cost of operating the system for 1 h at peak is

$$C_1(P_1) = 200 + 15.44 + 0.20 \cdot 44^2 = 1247 \, [£/h] \tag{12.17}$$

Similarly

$$C_2(P_2) = 3780 \, [£/h]$$
$$C_3(P_3) = 3195 \, [£/h] \tag{12.18}$$
$$C_4(P_4) = 3700 \, [£/h]$$

This gives a total of 11 922£/h. This gives an average cost of

$$a_C = \frac{11,922}{472.5} = 25.2 \, [£/MWh] \tag{12.19}$$

whereas the system marginal cost is 44£/MWh.

12.2 Fundamental Principles of Generation System Planning

When planning investment in generation systems, it is important to consider both *investment* and *operating costs*. Generally, generation planning time horizons may be decades to ensure that load growth is adequately met.

We will consider a static generation planning problem in order to highlight basic principles that are used to optimise generation technology mixes. In order to compare the cost associated with different generation technologies, operating costs are considered over a period of one year, while the investment costs are annuitized. The total annual operation and investment costs are then divided by the installed capacity expressed in £/kW.

Consider three generation technologies

i. *Base-load* generation, characterized by high capital (I_{BL}) and low per unit operating costs (O_{BL}), such as nuclear.
ii. *Mid-merit* generation, characterized by medium capital (I_{MM}) and per unit operating costs (O_{MM}), and
iii. *Peak-load* generation, characterized by low capital (I_{PL}) and high per unit operating costs (O_{PL}).

The objective of the generation planning exercise is to determine the capacities of individual generation technologies to be installed so that the total investment and operating costs are minimized. The choice of generation type depends on expected operating hours, which is illustrated in Figure 12.6. It should be

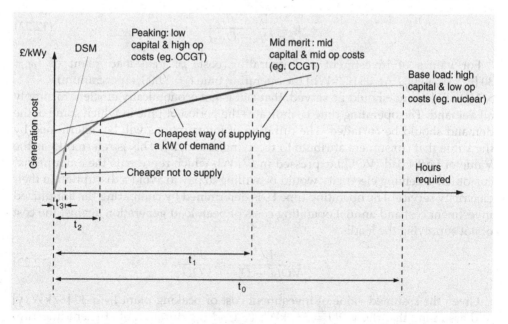

Figure 12.6 Determining operating hours of individual plant technologies

stressed that the expected operating times are a function of fixed and operating costs of generation only.

Base-load plant will be operating under all demand condition for the entire year ($t_0 = 8760$ h). The operating time for mid-merit plant is t_1 and for peaking-plant is t_2. For the duration of t_3, a portion of demand will be curtailed.

For operating time t_1 the per unit cost of investment and operation per year (£/kW/y) of mid-merit or base-load plant are equal:

$$I_{MM} + t_1 \cdot O_{MM} = I_{BL} + t_1 \cdot O_{BL} \tag{12.20}$$

In other words, we are indifferent if we invest in 1 MW of base load or mid-merit generation technology, if this unit will be operating for a period of t_1.

It is now straightforward to evaluate the operating time t_1 that determines the annual duration of operation of mid-merit plant:

$$t_1 = \frac{I_{BL} - I_{MM}}{O_{MM} - O_{BL}} \tag{12.21}$$

For values of annuitized investment costs of $I_{BL} = 170$ [£/kW/y] and $I_{MM} = 50$ [£/kW/y] and operating costs of $O_{BL} = 0.01$ [£/kWh] and $O_{MM} = 0.04$ [£/kWh] associated with base-load and mid-merit generation respectively, the operating time $t_1 = 4,000$ h (per annum).

Similarly, the operating time t_2, that determines the duration of the operation of peak-load generation, can be calculated by comparing annuitized investment cost and annual operating costs of mid-merit and peak-load generation:

$$t_2 = \frac{I_{MM} - I_{PL}}{O_{PL} - O_{MM}} \tag{12.22}$$

For values of investment and operating costs of peak-load plant of $I_{PL} = 30$ [£/kW/y] $O_{PL} = 0.06$ [£/kWh] the operating time $t_2 = 1,000$ h (per annum).

Not all demand should be served, that is it is not economically efficient to supply all demand. The operating time t_3, indicates the period of time in which some of the demand should be curtailed. The duration of interruptions will be determined by the value that consumers attribute to receiving electricity. This is referred to as the Value of Lost Load (*VOLL*) expressed in £/kWh which represents the amount that customers receiving electricity would be willing to pay to avoid a disruption in their electricity service. The operating time t_3, is determined by comparing the annuitized investment cost and annual operating costs of peak-load generation against the cost of not supplying the load:

$$t_3 = \frac{I_{PL}}{VOLL - O_{PL}} \approx \frac{I_{PL}}{VOLL} \tag{12.23}$$

Given the assumed value of investment cost of peaking plant $I_{PL} = 30$ [£/kW/y] and assuming that the $VOLL = 5$ £/kWh ($\gg O_{PL}$) the duration of efficient interruptions would be 6 hours per year.

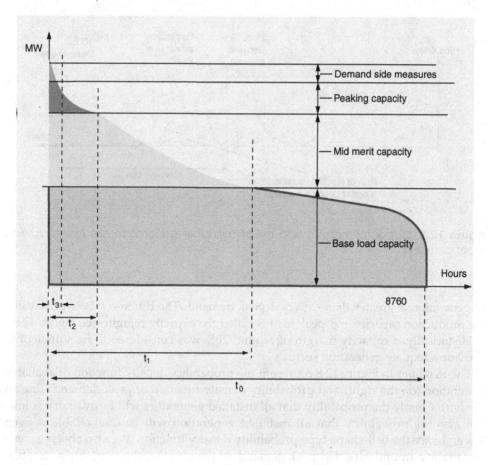

Figure 12.7 Determining the capacities of different generation plant technologies on the basis of operating hours calculated in Figure 12.6

Once the critical times are determined, we then use the annual load duration curve[2] (LDC) that links generating capacity requirements and capacity utilization, to determine the corresponding capacities of each generation type, as illustrated in Figure 12.7.

The plant mix determined by the capacities indicted in Figure 12.7, will minimize the overall generation investment and operating costs.

For reliable power supply sufficient production capacity has to be provided to cover the consumption and additional reserves required for reliable power system operation due to the unanticipated events, outages of generation or demand increases. In order to mitigate the effects of these uncertainties, the installed capacity

---

[2] The LDC is derived from a chronological load curve but the demand data is ordered in descending order of magnitude, rather than chronologically.

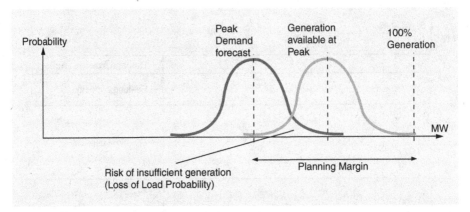

Figure 12.8 Capacity margins and risk of available generation not being able to meet demand

of generation is greater than expected peak demand. The difference between available production capacity and peak load is called the capacity margin (see Figure 12.8).

Historically, a capacity margin of around 20% was considered to be sufficient to provide adequate generation security.

The two lines in Figure 12.8 represent the probability density function of available generation (on the right) and probability density function of peak demand (line on the left). Clearly the probability that all installed generators will be available is low, but also the probability that all installed generation will be unavailable is even lower; hence the bell-shape type probability density function. We also observe some uncertainty around expected peak demand. In the case that the available generation is less than peak demand, then demand will need to be curtailed.

Generation adequacy is often assessed by determining the likelihood of there being insufficient generation to meet demand, or in other words, by calculating the risk that supply shortages will occur. Clearly, the larger the generation capacity margin the lower the risk of supply shortages. The risk of supply shortage can be quantified by the probability that peak demand will not be met by the available generation. This is called Loss of Load Probability (LOLP). The LOLP has been used as an index to measure generation capacity adequacy and risks of interruptions. The typical value of LOLP considered to be acceptable is around or below 10%. This is often interpreted to mean that demand curtailment, occurring once in 10 years on average, would be considered acceptable. Similarly, the index LOLE (Loss of Load Expectation) has also been used in many countries to assess generation capacity adequacy and risks of interruptions. The LOLE is the sum of LOLPs calculated for every hour of system operation during a year, multiplied by the corresponding number of hours in a year. Hence, LOLE represents the *expected* number of hours (in a year) in which demand may exceed available generation and hence some demand would be curtailed.

Example 12.3

Calculate the LOLP (at peak demand) and LOLE for the system in Figure 12.9.

In this example a simple three-generator system is considered. Their capacities (P_1, P_2 and P_3) and corresponding availabilities (A_1, A_2 and A_3) are indicated on Figure 12.9(a). The first step is to evaluate various State Probabilities and Cumulative Probabilities as indicated in Table 12.1.

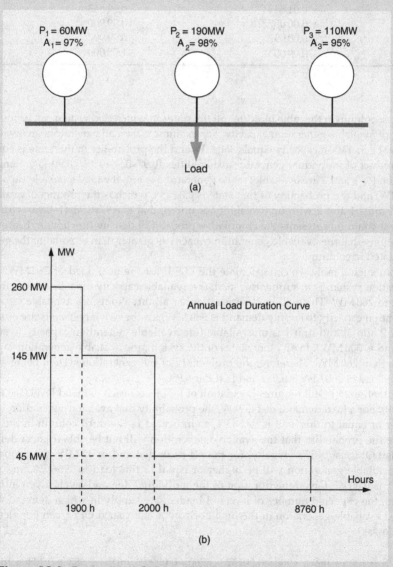

Figure 12.9 Evaluating LOLP and LOLE for a simple generation system

Table 12.1 State and cumulative probabilities for the system of Figure 12.9(a)

Available Generation (MW)	State Probability	Cumulative Probability Probability of available generation being greater than or equal to generation stated in column 1
360	0.90307	0.90307
300	0.02793	0.93100
250	0.04753	0.97853
190	0.00147	0.98000
170	0.01843	0.99843
110	0.00057	0.99900
60	0.00097	0.99997
0	0.00003	1.00000

The first column in the table indicates all the states in which this system can be found in terms of available generation capacity. For example, when all generators are available, the total generation capacity equals 360 MW and the probability of this state is equal to the product of individual generator availabilities $(0.97*0.98*0.95 = 0.90307)$. Similarly, if generator 1 and 2 are available, while generator 3 is not, the total available capacity is 250 MW, and the probability of that state is 0.04753, which is the product of availabilities of units 1 and 2 with unavailability of unit 3, that is $0.97*0.98*(1-0.95) = 0.04753$. The last column represents the cumulative probability density function, indicating the probability that the available generation capacity is greater than or equal to the generation stated in column 1.

From such a table we can calculate the LOLP for the peak load of 260 MW. If the generation system is to supply this load, the available capacity must be greater than or equal to 260 MW. This will be the case if either all three units are available (available generation capacity to supply demand is 360 MW), or the two larger units are available (units 2 and 3) and unit 1 is unavailable (the available generation capacity to supply demand is 300 MW). In all other states of the system, the available generation capacity is less than 260 MW. Therefore, the probability of the generation system being able to supply load of 260 MW is 0.931 and LOLP is 6.9%.

Evaluation of LOLE requires calculation of LOLPs for each demand level (see Figure 12.9(b)). For a load demand of 145 MW, the probability that available generation will be higher or equal to this load is 99.843%, as indicated in the third column in the table. Hence the probability that the available generation will not be able to meet demand, that is LOLP, is 0.157%. Finally, for the off-peak demand of 45 MW, the probability that available generation will be higher or equal to this load is 99.997%, and hence LOLP, is 0.003%. Given the duration of the individual demand levels given in Figure 12.9(b), the expected number of hours of inadequate supply in which demand would exceed available generation in the time horizon of one year (LOLE), can be calculated as follows:

$$LOLE = 100 * 0.069 + 1900 * 0.00157 + 6760 * 0.00003 = 6.9 + 2.983 + 0.203 = 10.09 \text{ h}$$

12.3 Economic Operation of Transmission Systems

Three characteristics of the transmission network determine its performance and associated operating costs: (i) level of congestion (ii) losses and (iii) interruptions caused by outages of transmission circuits. These factors may impact the optimal generation despatch discussed in Section 12.1. The discussion will start with the examination of the effect of the first characteristics.

The transmission network is generally characterized by significant power flows from exporting to importing areas. This is the case when generation cost in exporting area is lower than the generation cost in importing area. This can be interpreted as location arbitrage: the transport of power from low cost area to high cost area by transmission network facilities.

However, in order to keep the power flows over the transmission network within permissible limits it may be required that occasionally generators in exporting areas with lower marginal costs are constrained off, and generators in importing areas with higher marginal cost are constrained-on, which increases the cost of the operation of the generation system. In other words, the presence of transmission network constraints will alter the despatch discussed in Section 12.1. The process of optimising the operation of the generation system while maintaining the network flows within the prescribed limits is known as the Security Constrained Optimal Power Flow (SC-OPF) problem. The SC-OPF is distinguished from simple economic despatch as the solution obtained (generation outputs of individual generators) not only minimizes the total generation costs but also simultaneously respects the constraints of the transmission system including limits in both pre- and post-contingency states. This could also include optimisation of network control devices such as phase shifting transformers.

In the long term, the costs of network congestion are managed by appropriate investment in network capacity[3] as will be analyzed in the next section.

Example 12.4

To examine the impact of limited network capacity, Example 12.1 is considered again, with the two generators G1 and G2 connected to two different busbars (busbar 1 and busbar 2 respectively) that are connected by two identical transmission lines of 200 MW capacity each. Note that the total system load of 750 MW is split between the two busbars: 670 MW in busbar 1 and 80 MW in busbar 2, as indicated in Figure 12.10. Assuming that the economic despatch obtained in Example 12.1 (unconstrained economic despatch) can be implemented, the corresponding power flows are evaluated and presented in Figure 12.10 (losses are ignored in this exercise). Given that production in busbar of 2 of 476.7 MW and that local demand is 80 MW, power transfer from busbar 2 to busbar 1 is 396.7 MW. In busbar 1, local demand of 670 MW is supplied from local production of 273.3 MW and the contribution from busbar 2 of 396.7 MW.

---

[3] Through cost benefit based transmission network design, the costs of network congestion are to be balanced with the cost of network reinforcement.

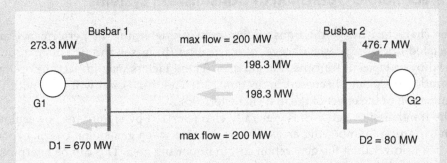

Figure 12.10 Implementing unconstrained economic despatch in Example 12.1

The resulting power flow from busbar 2 to busbar 1 of $F = 396.7\,\text{MW}$ appears to be acceptable given that the flows in individual circuits are within their capacities. However, modern transmission networks operate in accordance with N-1 security criteria, which requires that no transmission circuit should be overloaded (or cause any voltage and stability problems) in case of a single circuit outage. In this particular case, a fault on one of the lines would result in a significant overload on the remaining line, as the flow would increase from 198.3 MW to 396.7 MW, as illustrated in Figure 12.11.

In order to avoid this unacceptable operating state, generation despatch in the intact system (with both lines in service) will need to change.

This can be formally stated as the following optimisation problem:

$$\min_{P_1, P_2} C_1(P_1) + C_2(P_2)$$

$$st$$

$$P_1 + P_2 = 750 \qquad\qquad (12.24)$$

$$F \le 200$$

$$F = 670 - P_1 = P_2 - 80$$

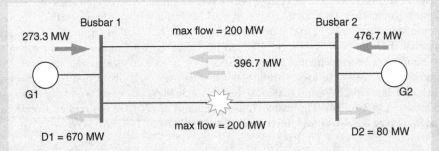

Figure 12.11 Post fault power flow leading to overload of the remaining circuit

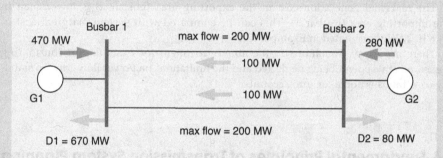

Figure 12.12 Security constrained economic despatch and secure transfer between busbars 1 and 2

The solution could however be found by inspection, as it is obvious that more efficient generator G2 in busbar 2 will be required to reduce its output. In order to re-establish power balance, the generator in busbar 1 will need to increase its output by exactly the same amount. The new, security constrained optimal despatch is shown in Figure 12.12.

Clearly, outage of one transmission circuit would not now overload the remaining line. The corresponding despatch is therefore called *security constrained economic despatch* and the resulting *secure power transfer* of $F = 200\,MW$ satisfies the N-1 operation rules. However, this new despatch is characterized by higher generation operating costs. Substituting the modified generation outputs (indicated in Figure 12.12) in expressions 12.3 and 12.4, the total generation costs of the security-constrained despatch can be calculated:

$$C_T^{SC} = C_1(470) + C_2(280) = 15\,631.2 \ [\text{£/h}] \tag{12.25}$$

The increase in costs above the cost of the unconstrained schedule given by Equation (12.12) is clearly driven by the limited capacity of the transmission network that connects export and import areas. The cost of constraints, also called cost of security or out-of-merit generation cost, can be calculated as the difference in costs between security-constrained (*SC*) and unconstrained (*U*) generation dispatches:

$$\Delta C_S = C_T^{SC} - C_T^{U} = 15\,631.2 - 14\,122.77 = 1\,508.43 \ [\text{£/h}] \tag{12.26}$$

Another important observation is that because of the network constraints, the two generators operate at different marginal costs. This is in contrast to Example 12.1 in which both generators operate at the same marginal cost. Hence in a constrained system, instead of a system marginal cost that would be applied to all network nodes, marginal costs are different in different locations:

$$\frac{dC_1(P_1)}{dP_1} = 10.66 + 0.052P_1 \bigg|_{P_1=470} = 35.1 \ [\text{£/MWh}] \tag{12.27}$$

$$\frac{dC_2(P_2)}{dP_2} = 12.48 + 0.026P_2 \bigg|_{P_2=280} = 19.76 \ [\text{£/MWh}]$$

The marginal generation cost in the exporting area (busbar 2) is lower than in the importing area (busbar 1). This can be compared with system marginal cost of 24.8 [£/MWh] calculated in Example 12.1.

There are many limitations in the above treatment for example not considering losses, reactive power being neglected and the limitations on power flows driven stability constraints is not taken into account.

12.4 Fundamental Principles of Transmission System Planning

Reinforcement of existing networks increases the amount of power that can be transported securely from exporting to importing areas. However, investments in new transmission equipment are costly and should therefore be undertaken only if they can be justified economically. In order to deliver maximum economic benefits, the electricity supply industry should follow the path of least-cost long-term development. This requires a coordinated approach to the optimization of the generation and transmission operation and development. Optimizing the transmission network in isolation from the generation resources would almost certainly not meet the above objective.

The cost-benefit approach to optimizing transmission investment is presented in Figure 12.13. Establishing the optimal level of network capacity that should be invested into should balance (i) the benefits associated with reduction of out-of-merit generation costs (cost of constraints) and costs of losses and (ii) cost of network investment that is dependent on the network capacity built. The optimal level of network capacity that should be built corresponds to the minimum of the total costs composed of cost of constraints and network investment.

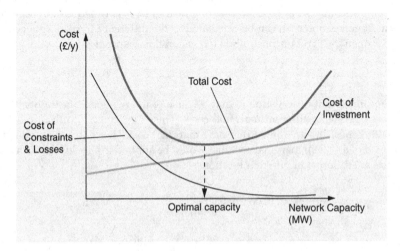

Figure 12.13 Cost-benefit approach to transmission planning

Example 12.5

To illustrate the basic principles for transmission expansion problem, we will consider the above example and discuss the economic impact of reinforcing the network between busbars 1 and 2, above the existing secure transfer capacity of 200 MW. This will include an analysis of the impact of network reinforcement that will provide additional secure transfer capability of ΔF on cost of security ΔC_S.

The out-of-merit generation cost for generation production of P_1 and P_2 in busbars 1 and 2 respectively, is given by:

$$\Delta C_S = C_T^{SC} - C_T^{U} = \left(143 + 10.66P_1 + 0.026P_1^2 + 221 + 12.48P_2 + 0.013P_2^2\right) - 14\,122.77 \tag{12.28}$$

By applying power balance equations to busbar 1 and busbar 2, generation productions can be expressed as a function of the additional power flow ΔF that can be facilitated through the network reinforcement:

$$P_1 + \Delta F + 200 = D_1 => P_1 = 470 - \Delta F$$
$$P_2 = 200 + \Delta F + D_2 = \Delta F + 280 \tag{12.29}$$

By substituting generation productions P_1 and P_2 given in (12.29) into equation (12.28), we finally obtain:

$$\Delta C_S = 0.039\Delta F^2 - 15.34\Delta F + 1508.4 \tag{12.30}$$

Adding secure transfer capacity of ΔF (on top of the existing 200 MW capacity), reduces network constraint (out-of-merit generation) costs, as shown in Figure 12.14.

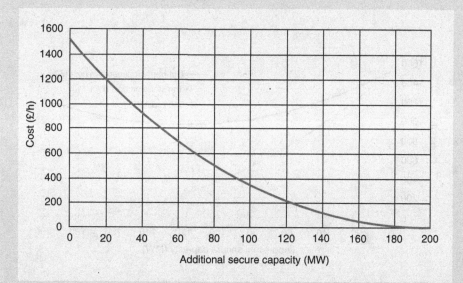

Figure 12.14 Cost of network constraints as a function of reinforced network capacity (ΔF)

Note that for the unconstrained economic despatch in Example 12.1, generation $P_1 = 273.33\,MW$ and $P_2 = 476.67\,MW$, constraint costs are zero ($\Delta C_S = 0$). As shown in Example 12.4, this despatch would result in the maximum power flow from busbar 2 to busbar 1 of 396.7 MW, and given that the existing network capacity is 200 MW, the network reinforcement should not exceed 196.7 MW ($\Delta F \leq 196.7\,MW$).

As indicated in Figure 12.13, the reduction in out-of-merit generation costs need to be considered together with the cost of network reinforcement that enables this increase in efficiency of system operation (reduction in out-of-merit generation costs).

For the sake of simplicity, it is assumed that the annuitized cost of reinforcing the transmission corridor is a linear function of additional secure capacity delivered ΔF:

$$C_{\Delta F} = 2 \cdot k \cdot L \cdot \Delta F \; [\pounds/y] \tag{12.31}$$

Where L is the length of the line in kilometers, k is the proportionality factor representing the annuitized marginal cost of reinforcing 1 km of the transmission line and its dimensions are $\pounds/(MW\cdot km\cdot year)$. The multiplication factor 2 signifies that two circuits would be reinforced. Dividing this annuitized cost by the number of hours in a year ($\tau_0 = 8760\,h$), the transmission investment costs are expressed per hour and can now be directly compared with hourly savings in generation costs. If we assume, for this example, that the line is 300 km long and that $k = 87.6$ $\pounds/(MW\cdot km\cdot year)$

$$C_{\Delta F} = C_{\Delta F}/8760 = 6 \cdot \Delta F \; [\pounds/h] \tag{12.32}$$

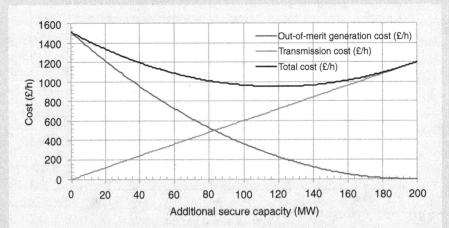

Figure 12.15 Finding optimal additional secure capacity by minimizing the sum of out-of-merit generation cost and transmission investment costs

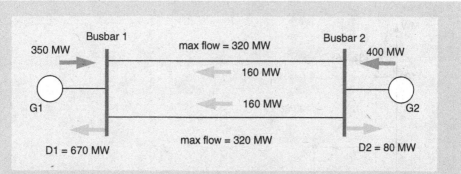

Figure 12.16 Generation despatch and flows after reinforcing the corridor and providing 120 MW of additional secure transfer

In order to determine the optimal capacity of network reinforcement, the sum of network costs and out-of-merit generation costs needs to be minimized:

$$\min\{(6\,\Delta F) + (0.039\Delta F^2 - 15.34\Delta F + 1508.4)\} \Rightarrow 0.078\Delta F - 9.34 = 0 \qquad (12.33)$$

Which gives optimal network reinforcement: $\Delta F = 120\,\text{MW}$

Changes in out-of-merit generation operating costs, transmission reinforcement cost and the total costs, as a function of additional capacity ΔF are presented in Figure 12.15.

Figure 12.16 shows the optimal despatch and secure power transfer over the transmission corridor after the circuit has been reinforced to provide an additional 120 MW of secure transfer capacity.

The transmission reinforcement, will impact the marginal costs are two locations:

$$\left.\frac{dC_1(P_1)}{dP_1}\right|_{P_1=360} = 28.9\,[\text{£/MWh}]$$

$$\qquad (12.34)$$

$$\left.\frac{dC_2(P_2)}{dP_2}\right|_{P_2=400} = 22.9\,[\text{£/MWh}]$$

We observe that the price difference between the two locations (28.9 [£/MWh] − 22.9 [£/MWh] is equal to the incremental investment costs of the line:

$$\frac{dC_{\Delta F}}{d\Delta F} = 6\,[\text{£/MWh}] \qquad (12.35)$$

12.5 Distribution and Transmission Network Security Considerations

Historically, the design and structure of electricity distribution and transmission networks was driven by an overall design philosophy developed to support large-scale generation technologies. Modern network design standards specify minimum requirements for transmission capacity, taking into account both network security and economics. Generally, the network must be able to continue to function after a

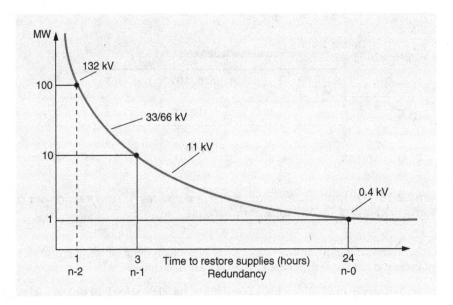

Figure 12.17 Philosophy of network design standards, level of redundancy increases with the amount of demand supplied

loss of a single circuit (or a double circuit on the same tower). After a loss of a circuit due to fault (e.g. lightning strike) the remaining circuits that take over the load of the faulty line, must not become overloaded.

In the majority of industrialized countries, the level of security in distribution and transmission networks is defined in terms of the time taken to restore power supplies following a predefined set of outages. Security levels in distribution systems are graded according to the total amount of peak power supplied and potentially being curtailed due to outages of network assets. In general, network redundancy has been specified according to a principle that the greater the amount of power which can be lost, the shorter the recommended restoration time. A simplified illustration of this network design philosophy, as implemented in the UK, is presented in Figure 12.17. For instance small demand groups, less than 1 MW peak, are provided with the lowest level of security, and have no redundancy (N-0 security). This means that any fault will cause an interruption and the supply will be restored only after the fault is repaired. It is expected that this could take up to about 24 hours.

- n-0 = no redundancy in security (must wait for repair of network)
- n-1 = one level of network redundancy; and
- n-2 = two levels of network redundancy

For demand groups with peak loads between 1 MW and 100 MW, although a single fault may lead to an interruption, the bulk of the lost load should be restored within 3 hours. This requires the presence of redundancy, as 3 hours is usually insufficient to implement repairs, but it does allow network reconfiguration activities.

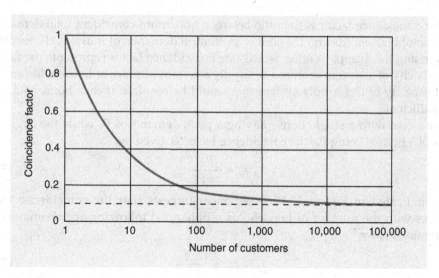

Figure 12.18 Load coincidence factor as a function of the number of typical households

Such network designs are often described as providing n-1 security. For demand groups with peak load larger than 100 MW, the networks should be able to provide supply continuity to customers following a single circuit outage (with no loss of supply) but also provide significant redundancy to enable supply restoration following a fault on another circuit superimposed on the existing outage, that is n-2 security.

This network design practice has effectively determined the characteristics regarding quality of service as experienced by end customers. The performance of 11 and 0.4 kV networks has a dominant effect on the overall quality of service seen by end customers. The vast majority of interruptions (about 90%) have their cause in these networks. This is primarily driven by the radial design of these networks, as any fault on a circuit leads to an interruption to some customers (unlike transmission and higher voltage level distribution networks that operate interconnected).[4]

When considering the design of network equipment to deal with peak demand conditions, it is important to consider diversity of demand associated with their use of different appliances. This is fully exploited both in system design and operation. The capacity of an electricity system supplying several thousand households would be only about 10% of the total capacity that would be required if each individual household were to be self sufficient (provide its own generation capacity). Distribution electricity networks are essential for achieving this significant benefit of load diversity. However, no material gains in the capacity of the electricity supply system would be made from increasing further the number of the households. This phenomenon is illustrated in Figure 12.18, which shows how the demand coincidence factor changes with the number of households.

---

[4] On average electricity consumers in Europe experience less than one interruption per year lasting about an hour. Consumers in urban areas have on the whole fewer interruptions than rural customers.

The coincidence factor is the ratio between maximum coincident total demand of a group of households and the sum of maximum demands of individual consumers comprising the group. In other words, the coincidence factor represents the ratio of the capacity of a system required to supply a certain number of households and the total capacity of the supply system that would be required if each household were self sufficient.

For a case with a single house having a peak demand of P_1 while the aggregate peak of n houses being P_n, the coincidence factor is given by:

$$j_n = \frac{P_n}{n \cdot P_1} \tag{12.36}$$

As indicated in Figure 12.18 experience suggests that the coincidence factor reduces with the number of households (appliances) following approximately the expression below:

$$j_n = j_\infty + \frac{1 - j_\infty}{\sqrt{n}} \tag{12.37}$$

where j_∞ represents the coincidence factor of infinite number of households (appliances).

Equation (12.37) shows that for $j_\infty = 0.1$, and for the number of households (appliances) being 100, the aggregate peak demand (diversified peak) will be only 20% of the sum of the individual peaks (non-diversified peak). Note that the diversity effect saturates for the number of households beyond 10 000.

12.6 Drivers for Change

As discussed, the location of generation relative to demand is the dominant factor driving the design and operation of electricity networks. Furthermore, the type of generation technology used, together with the pattern of usage, will make an impact on the actual network operation and development. Finally, advances in technology may open up new opportunities for achieving further improvement in efficiency of operation and investment in transmission and distribution networks.

With this in mind, there are four main drivers that may change the conventional philosophy system operation and development.

- First, generation, transmission and distribution systems in most developed countries were expanded considerably in the late 1950s and early 1960s. These assets are now approaching the end of their useful life. It is expected that a significant proportion of these assets will need to be replaced in the next two decades.
- Second, a number of countries are committed to respond to the climate change challenge and the energy sector, and in particular the electricity, will be required to deliver the changes necessary. In the last decade, many countries have supported deployment of distributed generation (DG) of various technologies (particularly renewables and CHP) to reduce carbon emissions and the need to improve system efficiency. These generation technologies range from kW size domestic PV

and micro CHP systems to several hundred MWs of wind generation connected to EHV (extra high voltage) distribution networks (132 kV). Very large windfarms are connected directly to transmission networks. This trend is likely to accelerate in the next decade and beyond as a key part of future energy policy.

- Third, there have recently been some major advances made in information and communication technologies (ICT), that in principle, could enable the development of significantly more sophisticated approaches to control and management of the system and hence increase the efficiency of operation and utilization of network investments.

- Finally, there has been significant research and development effort invested in the development of a number of control devices and concepts, such as FACTS (flexible AC transmission systems), storage and demand-side management which could be used to provide real time control of power flows in the network increasing utilization of transmission circuits. Similarly, greater automation of distribution network control could facilitate increase in utilization of network circuits.

In summary, the need to respond to climate change, improve efficiency of the system and increase fuel diversity and enhance security of supply, coupled with rapidly ageing assets and recent development in ICT, opens up the question of the strategy for infrastructure replacement in particular the design and investment in future electricity networks. This coincidence of factors presents an opportunity to re-examine the philosophy of the traditional approaches to system operation and design and develop a policy that will provide a secure, efficient and sustainable future energy supply. Although this requirement may not necessarily lead to an immediate radical change in the system, like-for-like replacement of assets is unlikely to be optimal.

Problems

12.1 The input-output curve of a coal-fired generating unit (with a maximum output of 550 MW) is given by the following expression:

$$H(P) = 126 + 8.9P + 0.0029P^2 [MJ/h]$$

If the cost of coal is $1.26\,£/MJ$, calculate the output of the unit when the system marginal cost is

a. 13 [£/MWh] and
b. 22 [£/MWh].

(Answer: (a) $P = 244.4\,MW$, (b) $P = 550\,MW$)

12.2 The incremental fuel costs of two units in a generating station are as follows:

$$\frac{dF_1}{dP_1} = 0.003P_1 + 0.7$$

$$\frac{dF_2}{dP_2} = 0.004P_2 + 0.5$$

marginal cost are in £/MWh and unit outputs are in MW.

Assuming continuous running with a total load of 150 MW calculate the saving per hour obtained by using the most economical division of load between the units as compared with loading each equally. The maximum and minimum operational loadings are the same for each unit and are 125 MW and 20 MW.

(Answer: $P_1 = 57$ MW, $P_2 = 93$ MW; saving £1.12 per hour)

12.3   What is the merit order used for when applied to generator scheduling?

A power system is supplied by three generators. The functions relating the cost (in £/h) to active power output (in MW) when operating each of these units are:

$$C_1(P_1) = 0.04P_1^2 + 2P_1 + 250$$

$$C_2(P_2) = 0.02P_2^2 + 3P_2 + 450$$

$$C_3(P_3) = 0.01P_3^2 + 5P_3 + 250$$

The system load is 525 MW. Assuming that all generators operate at the same marginal cost, calculate:

a. the marginal cost;
b. optimum output of each generator;
c. the total hourly cost of this despatch.

(Answer: (a) £10/MWh; (b) $P_1 = 100$ MW, $P_2 = 175$ MW, $P_3 = 250$ MW;
(c) £4562.5/h)

12.4   A power system is supplied from three generating units that have the following cost functions:

$$\text{Unit A} : 13 + 1.3\,P_A + 0.037\,P_A^2 \ [\$/h]$$

$$\text{Unit B} : 23 + 1.7\,P_B + 0.061\,P_B^2 \ [\$/h]$$

$$\text{Unit C} : 19 + 1.87\,P_C + 0.01\,P_C^2 \ [\$/h]$$

a. How should these units be dispatched if a load of 380 MW is to be supplied at minimum cost?
b. If, in addition to supplying a 380-MW load, the system can export (sell) energy to the neighbouring country in which the system marginal costs is 10.65 $/MWh. What is the optimal amount of power that should be exported?
c. Repeat problem (b) if the outputs of the generating units are limited as
   $$P_A^{MAX} = 110 \text{ MW}$$

follows: $P_B^{MAX} = 85$ MW

$$P_C^{MAX} = 266 \text{ MW}$$

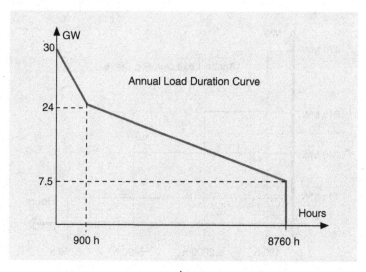

Figure 12.19 Load Duration Curve of Problem 12.5

(Answer:

a. $P_A = 77.60$ MW, $P_B = 43.79$ MW, $P_C = 258.61$ MW
b. $P_{ex} = 258.71$ MW
c. $P_{ex} = 69.36$ MW)

12.5 For the system with the load duration curve given in Figure 12.19 below, design a generation system to minimize the total investment and operating costs. The cost characteristics of different generation technologies are given below. Value of lost load is 8000 £/MWh.

Technology	Investment cost [£/kW/y]	Operating costs [£/MWh]
Base Load	250	5
Mid Merit	80	50
Peak Load	30	90

Calculate the amount of energy produced by each technology and the energy not served.

(Answer: base load 18.0 GW capacity and 131.3 TWh energy production, mid merit 5.3 GW and 13.3 TWh energy production, peak load generation 6.7 GW and 3.5 TWh energy production, expected energy unserved 47 MWh)

12.6 A power system contains five identical generators of capacity of 120 MW and availability of 96% that supply demand with the Load Duration Curve presented in Figure 12.20.
Calculate the LOLP at peak and LOLE.
(Answer: LOLP $= 1.48\%$, LOLE $= 4.1$ h)

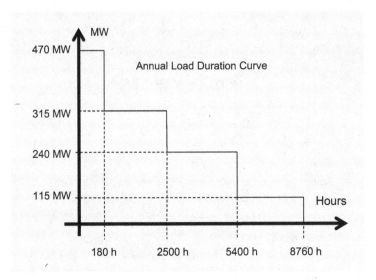

Figure 12.20 Load Duration Curve of Problem 12.6

12.7   A power system consists of two areas that are currently not connected and the national transmission operator is considering building an interconnector. Generation costs in the two areas are:

$$C_A(P_A) = 6P_A + 0.025P_A^2$$
$$C_B(P_B) = 10P_B + 0.11P_B^2$$

with the demand in the two areas being $DA = 600\,MW$ and $DB = 1000\,MW$.
   Calculate the benefits in enhancing the efficiency of the overall generation system operation that an interconnector of

a. 200 MW and
b. 400 MW

would create.

(Answer: (a) 33 400 £/h, (b) 56 000 £/h)

12.8   A power system of a small country is composed of two regions that are not connected. Generators 1 and 2 are located in the Northern Region while generators 3 and 4 are located in the Southern Region. The load in the Northern Region is 100 MW and the load in the Southern Region is 420 MW. Marginal costs of these generators are:

   Northern Region

$$MC_1 = 3 + 0.02P_1 \; [\text{£/MWh}]$$
$$MC_2 = 4 + 0.04P_2 \; [\text{£/MWh}]$$

Southern Region

$$MC_3 = 3.6 + 0.025P_3 [£/MWh]$$

$$MC_4 = 4.2 + 0.025P_4 [£/MWh]$$

a. Calculate the marginal costs in both regions and the corresponding generation dispatches.

b. The transmission company is considering building a 450km long transmission link between the two regions. Show that the optimal transmission capacity required to connect the two regions should be about 250 MW, when the annuitized investment cost of transmission is 37£/MW.km.year.

(Answer:

a. $MC_{north} = 4.67$ £/MWh; $P_1 = 83.33\,MW\,P_2 = 16.67\,MW$

   $MC_{south} = 9.15$ £/MWh, $P_3 = 222\,MW\,P_4 = 198\,MW$

12.9 Two areas, A and B, of a power system are linked by transmission link, with a secure capacity of 900 MW. System load is concentrated in area B. In winter, the load is 3,500 MW while summer load is 2000 MW. The cost of generation in the areas can be modelled by the following expressions, where P is in MW:

$$C(P_A) = 70 + P_A + 0.001\,P_A^2 \,[£/h] \text{ for area A, and}$$

$$C(P_B) = 50 + 2P_B + 0.002\,P_B^2 \,[£/h] \text{ for area B.}$$

a. Determine the optimal levels of generation in areas A and B for each of the winter and summer seasons neglecting the constraints of the transmission system. Calculate the marginal cost of production in each season.

b. If necessary, modify the levels of generation computed in (a) to take into consideration the capacity of the existing transmission link. What are the marginal cost prices in areas A and B in each of the seasons?

c. Assuming that the duration of the winter period is 2500 h, calculate the generation cost in each season. What are the total annual generation costs?

d. The transmission company is considering doubling the transmission capacity between the two areas in order to reduce generation cost. Assuming that the annuitized investment cost of the reinforcement is 1 000 000 £/year, determine if this proposed investment is justified.

(Answer:

a. Winter: $MC = 6$ £/MWh, $P_A = 2500\,MW$, $P_B = 1000\,MW$

   Summer: $MC = 4$ £/MWh, $P_A = 1500\,MW$, $P_B = 500\,MW$

b. Winter: $P_A = 900$ MW, $MC_A = 2.8$ £/MWh

     $P_B = 2600$ MW, $MC_B = 12.4$ £/MWh

     Summer: $P_A = 900\, MW$, $MC_A = 2.8$ £/MWh

     $P_B = 1100$ MW, $MC_B = 6.4$ £/MWh

c. Winter: $C_w = £35.85$ m, Summer: £33.61 m, Annual costs: £69.46 m

d. The case for proposed investment is justified.

12.10   A distribution substation will supply 160 households of two types. Large detached houses (45) with expected peak demand of 12 kW and smaller terraced houses (115) with peak demand of 8 kW. Coincidence factors for an infinite number of detached and terraced houses are 0.12 and 0.18 respectively. Assuming that peaks of both types of houses coincide, calculate the peak demand of the proposed 11 kV/0.4 kV substation.

(Answer: $P_{DS} = 230$ kW)

Appendix A

Synchronous Machine Reactances

Electric Power Systems, Fifth Edition. B.M. Weedy, B.J. Cory, N. Jenkins, J.B. Ekanayake and G. Strbac.
© 2012 John Wiley & Sons, Ltd. Published 2012 by John Wiley & Sons, Ltd.

Table A.1 Typical percentage reactances of synchronous machines at 50 Hz – British

Type and rating of machine	Positive sequence			Negative sequence X_2	Zero sequence X_0	Short-circuit ratio
	X''	X'	X_s			
11 kV Salient-pole alternator without dampers	22.0	33.0	110	2.0	6.0	–
11.8 kV, 60 MV, 75 MVA Turboalternator	12.5	17.5	201	13.5	6.7	0.55
11.8 kV, 56 MW, 70 MVA Gas-turbine turboalternator	10.0	14.0	175	13.0	5.0	0.68
11.8 kV, 70 MW, 87.5 MVA Gas-turbine turboalternator	14.0	19.0	195	16.0	7.5	0.55
13.8 kV, 100 MW, 125 MVA Turboalternator	20.0	28.0	206	22.4	9.4	0.58
16.0 kV, 275 MW, 324 MVA Turboalternator	16.0	21.5	260	18.0	6.0	0.40
18.5 kV, 300 MW, 353 MVa Turboalternator	19.0	25.5	265	19.0	11.0	0.40
22 kV, 500 MW, 588 MVA Turboalternator	20.5	28.0	255	20.0	6.0–12.0	0.40
23 kV, 600 MW, 776 MVA Turboalternator	23.0	28.0	207	26.0	15.0	0.50

Table A.2 Approximate reactance values of three-phase 60-Hz generating equipment – US (Values in per unit on rated kVA base)

| Apparatus | Positive sequence | | | | | | Negative sequence X_2 | | Zero sequence X_0 | |
| | Synchronous X_s | | Transient X' | | Subtransient X'' | | | | | |
	Average	Range	Average	Range	Average	Range	Average	Range	Average	Range
2-pole turbine generator (45 psig inner-cooled H_2)	1.65	1.22–1.91	0.27	0.20–0.35	0.21	0.17–0.25	0.21	0.17–0.25	0.093	0.04–0.14
2-pole turbine generator (30 psig H_2 cooled)	1.72	1.61–1.86	0.23	0.188–0.303	0.14	0.116–0.17	0.14	0.116–0.17	0.042	0.03–0.073
4-pole turbine generator (30 psig H_2 cooled)	1.49	1.36–1.67	0.281	0.265–0.30	0.19	0.169–0.208	0.19	0.169–0.208	0.106	0.041–0.1825
Salient-pole generator and motors (with dampers)	1.25	0.6–1.5	0.3	0.2–0.5	0.2	0.13–0.32	0.2	0.13–0.32	0.18	0.03–0.23
Salient-pole generator (without dampers)	1.25	0.6–1.5	0.3	0.2–0.5	0.3	0.2–0.5	0.48	0.35–0.65	0.19	0.03–0.24
Synchronous condensers (air-cooled)	1.85	1.25–2.20	0.4	0.3–0.5	0.27	0.19–0.3	0.26	0.18–0.4	0.12	0.025–0.15
Synchronous condensers (H_2 cooled at 1/2 psig rating)	2.2	1.5–2.65	0.48	0.36–0.6	0.32	0.23–0.36	0.31	0.22–0.48	0.14	0.03–0.18

With permission of Westinghouse Corp.

Table A.3 Principal data of 200–500 MW turbogenerators – Russia

	Units	Values of parameters of turbogenerators of various types					
		1	2	3	4	5	6
Power	MW/MVA	200/235	200/235	300/353	300/353	500/588	500/588
Cooling of winding							
(stator)		Water	Hydrogen	Water	Hydrogen	Water	Water
(rotor)		Hydrogen[c]	Hydrogen	Hydrogen	Hydrogen	Hydrogen	Water
Rotor diameter	m	1.075	1.075	1.075	1.120	1.125	1.120
Rotor length	m	4.35	5.10	6.1	5.80	6.35	6.20
Total weight	kg/kVA (N/kVA)	0.93 (9.1)	1.3 (12.7)	0.98 (9.6)	1.05 (10.3)	0.64 (6.26)	0.63 (6.17)
Rotor weight	kg/kVA (N/kVA)	0.18 (1.76)	0.205 (2.01)	0.16 (1.57)	0.158 (1.55)	0.11 (1.08)	0.1045 (1.025)
X_s^a	%	188.0	184.0	169.8	219.5	248.8	241.3
X'	%	27.5	29.5	25.8	30.0	36.8	37.3
X''	%	19.1	19.0	17.3	19.5	24.3	24.3
X_q	%	188.0	184.0	169.8	219.5	248.8	241.3
X_0	%	8.5	8.37	8.8	9.63	15.0	14.6
τ_1^b	s	2.3	3.09	2.1	2.55	1.7	1.63

[a]For all reactances, unsaturated values are given.
[b]Without the turbine.
[c]Hydrogen pressure for columns 1–4 is equal to 3 atm (304,000 N/m²) and for column 5, 4 atm (405 000 N/m²).
(Source: Glebov, I, A, C.I.G.R.E., 1968, Paper 11-07.)

Appendix B

Typical Transformer Impedances

Electric Power Systems, Fifth Edition. B.M. Weedy, B.J. Cory, N. Jenkins, J.B. Ekanayake and G. Strbac.
© 2012 John Wiley & Sons, Ltd. Published 2012 by John Wiley & Sons, Ltd.

Table B.1 Standard impedance limits for power transformers above 10000 kVA (60 Hz)

Highest voltage winding (BIL kV)	Low-voltage winding (BIL kV)	At kVA base equal to rating of largest capacity winding, 55°C							
		Self-cooled (OA), self-cooled/forced-air cooled (OA/FA) self-cooled/forced-air, forced-oil cooled (OA, FOA)				Forced-oil cooled (FOA and FOW)			
	For intermediate BIL use value for next higher BIL listed	Standard impedance (%)				Standard impedance (%)			
		Ungrounded neutral operation		Grounded neutral operation		Ungrounded neutral operation		Grounded neutral operation	
		Min.	Max.	Min.	Max.	Min.	Max.	Min.	Max.
110 and below	110 and below	5.0	6.25			8.25	10.5		
150	110	5.0	6.25			8.25	10.5		
200	110	5.5	7.0			9.0	12.0		
	150	5.75	7.5			9.75	12.75		
250	150	5.75	7.5			9.5	12.75		
	200	6.25	8.5			10.5	14.25		
350	200	6.25	8.5			10.25	14.25		
	250	6.75	9.5			11.25	15.75		
450	200	6.75	9.5	6.0	8.75	11.25	15.75	10.5	14.5
	250	7.25	10.75	6.75	9.5	12.0	17.25	11.25	16.0
	350	7.75	11.75	7.0	10.25	12.75	18.0	12.0	17.25
550	200	7.25	10.75	6.5	9.75	12.0	18.0	10.75	16.5
	350	8.25	13.0	7.25	10.75	13.25	21.0	12.0	18.0
	450	8.5	13.5	7.75	11.75	14.0	22.5	12.75	19.5
650	200	7.75	11.75	7.0	10.75	12.75	19.5	11.75	18.0
	350	8.5	13.5	7.75	12.0	14.0	22.5	12.75	19.5

(continued)

Table B.1 (*Continued*)

At kVA base equal to rating of largest capacity winding, 55 °C

Highest voltage winding (BIL kV)	Low-voltage winding (BIL kV) For intermediate BIL use value for next higher BIL listed	Self-cooled (OA), self-cooled/forced-air cooled (OA/FA) self-cooled/forced-air, forced-oil cooled (OA, FOA) Standard impedance (%)				Forced-oil cooled (FOA and FOW) Standard impedance (%)			
		Ungrounded neutral operation		Grounded neutral operation		Ungrounded neutral operation		Grounded neutral operation	
		Min.	Max.	Min.	Max.	Min.	Max.	Min.	Max.
750	450	9.25	14.0	8.5	13.5	15.25	24.5	14.0	22.5
	250	8.0	12.75	7.5	11.5	13.5	21.25	12.5	19.25
	450	9.0	13.75	8.25	13.0	15.0	24.0	13.75	21.5
	650	10.25	15.0	9.25	14.0	16.5	25.0	15.0	24.0
825	250	8.5	13.5	7.75	12.0	14.25	22.5	13.0	20.0
	450	9.5	14.25	8.75	13.5	15.75	24.0	14.5	22.25
	650	10.75	15.75	9.75	15.0	17.25	26.25	15.75	24.0
900	250			8.25	12.5			13.75	21.0
	450			9.25	14.0			15.25	23.5
	750			10.25	15.0			16.5	25.5
1050	250			8.75	13.5			14.75	22.0
	550			10.0	15.0			16.75	25.0
	825			11.0	16.5			18.25	27.5
1175	250			9.25	14.0			15.5	23.0
	550			10.5	15.75			17.5	25.5
	900			12.0	17.5			19.5	29.0
1300	250			9.75	14.5			16.25	24.0
	550			11.25	17.0			18.75	27.0
	1050			12	18.25			20.75	30.5

With permission of Westinghouse Corp.

Appendix C

Typical Overhead Line Parameters

Electric Power Systems, Fifth Edition. B.M. Weedy, B.J. Cory, N. Jenkins, J.B. Ekanayake and G. Strbac.
© 2012 John Wiley & Sons, Ltd. Published 2012 by John Wiley & Sons, Ltd.

Table C.1   Overhead-line parameters–50 Hz (British)

Parameter	275 kV		400 kV	
	$2 \times 113\,\mathrm{mm}^2$	$2 \times 258\,\mathrm{mm}^2$	$2 \times 258\,\mathrm{mm}^2$	$4 \times 258\,\mathrm{mm}^2$
$Z_1\,\Omega/\mathrm{km}$	$0.09 + j0.317$	$0.04 + j0.319$	$0.04 + j0.33$	$0.02 + j0.28$
$Z_0\,\Omega/\mathrm{km}$	$0.2 + j0.87$	$0.14 + j0.862$	$0.146 + j0.862$	$0.104 + j0.793$
$Z_{mo}\,\Omega/\mathrm{km}$	$0.114 + j0.487$	$0.108 + j0.462$	$0.108 + j0.45$	$0.085 + j0.425$
$Z_p\,\Omega/\mathrm{km}$	$0.127 + j0.5$	$0.072 + j0.5$	$0.075 + j0.507$	$0.048 + j0.45$
$Z_{pp}\,\Omega/\mathrm{km}$	$0.038 + j0.183$	$0.033 + j0.182$	$0.035 + j0.177$	$0.028 + j0.172$
$B_1\,\mu\mathrm{mho}/\mathrm{km}$	3.60	3.65	3.53	4.10
$B_0\,\mu\mathrm{mho}/\mathrm{km}$	2.00	2.00	2.00	2.32
$B_{m0}\,\mu\mathrm{tmho}/\mathrm{km}$	5.94	7.00	7.75	8.50

Z_1 Positive-sequence impedance.
Z_0 Zero-sequence impedance of a d.c. 1 line.
Z_{m0} Zero-sequence mutual impedance between circuits.
Z_p Self-impedance of one phase with earth return.
Z_{pp} Mutual impedance between phases with earth return.
B_1 Positive-sequence susceptance.
B_0 Zero-sequence susceptance of a d.c. 1 line.
B_{m0} Zero-sequence mutual susceptance between circuits.
Note: A d.c. 1 line refers to one circuit of a double-circuit line in which the other circuit is open at both ends.
The areas quoted are copper equivalent values based on $0.4\,\mathrm{in}^2$ and $0.175\,\mathrm{in}^2$.

Table C.2 Overhead line data – a.c. (60 Hz) and d.c. lines

Region:	500–550 kV–a.c.		700–750 kV	h.v.d.c.	
	Pacific	Canada	Canada	Mountain	Pacific
Utility:	So. California Edison Co.	Ontario Hydro	Quebec Hydroelectric Co.	U.S. Bureau of Reclamation	Los Angeles Dept. of W. & P.
Line name or number	Lugo-Eldorado	Pinard-Hanmer	Manics.-Levis	Oregon-Mead	Dalles-Sylmar
Voltage (nominal), kV; a.c. or d.c.	500; a.c.	500; a.c.	735; a.c.	750 (±375); d.c.	750 (±375); d.c.
Length of line, miles	176	228	236	560	560
Originates at	Lugo sub	Pinard TS	Manicouagan	Oregon border	Oregon border
Terminates at	Eldorado sub	Hanmer TS	Levis sub	Mead sub	near, Los Angeles
Year of construction	1967–69	1961–63	1964–65	1966–70	1969
Normal rating/cct, MVA	1000	–	1700	1350 MW	1350 MW
Structures					
(S)teel, (W)ood, (A)lum	S	S, A	S	S	S or A
Average number/mile	4.21	3.7	3.8	4.5	4.5
Type: (S)q, (H), guy (V) or (Y)	S (S-56)	V(A-51, S-51)	–; (S-71)	S, T (S-72, 73)	S, T (S-72, 73)
Min 60-Hz flashover, kV	850	1110	–	–	915
Number of circuits: Initial; Ultimate	1; 1	1; 1	1; 1	1; 1	1; 1
Crossarms (S)teel, (W)ood, (A)lum	–	S, A	S	S	S or A
Bracing (S)teel, (W)ood, (A)lum	–	–	S	S	S or A
Average weight/structure, lb	19 515	10 800; 4730	67 3000	11 000	–
Insulation in guys (W)ood, (P)orc, kV	No guys	Nil	–	None	–
Conductors					
Al, ACSR, ACAR, AAAC, 5005	ACSR 84/19	ACSR 18/7	ACSR 42/7	ACSR 96/19	ACSR 84/19
Diameter, in.; MCM	1.762; 2156	0.9; 583	–; 1361	–; 1857	1.82; 2300
Weight 1 cond./ft, lb	2.511	0.615	1.468	2.957	2.68
Number/phase; Bundle spacing, in	2; 18	4; 18	4; 18	2; 16	2; 18
Spacer (R)igid, (S)ring, (P)reform	S	R	R	S or P	S or A
Designed for-A/phase	2400	2500	–	1800	3380

(continued)

Table C.2 (Continued)

Region:	500–550 kV–a.c.		700–750 kV	h.v.d.c.	
	Pacific	Canada	Canada	Mountain	Pacific
Ph. config. (V)ertical, (H)orizontal, (T)riangular	H	H	H	H	H
Phase separation, ft	32	40	50	38–41	39–40
Span: Normal, ft; Maximum, ft	1500; 2764	1400; 2550	1400; 2800	1150; 1800	1175; –
Final sag, ft; at °F; Tension, 10^3 lb	46 at 130 F; 19.14	5.0 at 60 F; 3.4	226 at 120 F; 6.2	40 at 120 F; 12.4	35 at 60 F; 13.0
Minimum clearance: Ground, ft; Structure, ft	40; 12.5	33; 10.5	45; 18.3	35; 7.75	35; 7.8
Type armour at clamps	Preformed	Nil	None	–	None
Type vibration dampers	Stockbridge	Spacer damper	Tuned spacer	–	Stockbridge
Line altitude range, ft	1105–4896	800–1300	50–2500	2000–7000	400–8000
Max corona loss, kW/three-ph. mile	5	–	74 ($\frac{1}{2}$ in. snow/h)	–	–
Insulation					
Basic impulse level, kV	2080	1800	–	–	–
Tangent: (D)isk or (L)ine-post	D	D	D	D	D
Number strings in (V) or (P)arallel	2V	1 or 2P	4V	Single	Single
Number units; Size; Strength, 10^3 lb	27; $5^3\!/_4 \times 10$; 30	23; $5^3\!/_4 \times 10$; 25	35; 5×10; 15	–;–;–	–;–;–
Angle: (D)isk or (L)ine-post	D	D	D	D	D
Number strings in (V) or (P)arallel	2V	3P	4P	Single	Single
Number units; Size; Strength, 10^3 lb	25; $6^1\!/_4 \times 10^3\!/_8$; 40	26; $5^3\!/_4 \times 10$; 25	35; $6^1\!/_4$/10; 36	–;–;–	–;–;–
Insulators for struts	None	None	–	–	None
Terminations: (D)isk or (L)ine-post	D	D	–	D	D
Number strings in (P)arallel	2P	3P	–	–	4P
Number units; Size; Strength, 10^3 lb	25; $6^1\!/_4 \times 10^3\!/_8$; 40	26; $5^3\!/_4 \times 10$; 25	–;–;–	–;–;–	–;–;–
BIL reduction, kV or steps	3.5 steps	–	–	–	–
BIL of terminal apparatus, kV	1150–1425	1675–1900	2050–2200	–	1300

(continued)

Table C.2 (*Continued*)

Region:	500–550 kV–a.c.		700–750 kV	h.v.d.c.	
	Pacific	Canada	Canada	Mountain	Pacific
Protection					
Number shield wires; Metal and Size	2; 7 No. 6 AW	2; 5/16 in. galv.	2; 7/16 in. galv.	1; –	2; –
Shield angle, deg; Span clear, ft	24.5; 28	20; 48	20; 50	30; –	20; –
Ground resistance range, Ω		15			30 max.
Counterpose: (L)inear, (C)rowfoot	C	L	L		
Neutral gdg: (S)ol, (T)sf;	S; 0	S; –	S; –	S; –	S; –
Arrester rating, kV; Horn gap, in	420; None	480/432; None	636; –	–; –	–; –
Arc ring diameter, in.: Top; Bottom	None	None	–; –	–; –	–; –
Expected outages/100 miles/year	0.5	1			0.2
Type relaying	Ph. compar., carr.	Dir'l compar.			
Breaker time, cyc; Reclos, cycles	2; None	3; None	6; 24	–; –	–; –
Voltage Regulation					
Nominal, %	±10	–	–	Variable	Grid control
Synch, cond. MVA; Spacing, miles	None	None	–; –	–; –	None
Shunt capac. MVA; Spacing, miles	None	None	330; 236	–; –	On a.c. terminals
Series capac. MVA; Compens., %	None	None	None	–; –	–; –
Shunt reactor, MVA; Spacing, miles	–; –	–	–	–	–
Communication					
(T)elephone circuits on structures	None	None	None	None	None
(C)arrier; Frequency, kHz	C; –	C; 50–98	C; 42.0 and 50.0	None	None
(P)ilot wire	None	–	P	None	None
(M)icrowave; Frequency, MHz	M; 6700	–; –	M; 777.35	M; –	–

(*With permission of Edison Electric Insitute*)

Further Reading

This bibliography lists some of the many excellent books on power systems. A key is provided to indicate the most important books that should be looked at first if seeking more information or historic developments on topics in the denoted chapters. All books will have more references to consult.

Book	Chapter
Adkins, B. and Harley, R. (1978) *General Theory of Alternating Current Machines*, Chapman and Hall, London.	3[†]
Anderson, P.M. and Fouad, A.A. (1977) *Power System Control and Stability*, Iowa State University Press, Ames, Iowa.	8[†]
Arrillaga, J. (1983) *High Voltage Direct Current Transmission*, Peter Peregrinus on behalf of the Institution of Electrical, Engineers, UK.	9[†]
Arrillaga, J., Liu, Y.H., and Watson, N.R. (2007) *Flexible Power Transmission: The HVDC Options*, John Wiley & Sons.	9[†]
Arrillaga, J. and Arnold, C.P. (1990) *Computer Analysis of Power Systems*, John Wiley & Sons, Chichester.	6[†]
Berrie, T.W. (1990) *Power System Economics*, Peter Peregrinus on behalf of the Institution of Electrical, Engineers, UK.	12[*]
Berrie, T.W. (1992) *Electricity Economics and Planning*, Peter Peregrinus, on behalf of the Institution of Electrical, Engineers, UK.	12[*]
Bewley, L.W. (1961) *Travelling Waves on Transmission Systems*, Dover Books, New York.	10[**]
Bickford, J.P., Mullineux, N., and Reed, J.R. (1976) *Computation of Power System Transients*, Peter Peregrinus on behalf of the Institution of Electrical Engineers, UK.	10[†]
British Electricity International: Modern Power Station Practice (1991), Vols L. and K, Pergamon, Oxford.	1,11[†]

Electric Power Systems, Fifth Edition. B.M. Weedy, B.J. Cory, N. Jenkins, J.B. Ekanayake and G. Strbac.
© 2012 John Wiley & Sons, Ltd. Published 2012 by John Wiley & Sons, Ltd.

Debs, A.S. (1988) *Modern Power System Control and Operation*, Kluwer
Academic, Boston. — 4,5[†]

E.H.V. Transmission Line Reference Book (1968), Edison Electric Institute,
New York. — 1,10,11[†]

El Abiad, A.H. and Stagg, G.W. (1968) *Computer Methods in Power System
Analysis*, McGraw-Hill, New York. — 6,7[**]

Electrical Transmission and Distribution Reference Book (1964),
Westinghouse Electric Corporation, East Pittsburgh, Pennsylvannia. — 1,10,11[**]

Electricity Council (1991) *Power System Protection*, vols **1–3**, Peter
Peregrinus on behalf of the Institution of Electrical Engineers, UK. — 11[†]

Elgerd, O.I. (1983) *Electric-Energy Systems Theory—An Introduction*,
2nd edn, McGraw-Hill. — 3,4,5,6,7[†]

Fitzgerald, A.E., Kingsley, C., and Umans, S. (1990) *Electric Machinery*,
5th edn, McGraw-Hill. — 3[†]

Flurscheim, C.H. (ed.) (1987) *Power Circuit Breaker Theory and Design*,
Peter Peregrinus on behalf of the Institution of Electrical, Engineers,
UK. — 10,11[†]

Fouad, A.A. and Vittal, V. (1992) *Power System Transient Stability Analysis
Using the Transient Energy Function Method*, Prentice Hall, Englewood
Cliffs, New Jersey. — 8,10[†]

Giles, R.L. (1970) *Layout of E.H.V. Substations*, Cambridge University
Press, Cambridge. — 11[†]

Glover, J.D. and Sarma, M. (1994) *Power System Analysis and Design*,
2nd edn, PWS, Boston, Massachusetts. — 6,7[†]

Gonen, T. (1986) *Electric Power Distribution System Engineering*,
McGraw-Hill. — 1,3[†]

Grainger, J.J. and Stevenson, W.D. Jr (1994) *Power System Analysis*,
McGraw-Hill International. — 3,4,6,7,8[†]

Greenwood, A. (1971) *Electrical Transients in Power Systems*, Wiley-
Interscience. — 10[**]

Gross, C.A. (1986) *Power System Analysis*, 2nd edn, John Wiley & Sons,
New York. — 2,6[†]

Hindmarsh, J. (1984) *Electric Machines*, 4th edn, Pergamon, London. — 3[*]

HVDC: Connecting to the Future, (2010), Alstom Grid, Stafford. — 10[†]

Johns, A.T. and Salman, S.K. (1996) *Digital Protection for Power Systems*,
Peter Peregrinus on behalf of the Institution of Electrical Engineers,
UK. — 11[†]

Kirchmayer, L.K. (1958) *Economic Operation of Power Systems*, John Wiley
& Sons, New York. — 12[**]

Kirschen, D.S. and Strbac, G. (2004) *Fundamentals of Power System
Economics*, John Wiley and Sons, Chichester, UK. — 12[*]

Kundur, P. (1994) *Power System Stability and Control*, McGraw-Hill, New
York. — 8[†]

Lakervi, E. and Holmes, E.J. (1995) *Electricity Distribution Network Design,*   1,5[†]
   Peter Peregrinus on behalf of the Institution of Electrical Engineers,
   UK.

Lander, C.W. (1996) *Power Electronics,* 3rd edn, McGraw-Hill.   10[*]

Looms, J.S.T. (1990) *Insulators for High Voltages,* Peter Peregrinus on   11[*]
   behalf of the Institution of Electrical Engineers, UK.

Machowski, J., Bialek, J.W., and Bumby, J.R. (1997) *Power System*   8[†]
   Dynamics and Stability, John Wiley & Sons, Chichester.

Network Protection and Automation Guide, (2011), 5th edn, Alstom Grid,   11[†]
   Stafford UK.

Pai, M.A. (1981) *Power System Stability. Analysis by the Direct Method of*   8[†]
   Lyapunov, North Holland Publishing, Amsterdam.

Pai, M.A. (1989) *Energy Function Analysis for Power System Stability,*   8[†]
   Kluwer Academic, Boston, Massachusetts.

Pavella, M. and Murthy, P.J. (1994) *Transient Stability of Power Systems,*   8[†]
   Theory and Practice, John Wiley & Sons, Chichester.

Rudenberg, R. (1969) *Transient Performance of Electrical Power Systems,*   10[†]
   MIT Press, Boston.

Say, M.G. (1976) *Alternating Current Machines,* 4th edn, Pitman, London.   3[**]

Schweppe, F.C., Caramanis, M.C., Tabors, R.D., and Bohn, R.E. (1988)   12[**]
   Spot Pricing of Electricity, Kluwer Academic Publishers, Boston, MA.

Tleis, N.D. (2008) *Power Systems Modelling and Fault Analysis: Theory and*   7[†]
   Practice, Newnes, Oxford.

Transmission Line Reference Book, 345 kV and Above, (1975), Electric Power   3[†]
   Research Institute, Palo Alto, California.

Turvey, R. and Anderson, D. (1977) *Electricity Economics: Essays and Case*   12[**]
   Studies, Johns Hopkins University Press.

Weedy, B.M. (1980) *Underground Transmission of Electric Power,* John   3[*]
   Wiley & Sons, Chichester.

Wood, A.J. and Wollenberg, B.F. (1996) *Power Generation, Operation and*   8[†]
   Control, 2nd edn, John Wiley & Sons, New York.

Key: [†] consult for advanced study, [*] read for background knowledge, [**] classic book.

Index

Electric Power Systems, Fifth Edition. B.M. Weedy, B.J. Cory, N. Jenkins, J.B. Ekanayake and G. Strbac.
© 2012 John Wiley & Sons, Ltd. Published 2012 by John Wiley & Sons, Ltd.